Lecture Notes in Mathematics

Edited by A. Dold and B. Eckmann

Subseries: Fondazione C.I.M.E., Firenze
Adviser: Roberto Conti

996

Invariant Theory

Proceedings of the 1st 1982 Session of the Centro Internazionale Matematico Estivo (C.I.M.E.) Held at Montecatini, Italy, June 10–18, 1982

Edited by F. Gherardelli

Springer-Verlag
Berlin Heidelberg New York Tokyo 1983

Editor

Francesco Gherardelli
Istituto Matematico U. Dini, Università degli Studi
Viale Morgagni 67 A, 50134 Firenze, Italy

AMS Subject Classifications (1980): 14D20, 14D25

ISBN 3-540-12319-9 Springer-Verlag Berlin Heidelberg New York Tokyo
ISBN 0-387-12319-9 Springer-Verlag New York Heidelberg Berlin Tokyo

Printing and binding: Beltz Offsetdruck, Hemsbach/Bergstr.
2146/3140-543210

Math

Sup

C.I.M.E. Session on Invariant Theory

List of Participants

B. Ådlandsvik, Matematisk Institutt, Allégt. 55, 5000 Bergen, Norway

A. Albano, Istituto di Geometria "C. Segre", Via Principe Amedeo 8, 10123 Torino

G. Almkvist, PL 500, 24300 Höör, Sweden

L. Amodei, Istituto Matematico Università, Via F. Buonarroti 2, 56100 Pisa

F. Arnold, Willy-Andreas-Allee 5, 7500 Karlsruhe, West Germany

E. Ballico, Scuola Normale Superiore, Piazza dei Cavalieri 7, 56100 Pisa

R. Benedetti, Istituto Matematico "L. Tonelli", Via F. Buonarroti 2, 56100 Pisa

J. Bertin, 18 rue du Sénéchal, 31000 Toulouse, France

C. Blondel, UER de Math. 45-55, Univ. Paris VII, 2 Pl. Jussieu, 75221 Paris, France

J. F. Boutot, 15 rue Erard, 75012 Paris, France

M. Brion, E.N.S., 45 rue d'Ulm, 75005 Paris, France

F. Catanese, Istituto Matematico Università, Via F. Buonarroti 2, 56100 Pisa

A. Collino, Istituto di Geometria, Università, Via Principe Amedeo 8, 10123 Torino

P. Cragnolini, Istituto Matematico Università, Via F. Buonarroti 2, 56100 Pisa

M. Dale, Matematisk Institutt, Allégt. 55, 5000 Bergen, Norway

C. De Concini, Istituto Matematico Università, Via F. Buonarroti 2, 56100 Pisa

A. Del Centina, Istituto Matematico Università, Viale Morgagni 67/A, 50134 Firenze

G. D'Este, Istituto di Algebra e Geometria, Via Belzoni 7, 35100 Padova

D. Dikranjan, Istituto Matematico Università, Via F. Buonarroti 2, 56100 Pisa

G. Elencwajg, IMSP-Mathématiques, Parc Valrose, 06034 Nice-Cedex, France

J. Eschgfäller, Istituto Matematico Università, Via Machiavelli 35, 44100 Ferrara

P. Gerardin, UER de Mathématiques, Université Paris VII, 2 Place Jussieu, 75231 Paris-Cedex, France

F. Gherardelli, Istituto Matematico Università, Viale Morgagni 67/A, 50134 Firenze

P. Gianni, Dipartimento di Matematica, Via F. Buonarroti 2, 56100 Pisa

D. Gieseker, Department of Mathematics, UCLA, Los Angeles, Cal. 90024, USA

A. Gimigliano, Viale della Repubblica 85, 50019 Sesto Fiorentino, Firenze

K. Girstmair, Institut für Mathematik, Universität Innsbruck, Innrain 52, A-6020 Innsbruck, Austria

A. Helversen-Pasotto, Laboratoire associé du CNRS n. 168, Départment de Mathématique, IMSP, Université de Nice, Parc Valrose, 06034 Nice-Cedex, France

T. Johnsen, Lallakroken 8 C, Oslo 2, Norway

V. Kac, Department of Mathematics, MIT, Cambridge, Mass. 02139, USA

M. Laglasse-Decauwert, USMG Institut Fourier, Mathématiques Pures, B.P. 116, 38402 Saint Martin D'Heres, France

A. Lanteri, Istituto Matematico "F. Enriques", Via C. Saldini 50, 20133 Milano

A. Lascoux, L.I.T.P., UER Maths Paris VII, 2 Place Jussieu, 75251 Paris Cedex 05, France.

D. Luna, 12 Rue Fracy, 38700 La Tronche, France

V. Mehta, Department of Mathematics, The University of Bombay, Bombay, India

M. Meschiari, Via Baraldi 12, 41100 Modena

D. Montanari, Via Asmara 38, 00199 Roma

J. Oesterlé, E.N.S., 45 rue d'Ulm, 75005 Paris, France

P. Oliverio, Scuola Normale Superiore, Piazza dei Cavalieri 7, 56100 Pisa

M. Palleschi, Via Bergognone 27, 20144 Milano

F. Pauer, Institut für Mathematik, Universität Innsbruck, Innrain 52, A-6020 Innsbruck, Austria

E. Previato, Department of Mathematics, Harvard University, Cambridge,Mass.02138, USA

M. Roberts, 9 Pelham Grove, Liverpool 17, England

N. Rodinò, Via di Vacciano 87, 50015 Grassina, Firenze

S. Rosset, School of Mathematics, University of Tel-Aviv, Tel-Avis, Israele

P. Salmon, Via Rodi 14/9, 16145 Genova

W. K. Seiler, Mathematisches Institut II, Englerstrasse 2, 7500 Karlsruhe 1, West Germany

M. Seppälä, Fakultat für Mathematik, Universität Bielefeld, BRD-4800 Bielefeld 1, Germania Occ.

R. Smith, Univ. of Georgia, Graduate Students Res. Center, Athens,Ga 30602, USA

C. Traverso, Dipartimento di Matematica, Via F. Buonarroti 2, 56100 Pisa

C. Turrini, Istituto Matematico "F. Enriques", Via C. Saldini 50, 20133 Milano

L. Verdi, Istituto Matematico Università, Viale Morgagni 67/A, 50134 Firenze

A. Verra, Via Assarotti 16, 10122 Torino

G. Weill, 300 E 33rd Street, New York, N.Y. 10016, USA

INDEX

C. DE CONCINI and C. PROCESI, *Complete symmetric varieties* 1

D. GIESEKER, *Geometric invariant theory and applications to moduli problems* .. 45

V. G. KAC, *Root systems, representations of quivers and invariant theory* 74

G. ALMKVIST, *Invariant of* $\mathbb{Z}/p\mathbb{Z}$ *in characteristic* p 109

A. LASCOUX and M.P. SCHÜTZENBERGER, *Symmetric and flag manifolds* 118

V. B. MEHTA, *On some restriction theorems for semistable bundles* 145

COMPLETE SYMMETRIC VARIETIES

by

C. De Concini and C. Procesi

Università di Roma II — Università di Roma

Nur der Philister schwärmt für absolute Symmetrie

H. Seidel, ges. w. 1,70 *

INTRODUCTION

In the study of enumerative problems on plane conics the following variety has been extensively studied ([6],[7],[15],[17],[18],[19],[20],[23],[25]).

We consider pairs (C,C') where C is a non degenerate conic and C' its dual and call X the closure of this correspondence in the variety of pairs of conics in $\mathbb{P}^2$ and $\check{\mathbb{P}}^2$.

On this variety acts naturally the projective group of the plane and one can see that X decomposes into 4 orbits: X_o open in X; X_1, X_2 of codimension 1 and $X_3 = \bar{X}_1 \cap \bar{X}_2$ of codimension 2. All orbit closures are smooth and the intersection of $\bar{X}_1$ with $\bar{X}_2$ is transversal. This theory has been extended to higher dimensional quadrics ([1],[15],[17],[21]) and also carried out in the similar example of collineations ([16]).

The renewed interest in enumerative geometry (see e.g. [11]) has brought back some interest in this class of varieties ([22],[5] , cf.§6).

In this paper we will study closely a general class of varieties, including the previous examples, which have a significance for enumerative problems.

Let $\bar{G}$ be a semisimple adjoint group, $\sigma: \bar{G} \to \bar{G}$ an automorphism of order 2 and $\bar{H} = \bar{G}^\sigma$. We construct a canonical variety X with an action of $\bar{G}$ such that

1) X has an open orbit isomorphic to $\bar{G}/\bar{H}$
2) X is smooth with finitely many $\bar{G}$ orbits
3) The orbit closures are all smooth
4) There is a 1-1 correspondence between the set of orbit closures and the family of subsets of a set I_ℓ with ℓ elements. If $J \subseteq I_\ell$ we denote by S_J the corresponding orbit closure
5) We have $S_I \cap S_J = S_{I \cup J}$ and codim S_I = card I

* We thank the "Lessico intellettuale europeo" for supplying the quotation.

6) Each S_I is the transversal complete intersection of the $S_{\{u\}}$, $u \in I$

7) For each S_I we have a $\bar{G}$ equivariant fibration $\pi_I : S_I \to G/P_I$ with P_I a parabolic subgroup with semisimple Levi factor L, σ stable, and the fiber of π_I is the canonical projective variety associated to L and $\sigma|L$

Using results of Bialynicki Birula [2] we give a paving of X by affine spaces and compute its Picard group. We describe the positive line bundles on X and their cohomology in a fashion similar to that of "Flag varieties".

Next we give a precise algorithm which allows to compute the so called characteristic numbers of basic conditions (in the classical terminology) in all cases. The computation can be carried out mechanically although it is very lengthy.

As an example we give the classical application due to H.Schubert [14] for space quadrics and compute the number of quadrics tangent to nine quadrics in general position.

We should now make three final remarks. First of all our method has been strongly influenced by the work of Semple [15], we have in fact interpeted his construction in the language of algebraic groups. The second point will be taken in a continuation of this work.Briefly we should say that a general theory of group embeddings due to Luna and Vust [13] has been used by Vust to classify all projective equivariant embeddings of a symmetric variety of adjoint type and in particular the ones which have the property that each orbit closure is smooth. We call such embeddings wonderful. It has been shown by Vust that such embeddings are all obtained in most cases from our variety X by successive blow ups, followed by a suitable contraction.

This is the reason why we sometimes refer to X as the minimal compactification, in fact it is minimal only among this special class.

The study of the limit provariety obtained in this way is the clue for a general understanding of enumerative questions on symmetric varieties as we plan to show elsewhere.

Finally we have restricted our analysis to characteristic 0 for simplicity. Many of our results are valid in all characteristics (with the possible exception of 2) and some should have a suitable characteristic free analogue. Hopefully an analysis of this theory may have same applications to representation theory also in positive characteristic.

The first named author wishes to thank the Tata institute of Fundamental research and the C.N.R. for partial financial support during the course of this research. Special thank go to the C.I.M.E.

which allowed him to lecture on the material of this paper at the meeting on the "Theory of Invariants" held in Montecatini in the period June 10-18, 1982.

The second named author aknowledges partial support from Brandeis University and grants from N.S.F. and C.N.R. during different periods of the development of this research.

1. PRELIMINARIES

In this section we collect a few more or less well known facts.

1.1. Let G be a semisimple simply connected algebraic group over the complex numbers. Let $\sigma: G \to G$ be an automorphism of order 2 and $H=G^\sigma$ the subgroup of G of the elements fixed under σ. The homogeneous space G/H is by definition a symmetric variety and more generally, if G' is a quotient of G by a (finite) σ stable subgroup of the center of G, the corresponding G'/H' will again be a symmetric variety.

Let $\underline{g}$, $\underline{h}$ denote the Lie algebras of G, H respectively. σ induces an automorphism of order 2 in $\underline{g}$ which will again be denoted by σ and $\underline{h}$ is exactly the +1 eigenspace of σ.

We recall a well known fact:

PROPOSITION. Every σ-stable torus in G is contained in a maximal torus of G which is σ stable.

If T is a σ stable torus and $\underline{t}$ its Lie algebra, we can decompose $\underline{t}$ as $\underline{t} = \underline{t}_o \oplus \underline{t}_1$ according to the eigenvalues +1, -1 of σ. $\underline{t}_o$ is the Lie algebra of the torus $T_o = T^\sigma$ while $\underline{t}_1$ is the Lie algebra of the torus $T_1 = \{t \in T | t^\sigma = t^{-1}\}$ such a torus is called anisotropic. The natural mapping $T_o \times T_1 \to T$ is an isogeny, it is not necessarily an isomorphism since the character group of T need not decompose under σ into the sum of the subgroups relative to the eigenvalues ± 1. We indicate still by σ the induced mapping on $\underline{t}^*$ and can easily verify in case T is a maximal torus and $\Phi \subseteq \underline{t}^*$ the root system:

i) If $\underline{t} \oplus \sum_{\alpha\in\Phi} \underline{g}_\alpha$ is the root space decomposition of $\underline{g}$ then

$\sigma(\underline{g}_\alpha) = \underline{g}_{\alpha\sigma}$, hence $\sigma(\Phi) = \Phi$.

(ii) σ preserves the Killing form.

We want now to choose among all possible σ stable tori one for which dim T_1 is maximal and call this dimension the rank of G/H, indicated by ℓ.

1.2. Having fixed T and so the root system Φ we proceed now to fix the positive roots in a compatible way.

LEMMA. One can choose the set Φ^+ of positive roots in such a way that: If $\alpha \in \Phi^+$ and $\alpha \not\equiv 0$ on $\underline{t}_1$ then $\alpha^\sigma \in \Phi^-$.

PROOF. Decompose $\underline{t}^* = \underline{t}_o^* \oplus \underline{t}_1^*$; every root α is then written $\alpha = \alpha_o + \alpha_1$ and $\alpha^\sigma = \alpha_o - \alpha_1$. Choose two R-linear forms ϕ_o and ϕ_1 on $\underline{t}_o^*$ and $\underline{t}_1^*$ such that ϕ_o and ϕ_1 are non zero on the non zero components of the roots. We can replace ϕ_1 by a multiple if necessary so that, if $\alpha =$ $= \alpha_o + \alpha_1$ and $\alpha_1 \neq 0$ we have $|\phi_1(\alpha_1)| > |\phi_o(\alpha_o)|$. Consider now the R-linear form $\phi = \phi_o \oplus \phi_1$, we have that $\phi(\alpha) \neq 0$ for every root α ; moreover if $\alpha \not\equiv 0$ on t_1, i.e. $\alpha = \alpha_o + \alpha_1$ with $\alpha_1 \neq 0$ the sign of $\phi(\alpha)$ equals the sign of $\phi_1(\alpha_1)$. Thus, setting $\Phi^+ = \{\alpha \in \Phi | \phi(\alpha) > 0\}$ we have the required choice of positive roots. Let us use the following notations

$$\Phi_o = \{\alpha \in \Phi |\ \alpha | t_1 = 0\}, \quad \Phi_1 = \Phi - \Phi_o .$$

Clearly $\Phi_o = \{\alpha \in \Phi | \alpha^\sigma = \alpha\}$ while by the previous lemma σ interchanges Φ_1^+ with Φ_1^- .

Having fixed Φ^+ as in the above lemma we denote by $B \subset G$ the corresponding Borel subgroup and by B^- its opposite Borel subgroup.

1.3. It is now easy to describe the Lie algebra $\underline{h}$ in terms of the root decomposition. We have already noticed that $\sigma(\underline{g}_\alpha) = \underline{g}_{\alpha^\sigma}$.

LEMMA. If $\alpha \in \Phi_o$, σ is the identity on g_α .

PROOF. Let x_α , y_α , h_α be the standard sl_2 triple associated to α. Since $\alpha^\sigma = \alpha$ we have $\sigma(h_\alpha) = h_\alpha$. On the other hand since $\sigma(g_{\pm\alpha}) = g_{\pm\alpha}$ we have $\sigma(x_\alpha) = \pm x_\alpha$. Now if $\sigma(x_\alpha) = -x_\alpha$ we must have also $\sigma(y_\alpha) = -y_\alpha$ since $h_\alpha = [x_\alpha , y_\alpha]$. Now if we consider any element $s \in \underline{t}_1$ we have $[x_\alpha , s] = [y_\alpha , s] = 0$ since α vanishes on $\underline{t}_1$ by hypothesis. This implies, setting $t = x_1 + y_1$, that $\underline{t}_1 + Ct$ is a Toral subalgebra on which σ acts as -1. Since we can enlarge this to a maximal Toral subalgebra, we contradict the choice of T maximizing the dimension of T_1.

PROPOSITION. $\underline{h} = \underline{t}_o + \sum_{\alpha \in \Phi_o} g_\alpha + \sum_{\alpha \in \Phi_1} C(x_\alpha + \sigma(x_\alpha))$.

PROOF. Trivial from the previous lemma.

We may express a consequence of this, the so called Iwasawa decomposition: The subspace $\underline{t}_1 + \sum_{\alpha \in \Phi_1^+} Cx_\alpha$ is a complement to $\underline{h}$ and so it projects isomorphically onto the tangent space of G/H at H, in particular since Lie B $\supset t_1 + \sum_{\alpha \in \Phi_1^+} \mathbf{C}x_\alpha$, BH $\subset$ G is dense in G.

COROLLARY. $\dim G/H = \dim \underline{t}_1 + 1/2|\Phi_1|$.

1.4. If $\Gamma \subset \Phi_+$ is the set of simple roots, let us denote $\Gamma_o = \Gamma \cap \Phi_o$, $\Gamma_1 = \Gamma \cap \Phi_1$ explicitely:

$$\Gamma_o = \{\beta_1,\ldots,\beta_k\}; \qquad \Gamma_1 = \{\alpha_1,\ldots,\alpha_j\}.$$

LEMMA. For every $\alpha_i \in \Gamma_1$ we have that α_i^σ is of the form $-\alpha_k - \Sigma n_{ij}\beta_j$ for some $\alpha_k \in \Gamma_1$ and some non negative integers n_{ij}. Moreover, $\alpha_k^\sigma = -\alpha_i - \Sigma n_{ij}\beta_j$.

PROOF. By Lemma 1.2 we know that $\alpha_i^\sigma \in \Phi^-$ hence we can write $\alpha_i^\sigma = -(\Sigma m_{ik}\alpha_k + \Sigma n_{ij}\beta_j)$ where m_{ik} , n_{ij} are non negative integers. Thus $\alpha_i = \alpha_i^{\sigma\sigma} = \sum_k m_{ik}(\sum_t m_{kt}\alpha_t) + \Sigma m_{ik}\Sigma n_{kj}\beta_j - \Sigma n_{ij}\beta_j$. Since the simple roots are a basis of the root lattice we must have in particular $\Sigma m_{ik}m_{kt} = 0$ for $t \neq i$ and $\Sigma m_{ik}m_{ki} = 1$. Since the m_{ij}'s are non negative integers it follows that only one m_{ik} is non zero and equal to 1 and the m_{ki} is also equal to 1.

Now consider the fundamental weights. Since they form a dual basis of the simple coroots we also divide them:

$$\omega_1,\ldots,\omega_j, \qquad \zeta_1,\ldots,\zeta_k \qquad \text{where:}$$

$$(\omega_i , \check\beta_j) = 0, \qquad (\omega_i , \check\alpha_j) =: \delta_j^i \quad \text{and similarly for the } \zeta_j\text{'s.}$$

Since σ preserves the Killing form we have:

$$(\omega_i^\sigma , \check\beta_j^\sigma) = (\omega_i^\sigma , \check\beta_j) = 0$$

$$\delta_j^i = (\omega_j^\sigma , \check\alpha_i^\sigma) = (\omega^\sigma , \frac{2}{(\alpha_i , \alpha_i)}(-\alpha_k - \Sigma n_{ij}\beta_j))$$

$$= -(\omega_j^\sigma , \frac{2\alpha_k}{(\alpha_i , \alpha_i)}) = \frac{(\alpha_k , \alpha_k)}{(\alpha_i , \alpha_i)} (\omega_j^\sigma , \check\alpha_k)$$

We deduce that

$$\omega_i^\sigma = - \frac{(\alpha_k , \alpha_k)}{(\alpha_i , \alpha_i)} \omega_k.$$

Now ω_i^σ must be in the weight lattice so $\frac{(\alpha_k , \alpha_k)}{(\alpha_i , \alpha_i)}$ is an integer. Reversing the role of i and k we set that it must be 1 so

$$\omega_i^\sigma = -\omega_k .$$

We can summarize this by saying that we have a permutation $\tilde{\sigma}$ of order 2 in the indices 1,2,...,j such that $\omega_i^\sigma = -\omega_{\tilde{\sigma}(i)}$.

DEFINITION. A dominant weight is special if it is of the form $\Sigma n_i \omega_i$ with $n_i = n_{\tilde{\sigma}(i)}$. A special weight is regular if $n_i \neq 0$ for all i.

Thus we have that a weight λ is special iff $\lambda^\sigma = -\lambda$.

1.5.

LEMMA. Let λ be a dominant weight and let V_λ the corresponding irreducible representation of G with highest weight λ. Then if V_λ^H denotes the subspace of V_λ of H-invariant vectors dim $V_\lambda^H \leq 1$ and if $V_\lambda^H \neq 0$ λ is a special weight.

PROOF. Recall that $BH \subset G$ is dense in G so that H has a dense orbit in G/B. Also $V_\lambda \underset{G}{\cong} H^0(G/B,L)$ for a suitable line bundle L on G/B. So if $s_1, s_2 \in V_\lambda^H - \{0\}$, we have that $\frac{s_1}{s_2}$ is a meromorphic function on G/B constant on the dense H orbit, hence s_1 is a multiple of s_2 and our first claim follows.
Now assume $V_\lambda^H \neq 0$ and let $h \in V_\lambda^H - \{0\}$. Fix an highest weight vector $v_\lambda \in V_\lambda$ and let $U \subset V_\lambda$ be the unique T-stable complement to v_λ. Clearly U is B^- stable and $B^- H \subset G$ is dense in G. Then assume $h \in U$ but an the other hand $B^- Hh$ spans V_λ a contradiction. Hence

$$h = av_\lambda + u \quad , \quad a \in \mathbf{C} - \{0\},\ u \in U$$

Since $T_o \subset H$ and h is H invariant this implies $\lambda | T_o = id$ hence λ is special.

1.6. If λ is any integral dominant weight and V_λ the corresponding irreducible representation of G with highest weight λ, we define V_λ^σ to be the space V_λ with G action twisted by σ (i.e. we set $g \circ v$ in V_λ^σ to be $\sigma(g)v$, in V_λ).

LEMMA. If λ is a special weight then V_λ^σ is isomorphic to V_λ^*.

PROOF. V_λ^* can be characterized as the irreducible representation of G having $-\lambda$ as lowest weight. Now let $v_\lambda \in V_\lambda$ be a vector of weight λ, let P be the parabolic subgroup of G fixing the line through v_λ. P is generated by the Borel subgroup B and the root subgroups relative to the negative roots $-\alpha$ for which $\langle \alpha,\lambda \rangle = 0$. Thus the parabolic subgroup P^σ, transformed of P via σ, contains the root subgroups relative to the roots $\pm\beta_i$ and also to the roots α^σ, $\alpha \in \phi_1^+$. Now $\sigma(\phi_1^+) = \phi_1^-$ hence P^σ contains the opposite Borel subgroup B^-. Clearly $v_\lambda \in V_\lambda^\sigma$ is stabilized by P^σ hence v_λ is a minimal weight vector and its weight is $-\lambda$. This proves the claim.

1.7. We have just seen that, if λ is an integral dominant special weight V_λ is isomorphic, in a σ-linear way, to V_λ^*. Under this isomorphism the highest weight vector v_λ is mapped into a lowest weight vector in V_λ^*. We normalize the mapping as follows: In V_λ the line $\mathbb{C}v_\lambda$ has a unique T-stable complement $\bar{V}_\lambda$ we define $v^\lambda \in V_\lambda^*$ by: $\langle v^\lambda, v_\lambda \rangle = 1$, $\langle v^\lambda, \bar{V}_\lambda \rangle = 0$. v^λ is easily seen to be a lowest weight vector in V_λ^*.We thus define $h: V_\lambda^* \to V_\lambda$ to be the (unique) σ-linear isomorphism such that $h(v^\lambda) = v_\lambda$.

REMARK. If $V = \oplus V_{\lambda_i}$ is a G-module, the action of G on $\mathbb{P}(V)$ factors through $\bar{G}$ if and only if the center of G acts on each V_{λ_i} with the same character. This applies in particular when V is a tensor product of irreducible G-modules.

We now analyze the stabilizer in G, $\tilde{H}$; of the line generated by h.

LEMMA. i) $\tilde{H}$ equals the normalizer of H.

ii) We have an exact sequence $H \hookrightarrow \tilde{H} \twoheadrightarrow C$, where C is the subgroup of the center of G formed by the elements expressible as $g\sigma(g^{-1})$ for some $g \in G$.

iii) The stabilizer of the line generated by h in $\bar{G}$ is the subgroup fixed by the order two automorphism induced by σ on $\bar{G}$.

PROOF. Assume ${}^g h = \alpha h$, α a scalar. Since h is σ linear, ${}^g h = ghg^{-1} = g\sigma(g^{-1})h$. Therefore $g\sigma(g^{-1})$ acts on V_λ as a scalar. Since V_λ is irreducible this implies $g\sigma(g^{-1})$ lies in the center of G. Conversely if $g\sigma(g^{-1})$ lies in the center of G, $g \in \tilde{H}$. We claim $g \in N(H)$. In fact putting $\zeta = g\sigma(g^{-1})$ we get for each $u \in H$

$$\sigma(g^{-1}ug) = \sigma(g^{-1})u\sigma(g) = \sigma(g^{-1})\zeta^{-1}u\zeta\sigma(g) = g^{-1}ug.$$

Now assume $g \in N(H)$. To see that $g \in \tilde{H}$ it is sufficient to show that $g\sigma(g^{-1})$ lies in the center of G or equivalently that it acts trivially on $\underline{g}$ = Lie G via the adjoint representation. Decompose $\underline{g} = \underline{h} \oplus \underline{g}_1$. And

consider the subgroup K in Aut($\underline{g}$) generated by adN(H) and σ. Since adN(H) is reductive and has at most index 2 in K(N(H) is clearly σ stable) also K is reductive. We claim that both $\underline{h}$ and $\underline{g}_1$ are K stable. In fact $\underline{h}$ is clearly K stable and the reductivity of K implies that it has a K-stable complement in $\underline{g}$, but the unique σ stable complement of $\underline{h}$ is $\underline{g}_1$ so $\underline{g}_1$ is also K stable.
Now notice that since $g \in N(H)$, for each $u \in H$

$$g^{-1}ug = \sigma(g^{-1})u\sigma(g)$$

so that $g\sigma(g^{-1})$ commutes with H and acts trivially on $\underline{h}$. On the other-hand, if $x \in \underline{g}_1$, we have $\mathrm{ad}g^{-1}(x) \in \underline{g}_1$, since $\underline{g}_1$ is K stable, so

$$-\mathrm{ad}g^{-1}(x) = \sigma(\mathrm{ad}g^{-1}(x)) = -\mathrm{ad}\sigma(g^{-1})(x)$$

and hence $\mathrm{ad}g\sigma(g^{-1})(x) = x$ so $g\sigma(g^{-1})$ acts trivially also on $\underline{g}_1$, and so on $\underline{g}$. This proves i).
ii) is clear from the above.
To see iii) notice that the subgroup fixing the line generated by h in $\bar{G}$ is the image in $\bar{G}$ of $\tilde{H}$. Hence if we denote by σ' the automorphism induced by σ on $\bar{G}$ it consists of the elements such that $g\sigma'(g^{-1}) = id$ which are the elements fixed by σ'.

REMARKS. a) H has finite index in $\tilde{H}$..
b) $\tilde{H}$ is the largest subgroup of G with Lie$\tilde{H}$ = $\underline{h}$.

PROOF. a) follows from part ii) of the previous lemma and b) from the fact that H is connected (cf. [28]).

We complete v_λ to a basis $\{v_\lambda, v_1, v_2, \ldots, v_m\}$ of weight vectors and consider the dual basis $\{v^\lambda, v^1, v^2, \ldots, v^m\}$ in V_λ^*. We have $h(v^\lambda) = v_\lambda$ and, if χ_i is the weight of v_i we have $-\chi_i$ as weight of v^i and so $-\chi_i^\sigma$ as weight of $w_i = h(v^i)$. If we identify $\hom(V_\lambda^*, V_\lambda)$ with $V_\lambda \otimes V_\lambda$ we see that h is identified with the tensor

$$h = v_\lambda \otimes v_\lambda + \sum_{i=1}^{m} w_i \otimes v_i.$$

$v_\lambda \otimes v_\lambda$ has weight 2λ while $w_i \otimes v_i$ has weight $\chi_i - \chi_i^\sigma$.

The fact that h is σ-linear implies in particular that it is an H isomorphism. This in turn means that $\bar{h}$ is fixed under H.
Recall that $v_\lambda \otimes v_\lambda$ generates in $V_\lambda \otimes V_\lambda$ the irreducible module $V_{2\lambda}$.
Now order $\alpha_1, \ldots, \alpha_j$ so that $\alpha_s - \alpha_s^\sigma$ are mutually distinct for $s \leq \ell$ (and of course by 1.4 if $j > \ell$, for each $i > \ell$ there is an index $s \leq \ell$ such that $\alpha_s - \alpha_s^\sigma = \alpha_i - \alpha_i^\sigma$). Call $\bar{\alpha}_s = \frac{1}{2}(\alpha_s - \alpha_s^\sigma)$ $s \leq \ell$ the restricted simple roots.

PROPOSITION. i) If λ is a special weight then $V_{2\lambda}$ contains a non zero element h' fixed under H.
ii) h' is unique up to scalar multiples and can be normalized to be

$$h' = v_{2\lambda} + \sum z_i$$

with $v_{2\lambda}$ a highest weight vector of $V_{2\lambda}$ and the z_i's weight vectors having distinct weights whose weight is of the form $2(\lambda - \sum_{s=1}^{\ell} n_s\bar{\alpha}_s)$, n_i non negative integers.
iii) if λ is a regular special weight then we can assume that the vectors $z_1,\dots,z_\ell$ have weight $2(\lambda-\bar{\alpha}_1),\dots,2(\lambda-\bar{\alpha}_\ell)$.

PROOF. If we put h' equal to the image of $\bar{h}$ under the unique G-equivariant projection $V_\lambda \otimes V_\lambda \to V_{2\lambda}$, i) ii) follow from the expression of $\bar{h}$ as a linear combination of weight vectors given above. To see iii) assume λ (and hence 2λ) is a regular special weight. Since h' is fixed under H, $xh' = 0$ for any $x \in \underline{h} = \text{LieH}$. In particular if we let $\bar{\alpha}_s$ be a simple restricted root and $\alpha_s \in \Gamma_1$ be such that $\bar{\alpha}_s = \frac{1}{2}(\alpha_s - \alpha_s^\sigma)$ we have (cf. 1.3)

$$(x_{-\alpha_s} + \sigma(x_{-\alpha_s}))h' = 0, \quad x_{-\alpha_s} \in g_{-\alpha_s}.$$

But

$$(x_{-\alpha_s} + \sigma(x_{-\alpha_s}))v_{2\lambda} = x_{-\alpha_s}v_{2\lambda}$$

since $\sigma(x_{-\alpha_s}) \in g_{-\alpha_s^\sigma}$ and $-\alpha_s^\sigma \in \Phi_1^+$. Also by the regularity of 2λ $x_{-\alpha_s}v_\lambda$ is a non zero weight vector of weight $2\lambda - \alpha_s$. It follows that for some z_i, $\sigma(x_{-\alpha_s})z_i = -x_{-\alpha_s}v_{2\lambda}$ so that z_i has weight $2(\lambda-\bar{\alpha}_s)$ proving the claim.

The analysis just performed does not exclude that V_λ itself may contain a non zero H-fixed vector h_λ. In this case we have seen that we can normalize h_λ : $h_\lambda = v_\lambda + \sum u_i^1$, u_i^1 lower weight vectors. It follows that $h_\lambda \otimes h_\lambda$ must project to h in $V_{2\lambda}$ (by uniqueness of h).

Now the dominant λ's for which $\dim V_\lambda^H = 1$ have been determined completely [9],[24], the result is as follows: Let us indicate Λ^1 such set.

Consider the Killing form restricted to $\underline{t}_1$ and thus to $\underline{t}_1^*$. We look at the restriction of Φ_1 to $\underline{t}_1$, if $\alpha \in \Phi$, let us indicate $\bar{\alpha}$ the restriction of α to $\underline{t}_1$.

If $\mu \in \underline{t}_1^*$ let us indicate by $\tilde{\mu}$ its extension to $\underline{t}$ by setting it 0 to $\underline{t}_0$.
Then the theorem in [9] is:
Consider the set of $\mu \in \underline{t}_1^*$ such that

$\frac{(\mu,\bar{\alpha})}{(\bar{\alpha},\bar{\alpha})}$ is a positive integer for all $\alpha \in \phi$

Then the set of weights $\tilde{\mu}$ of $\underline{t}$ so obtained is exactly the set Λ^1 of λ for which $\dim V_\lambda^H = 1$. One can understand this theorem in a more precise way. If $\alpha \in \phi$, then $\tilde{\bar{\alpha}}$ is exactly $\frac{1}{2}(\alpha - \alpha^\sigma)$, and $(\bar{\alpha};\bar{\alpha}) = (\tilde{\bar{\alpha}},\tilde{\bar{\alpha}})$. Now also a weight ω is of the form $\tilde{\mu}$ if and only if $\omega = \frac{1}{2}(\omega - \omega^\sigma)$. For such weights of course $(\omega,\beta_j) = 0$. Thus we see immediately that Λ^1 is contained in the positive lattice generated by the weights ω_i if $\sigma(i) = i$ and $\omega_i - \omega_{\tilde{\sigma}(i)}$ if $\tilde{\sigma}(i) \neq i$.

To understand exactly the nature of Λ^1 we must see if

$$\frac{(\omega_i,\bar{\alpha})}{(\bar{\alpha},\bar{\alpha})} \qquad (\text{resp. } \frac{(\omega_i - \omega_{\tilde{\sigma}(i)},\bar{\alpha})}{(\bar{\alpha},\bar{\alpha})})$$

is an integer.

Since in any case for such special weights λ we have $2\lambda \in \Lambda^1$ one knows at least that these numbers are half integers. It follows in any case that Λ^1 is the positive lattice generated by the previous weights or their doubles. i.e.

$$\Lambda^1 = \sum_{i=1}^{\ell} n_i\mu_i, \quad n_i \geq 0 \quad \text{and} \quad \mu_i = \omega_i \quad \text{or}$$

$2\,\omega_i$ (resp. $\omega_i - \omega_{\tilde{\sigma}(i)}$ or $2(\omega_i - \omega_{\tilde{\sigma}(i)})$). Recall that $\ell = \text{rk}\,\Lambda^1$ is also the rank of the symmetric space.

2. THE BASIC CONSTRUCTION

2.1. We consider now a regular special weight λ and all the objects of the previous paragraph V_λ, $h' \in V_{2\lambda}$. Let now $\mathbb{P}_{2\lambda} = \mathbb{P}(V_{2\lambda})$ be the projective space of lines in $V_{2\lambda}$ and $\tilde{h} \in \mathbb{P}_{2\lambda}$ be the class of h'. The basic object of our nalysis is the orbit $G \cdot \tilde{h}$ of $\tilde{h}$ in $\mathbb{P}_{2\lambda}$ and its closure $\underline{\bar{X}} = \overline{G \cdot \tilde{h}}$. By construction $\underline{\bar{X}}$ is a G-equivariant compactification of the homogeneous space $G \cdot \tilde{h}$, furthermore the stabilizer $\tilde{H}$ of $\tilde{h}$ is a group containing the subgroup H.

We will analyze in detail $\tilde{H}$ and in particular will see that H has finite index in $\tilde{H}$. For the moment we concentrate our attention to $\underline{\bar{X}}$. Since $\underline{\bar{X}}$ is closed in $\mathbb{P}_{2\lambda}$ and G stable it contains the unique closed orbit of G acting on $\mathbb{P}_{2\lambda}$, i.e. the orbit of the highest weight vector $v_\lambda \otimes v_\lambda$. Now the following general lemma is of trivial verification:

LEMMA: If X is a G variety with a unique closed orbit Y and V is an open set in X with $Y \cap V \neq \phi$ then $X = \bigcup_{g \in G} gV$.

The use of this lemma for us is in the fact that it allows us to study the singularities of X locally in V.

2.2. Let λ be a regular special weight. Consider a G module $W \simeq V_{2\lambda} \oplus \sum V_{\mu_i}$ with $\mu_i = 2\lambda - \sum n_i 2\bar{\alpha}_i$ some $n_i > 0$. Let $h \in \underset{\sim}{V}$ be an H invariant with component h' in $V_{2\lambda}$. Decompose $V_{2\lambda} = \mathbf{C}v_{2\lambda} \otimes \tilde{V}_{2\lambda}$ in a T stable way and consider the open affine set $A = v_{2\lambda} \otimes \tilde{V}_{2\lambda} \otimes \sum V_{\mu_i} \subseteq \mathbb{P}(W)$. Notice that $h \in A$ and A is B^- stable.

LEMMA: The closure in A of the T^1 orbit T^1h is isomorphic to ℓ dimensional affine space $\mathbb{A}^\ell$. The natural morphism $T^1 \to T^1h \hookrightarrow \mathbb{A}^\ell$ has coordinates $t \to (t^{-2\bar{\alpha}_1}, t^{-2\bar{\alpha}_2}, \ldots, t^{-2\bar{\alpha}_\ell})$. T^1h is identified with the open set of $\mathbb{A}^\ell$ where all coordinates are non zero.

PROOF: By prop. 1.7 we can write $h = v_{2\lambda} + \sum_i z_i'$ with z_i' weight vectors of weights $\chi_i = 2\lambda - \sum m_j^{(i)} 2\bar{\alpha}_j$ (some $m_j > 0$) and $z_1', \ldots, z_\ell'$ of weights $2\lambda - 2\bar{\alpha}_1, \ldots, 2\lambda - 2\bar{\alpha}_\ell$. Let us apply an element $t \in T'$ to h we get $th = t^{2\lambda} v_{2\lambda} + \sum t^{\chi_i} z_i'$ which, in affine coordinates, is

$$v_{2\lambda} + \sum t^{\chi_i - 2\lambda} z_i'.$$

From the previous formula $\chi_i - 2\lambda = \sum_j m_j^{(i)} (-2\bar{\alpha}_j)$, this means that the coordinates of th are monomials in the first ℓ coordinates.

This means that T^1 maps to a closed subvariety of A, isomorphic to affine space $\mathbb{A}^\ell$, via the coordinates $(t^{-2\bar{\alpha}_1}, \ldots, t^{-2\bar{\alpha}_\ell})$. Since the restricted simple roots are linearly independent the orbit T^1h is the open dense subset of $\mathbb{A}^\ell$ consisting of the elements with non zero coordinates.

REMARK. The stabilizer of h in T^1 is the finite subgroup of the elements $t \in T^1$ with $t^{2\bar{\alpha}_i} = 1$.

2.3. Let us go back to $\underline{\bar{X}} \subseteq P_{2\lambda}$. Consider the open affine set $A = v_{2\lambda} \oplus \tilde{V}_{2\lambda} \subseteq P_{2\lambda}$ and set $V = A \cap \underline{\bar{X}}$. Remark that V is B^- stable, it contains $\tilde{h}$ and so also $\mathbb{A}^\ell$, the closure of $T^1\tilde{h}$ in A, hence $v_{2\lambda} \in V$ and therefore V has a non empty intersection with the unique closed orbit or G in $P_{2\lambda}$.

Let U be the unipotent group generated by the root subgroups X_α, $\alpha \in \phi_1^-$. Since U acts on V we have a well defined map $\phi\colon U \times \mathbb{A}^\ell \to V$ by the formula $\phi(u,x) = u \cdot x$.

PROPOSITION: $\phi : U \times \mathbf{A}^{\ell} \to V$ is an isomorphism.

PROOF. We first will construct a map $\psi: V \to U$ such that $\psi\phi(u,x) = u$, and prove that Im ϕ is dense in V. From this the claim follows; in fact consider the map $\zeta: V \to V$ given by $\zeta(v) = \psi(v)^{-1}v$, clearly $\zeta\phi(u,x) = x$ hence ζ maps V in $\mathbf{A}^{\ell}$ and setting $\phi': V \to U \times \mathbf{A}^{\ell}$ by $\phi'(v) = (\psi(v),\zeta(v))$ we have $\phi' \cdot \phi = 1_{U\times\mathbf{A}^{\ell}}$. Since $\phi(U \times \mathbf{A}^{\ell})$ is dense in V and $\phi\cdot\phi'$ is the identity we also have $\phi\cdot\phi' = 1_V$.

2.4. From now on we make the necessary steps for the construction of ψ.

Since 2λ is special we have, by our considerations of 1.6, that $V_{2\lambda}$ is isomorphic to $V_{2\lambda}^*$ in a σ-linear way. This isomorphism defines a non degenerate bilinear form $\langle\, ,\rangle$ on $V_{2\lambda}$ which is symmetric and satisfies the following properties:

$$\langle gu,v\rangle = \langle u,\sigma(g^{-1})v\rangle \quad \text{for each } g \in G,\ u,v \in V_{2\lambda}$$

$$\langle xu,v\rangle = -\langle u,\sigma(x)v\rangle \quad \text{for each } x \in \underline{g},\ u,v \in V_{2\lambda}$$

Remark that the tangent space τ in $v_{2\lambda}$ to the orbit $U\cdot v_{2\lambda}$ has as basis the elements $x_{-\alpha}v_{2\lambda}$, $\alpha \in \phi_1^+$ (since the opposite unipotent group of U is the unipotent radical of the parabolic subgroup P stabilizing the line through $v_{2\lambda}$). Let τ^o be the subspace generated by τ and $v_{2\lambda}$.

LEMMA: i) The form $\langle\, ,\rangle$ restricted to τ^o is non degenerate.
ii) τ^o is stable under P.
iii) The orthogonal $\tau^{o\perp}$ (relative to the given form) is stable under $\sigma(P)$.

PROOF: i) First of all remark that if $v_1,v_2 \in V_{2\lambda}$ are weight vectors of weights χ_1,χ_2 respectively and $\langle v_1,v_2\rangle \neq 0$ we have, for $t \in T$,
$t^{\chi_1}\langle v_1,v_2\rangle = \langle tv_1,v_2\rangle = \langle v_1,\sigma(t^{-1})v_2\rangle = t^{-\chi_2^{\sigma}}\langle v_1,v_2\rangle$ and so $\chi_1 = -\chi_2^{\sigma}$.
This implies that $v_{2\lambda}$ is orthogonal to $\tilde{V}_{2\lambda}$ and $\langle v_{2\lambda},v_{2\lambda}\rangle \neq 0$.
It remains to verify that on τ the form is non degenerate. Using our previous remark $\langle x_{-\alpha}v_{2\lambda},x_{-\beta}v_{2\lambda}\rangle = 0$ unless $\beta = -\alpha^{\sigma}$. In this case $\langle x_{-\beta^{\sigma}}v_{2\lambda},x_{-\beta}v_{2\lambda}\rangle = -c\langle v_{2\lambda},x_{\beta}x_{-\beta}v_{2\lambda}\rangle$ $(c \neq 0)$ and $\langle v_{2\lambda},x_{\beta}x_{-\beta}v_{2\lambda}\rangle =$
$= \langle v_{2\lambda},[x_{\beta},x_{-\beta}]v_{2\lambda}\rangle$ since $x_{\beta}v_{2\lambda} = 0$ this is $(2\lambda,\beta)\,\langle v_{2\lambda},v_{2\lambda}\rangle \neq 0$.
Since the map $\alpha \to -\alpha^{\sigma}$ is an involution of ϕ_1^+ the first claim follows.
ii) It is sufficient to show that τ^o is stable under the action of the Lie algebra of P. Since τ^o is stable under the torus T it is enough to show the stability of τ^o with respect to the elements x_{α} with $\alpha \in \phi_o \cup \phi_1^+$. Now $x_{\alpha}x_{-\beta}v_{2\lambda} = [x_{\alpha},x_{-\beta}]v_{2\lambda} + x_{-\beta}x_{\alpha}v_{2\lambda}$, if $\alpha \in \phi_o \cup \phi_1^+$ we have $x_{\alpha}v_{2\lambda} = 0$.
iii) This is clear from the properties of the form.

2.5.

LEMMA. $\mathbb{A}^{\ell} \subseteq v_{2\lambda} + \tau^{o\perp}$.

PROOF. We must show that, if $h' = v_{2\lambda} + \sum_i z_i$, each $z_i \in \tau^{o\perp}$. The weight of z_i is $\chi_i = 2\lambda - \sum n_j^{(i)} 2\bar{\alpha}_j$ so the only case to verify is when $-\sum n_j^{(i)} 2\bar{\alpha}_j = -\beta$ for some $\beta \in \phi_1^+$. Suppose this happens for z_{i_o}, since h' is H stable we have $(x_\beta + \sigma(x_\beta))h' = 0$; but $(x_\beta + \sigma(x_\beta))h' = x_\beta z_{i_o} +$ terms of weight different from 2λ, thus $x_\beta z_{i_o} = 0$. By the same weight considerations the only possible non zero scalar product between z_{i_o} and the elements of the basis of τ^o is the one with $x_{-\beta} v_{2\lambda}$, for this we have $\langle x_{-\beta} v_{2\lambda}, z_{i_o} \rangle = -\langle v_{2\lambda}, \sigma(x_{-\beta}) z_{i_o} \rangle = 0$, ($\sigma(x_{-\beta}) = c x_\beta$ some c).

2.6. Now we consider the projection π of $V_{2\lambda}$ onto $V_{2\lambda}/\tau^{o\perp}$, since $U \subset \sigma(P)$ we have a U action on $V_{2\lambda}/\tau^{o\perp}$ and the projection is equivariant. Let $K = \pi(v_{2\lambda} + \tilde{V}_{2\lambda})$, K is an affine hyperplane in $V_{2\lambda}/\tau^{o\perp}$ and it is U stable.

LEMMA. The map $j: U \to K$ defined by $j(u) = \pi(uv_{2\lambda})$ is a U equivariant isomorphism.

PROOF. From 2.4 we know that τ is the tangent space of $Uv_{2\lambda}$ in $v_{2\lambda}$. This implies that j is smooth at 1. Since j is U equivariant it is everywhere smooth. Now U has no finite subgroups and $\dim U = \dim K$ so j is an open immersion. It is a well known fact that an open immersion j of affine space $\mathbb{A}^n$ into another affine space $\bar{\mathbb{A}}^n$ of the same dimension is necessarily an isomorphism, we recall the proof: It the complement of $j(A^n)$ is non emty it is a divisor which has an equation f, this is a unit a A^n and hence a constant, giving a contradiction.
We can now construct ψ as required in 2.3, setting $\psi(v) = j^{-1}(\pi(v))$ for any $v \in V$, the fact that $\psi\phi(u,x) = u$ follows from the U equivariance of π and j and lemma 2.5.

2.7.

LEMMA. The image of ϕ is dense in V.

PROOF. The tangent space to $\mathbb{A}^{\ell}$ in $v_{2\lambda}$ is orthogonal to τ (cf. 2.5). This implies that the differential of ϕ in the point (1,0) is injective and so $\dim(\overline{\mathrm{Im}\phi}) = \dim(U \times \mathbb{A}^{\ell})$; now $\dim V = \dim \underline{\bar{X}} \leq \dim G/H = \dim(U \times \mathbb{A}^{\ell})$. Since V is irreducible we get that $V = \overline{\mathrm{Im}\phi}$.

PROPOSITION. The stabilizer of $\tilde{h}$ is $\tilde{H}$.

PROOF. We have shown in the previous lemma that $\dim X = \dim G/H$ hence the subgroup H has finite index in the stabilizer of $\tilde{h}$. From 1.7 the proposition follows.

2.8. Using proposition 2.3 we identify V with the affine space $U \times \mathbb{A}^\ell$.

PROPOSITION. The intersection between the orbit $G\tilde{h}$ and $U \times \mathbb{A}^\ell$ is the open set where the last ℓ coordinates are non zero.

PROOF. We go back to $h \in \hom(V_\lambda^*, V_\lambda) \simeq V_\lambda \otimes V_\lambda$ (cf. 1.7) and proceed as in 2.1, 2.2. Let $h^\#$ be the class of h in $\mathbb{P}(\hom(V_\lambda^*, V_\lambda)) = \mathbb{P}(V_\lambda \otimes V_\lambda)$ and $\bar{X}^\# = \overline{G \cdot h^\#}$. Setting $V_\lambda \otimes V_\lambda = V_{2\lambda} \oplus Z$, the decomposition in G submodules, we consider the affine space $A^\# = v_{2\lambda} + \tilde{V}_2 \oplus Z$ and the G equivariant projection $\rho\colon \mathbb{P}(V_\lambda \otimes V_\lambda) \to \mathbb{P}(V_{2\lambda})$ from $\mathbb{P}(Z)$, ρ is defined in the open set $\mathbb{P}(V_\lambda \otimes V_\lambda) - \mathbb{P}(Z)$, hence in particular in $V^\# = \bar{X}^\# \cap A^\#$.

From the analysis of 2.2 the closure in $A^\#$ of the orbit $T^1h^\#$ projects under ρ isomorphically onto $\mathbb{A}^\ell$ hence the isomorphism $\phi\colon U \times \mathbb{A}^\ell \to V$ factors through $\phi\colon U \times \mathbb{A}^\ell \xrightarrow{\phi^\#} V^\# \xrightarrow{\rho} V$. We know that $\dim V^\# = \dim X^\# = \dim G/\tilde{H}$ (cf. 1.7) so $\operatorname{Im}\phi^\#$ is dense in $V^\#$ and as in 2.3 this implies that $\phi^\#$ is an isomorphism. We now have that the union of the translates of $V^\#$ under G is an open dense subset in $X^\#$ isomorphic, under ρ, to $\bar{X}$; since $\bar{X}$ is complete this open set must be $\bar{X}^\#$. We can now prove the proposition working with $V^\#$, $\bar{X}^\#$ and $Gh^\#$. The points in $U \times \mathbb{A}^\ell$ where the last ℓ coordinates are non zero are in the B^- orbit of $h^\#$ hence in $Gh^\#$, we show now that the remaining points cannot be in $Gh^\#$. In order to do this we interpret such points as maps from V_λ^* to V_λ and show that an element of $\mathbb{A}^\ell$ with a zero coordinate is not of maximal rank, this is clear from the analysis of 1.7. Since every point in $V^\#$ is in the U orbit of a point in $\mathbb{A}^\ell$ the proposition follows.

3. THE MINIMAL COMPACTIFICATION

3.1. We can now completely describe the structure of the variety $\bar{X}$.

THEOREM.

i) $\bar{X}$ is smooth.

ii) $\bar{X} - G \cdot \tilde{h}$ is a union of ℓ smooth hypersurfaces S_i which cross transversely.

iii) The G orbits of $\bar{X}$ correspond to the subsets of the indeces $1,2,\dots,\ell$ so that the orbit closures are the intersections $S_{i_1} \cap S_{i_2} \cap \dots \cap S_{i_k}$.

iv) The unique closed orbity $Y \simeq G/P$ is $\bigcap_{i=1}^{\ell} S_i$.

PROOF. We have seen that the complement of $G \cdot \tilde{h} \cap V$ in V is the union of ℓ hypersurfaces which are in fact coordinate hyperplanes, since $V \simeq U \times \mathbb{A}^\ell$ and the ℓ hypersurfaces Σ_i are given by the equations $x_i = 0$ for the last ℓ coordinates. Furthermore, the description of the torus action of T_1 on $\mathbb{A}^\ell$ shows that, two points in V are in the same $U \times T_1$

orbit if and only if they lie in the same set of hyperplanes $\sum_i$. Now we claim that the hypersurfaces S_i are just the closure of the $\sum_i$ in $\bar{X}$. In fact, let S_i be any irreducible component of $S - G\cdot\tilde{h}$, necessarily S_i is G stable, since G is connected. Hence, $S_i \underset{\neq}{\supset} Y$ (the unique closed orbit) and $S_i \cap V$ is thus a component of $V - G\cdot h$. Hence, $S_i \cap V = \sum_i$ (up to reordering the indeces). Hence, $S_i = \bar{\sum}_i$ and conversely by the same argument, $\bar{\sum}_i$ is an irreducible component of $X - G\cdot\tilde{h}$, hence, it is G-stable.

To finish it is only necessary to remark that, since any point is G-conjugate to a point in V, the statement that two points in $\bar{X}$ are in the same orbit if and only if they are contained in the same S_i's follows from the similar statement relative to $U \times T_1$ in V.

3.2. Summarizing, we have found ℓ hypersurfaces S_i which are smooth. The orbits are just

$$O_{i_1,\ldots,i_k} = S_{i_1} \cap \ldots \cap S_{i_k} - \bigcup_{i\neq i_1,\ldots,i_k} S_{i_1} \cap \ldots \cap S_{i_k} \cap S_i$$

and $\bar{O}_{i_1,\ldots,i_k} = S_{i_1} \cap \ldots \cap S_{i_k}$ is smooth.

These are the only irreducible, closed G-stable subsets of $\bar{X}$. Their inclusion relations are, therefore, opposite to those of the faces of the simplex on the indeces $1,2,\ldots,\ell$. The statement iv) is then clear.

3.3. We have just seen that, given a regular special weight λ we can describe the structure of the variety $\bar{X} = \overline{G\,\tilde{h}} \subset \mathbb{P}(V_{2\lambda})$. Assume now that V_λ itself contains a non zero H-invariant line generated by h' and consider $\bar{X}' = \overline{G\cdot\tilde{h}'} \subset \mathbb{P}(V_\lambda)$.

PROPOSITION. There is a natural G-isomorphism $\psi: \bar{X}' \to \bar{X}$.

PROOF. Let us consider the map $\phi: V_\lambda \to V_{2\lambda}$ which is the composition of the map $f: V_\lambda \to V_\lambda \otimes V_\lambda$ defined by $f(v) = v \otimes v$ and of the G-equivariant projection $\pi: V_\lambda \otimes V_\lambda \to V_{2\lambda}$. Clearly ϕ is G-equivariant and we can normalize h' so that $\phi(h') = \bar{h}$. If we identify V_λ (resp. $V_{2\lambda}$) with $H^0(G/B,L_\lambda)$ (resp. $H^0(G/B,L_{2\lambda})$ (where L_μ is the line bundle relative to the dominant weight μ), we see that ϕ is the map taking a section into its square. Since G/H is irreducible, we then have that ϕ induces an embedding $\bar{\phi}: \mathbb{P}(V_\lambda) \to \mathbb{P}(V_{2\lambda})$ which is G-equivariant (and an isomorphism of $\mathbb{P}(V_\lambda)$ onto its image). Clearly $\bar{X}$ is contained in the image of $\bar{\phi}$ and is the image of $\bar{X}'$. Thus $\bar{\phi}$ induces the required isomorphism ψ.

3.4. We should remark that in the special case of a group G, considered as symmetric variety over G × G, one can more simply describe the construction ad follows. If λ is a regular dominant weight of G and V_λ the corresponding irreducible representation, we consider End $(V_\lambda) = V_\lambda \otimes V_\lambda^*$ as G × G module. G is then thought as the orbit of the identity $1 \in$ End (V_λ) and the compactification $X = \overline{G \cdot \bar{1}}$ can thus be thought as "degenerate" projective transformation of the flag variety. We will refer to this case as the "compactification of $\bar{G}$".

4. INDEPENDENCE ON λ

4.1. A priori the construction performed in §2 depends on the regular weight λ, we want to show now a different construction of $\underline{\bar{X}}$ which shows its independence on λ. Consider again the permutation $\tilde{\sigma}$ considered in 1.3. Each orbit of $\tilde{\sigma}$ consists of either one or two indices. Indexing the orbit by the indeces $\{1,\dots,\ell\}$, for each such index j we let λ be the sum of the fundamental weights (one or two) in the corresponding orbit. Thus a special weight is just a positive integral combination $\sum n_j \cdot \lambda_j$ while a regular one has the condition $n_j \neq 0$ for all j.

For each j we have $V_{\lambda_j} \simeq V_{\lambda_j}^*$ and a corresponding element $h_j \in V_{2\lambda_j}$. Consider then $\bar{h}_j \in \mathbb{P}(V_{2\lambda_j})$ and $\bar{h}' = (h_1,\dots,h_\ell) \in \Pi\, \mathbb{P}(V_{2\lambda_j})$. We claim that $\underline{\bar{X}}$ is isomorphic to $\overline{G \cdot \bar{h}'} \subseteq \Pi\, \mathbb{P}(V_{2\lambda_j})$. In fact, consider $\lambda = \sum n_j \lambda_j$ and $\bigotimes_j V_{\lambda_j}^{\otimes n_j} = \mathbb{Q}$. Clearly $\mathbb{Q} = V_\lambda \oplus \mathbb{Q}'$ with $\mathbb{Q}'$ a sum of represen̲tations with lower highest weights. The element

$$\bigotimes_j h_j^{\otimes n_j} : \bigotimes_j V_{\lambda_j}^{\otimes n_j} \to \bigotimes_j V_{\lambda_j}^{*\otimes n_j}$$

and in particular it maps V_λ in V_λ^* and by the uniqueness of h it coincides with h on V_λ. Now we have clearly a mapping $\pi\, \mathbb{P}(V_{2\lambda_j}) \to \mathbb{P}(\otimes V_{2\lambda_j}^{\otimes n_j})$ sending $\bar{h}'$ to $\otimes h_j^{\otimes n_j}$ and so $\overline{G \cdot \bar{h}'}$ is identical to the closure of the orbit of $\otimes h_j^{\otimes n_j}$. Let $\underline{\bar{X}}'$ be $G \cdot \otimes h_j^{\otimes n_j} \subseteq \mathbb{P}(\bigotimes_j V_{2\lambda_j}^{\otimes n_j})$. We wish to project $\underline{\bar{X}}'$ to X proving that they are isomorphic. In fact, we prove a more general statement which will be used later. Let us give a regular special weight λ and a representation W, with a line $\mathbb{C}h_W$ fixed under H, such that its T_1 weights are all of the form $\lambda - \sum n_i 2\bar{\alpha}_i$.

Suppose $h_\lambda \in V_\lambda$ is an H-invariant non zero vector and set $h = h_\lambda + h_W \in V_\lambda \oplus W$ and $\underline{\bar{X}}' = \overline{G\tilde{h}} \subseteq \mathbb{P}(V_\lambda \oplus W)$. If we project $\mathbb{P}(V_\lambda \oplus W)$ to $\mathbb{P}(V_\lambda)$ from $\mathbb{P}(W)$ we have

LEMMA. The projection is defined on $\underline{\bar{X}}'$ and establishes an isomorphism between $\underline{\bar{X}}'$ and $\underline{\bar{X}} = \overline{G\tilde{h}_\lambda}$.

PROOF. We can assume $W = \oplus V_i$, each V_i irreducible and containing a H fixed line $\mathbf{C}h_i$ so that the projection $\Pi_i: W \to V_i$ with kernel $\bigoplus_{j\neq i} V_i$ has the property $\Pi_i(h_W) = h_i$.

By reasoning as in 3.3 we can double all weights and assume $\lambda = 2\lambda'$ and V_i has weight $2\mu_i$. In this situation we can define in $\underline{\bar{X}}'$ the affine set V' as in 2.2 and carry out the same analysis verbatim due to the structure of the weights of h_W. Then we see that under the given map $\overset{\circ}{\underline{\bar{X}}}{}' = \bigcup_{g\in G} V'^g$ in $\underline{\bar{X}}'$ projects isomorphically onto $\underline{\bar{X}}$. Since $\underline{\bar{X}}$ is complete, it follows that $\overset{\circ}{\underline{\bar{X}}}{}'$ is also complete and hence $\overset{\circ}{\underline{\bar{X}}}{}' = \underline{\bar{X}}'$ as desired.

5. THE STABLE SUBVARIETIES

5.1. We have seen that in $\underline{\bar{X}}$ the only G stable subvarieties are of the form $W_{i_1,\ldots,i_k} = S_{i_1} \cap S_{i_2} \cap \ldots \cap S_{i_k}$ for a subset of the indices $1,2,\ldots,\ell$. We wish now to describe geometrically such a subvariety. Let us then consider the weights λ_j, $j = 1,2,\ldots,\ell$ defined in 4.1 and the two weights $\lambda_1 = \lambda_{i_1} + \lambda_{i_2} + \ldots + \lambda_{i_k}$ and $\lambda_2 = \lambda_{j_1} + \ldots + \lambda_{i_{\ell-k}}$ where $j_1,\ldots,j_{\ell-k}$ are the complement of $i_1,i_2,\ldots,i_k$ in $i,2,\ldots,\ell$. We can, as before, consider $\underline{\bar{X}}$ embedded in $\mathbb{P}(V_{2\lambda_1}) \times \mathbb{P}(V_{2\lambda_2}) \subseteq \mathbb{P}(V_{2\lambda_1} \otimes V_{2\lambda_2})$ and we can project $\underline{\bar{X}}$ to $\mathbb{P}(V_{2\lambda_1})$. Let us call Π_1 this projection which is clearly G equivariant and maps onto the closure of the orbit $\underline{\bar{X}}_1 = \overline{G \cdot \tilde{h}_{2\lambda_1}}$.

LEMMA. $\Pi_1(W_{i_1,\ldots,i_k})$ equals the unique closed orbit in $\underline{\bar{X}}_1$ (i.e. G/P_1, P_1 the parabolic, stabilizing the line through a highest weight vector in $V_{2\lambda_1}$).

PROOF: We may analyze the projection locally in V and in fact, since $V = U \cdot \mathbf{A}^\ell$, it is enough to study $\Pi_1(A^\ell \cap W_{i_1,\ldots,i_k}) = \Pi_1(A^\ell_{i_1,\ldots,i_k})$. We know that the intersection $\mathbf{A}^\ell \cap W_{i_1,\ldots,i_k}$ is that part $A_{i_1,\ldots,i_k}$ of $\mathbf{A}^\ell$ where the coordinates x_i (corresponding to $t^{-2\alpha_i}$) vanish, for $i = i_1,i_2,\ldots,i_k$. The weights of the representation $V_{2\lambda_1}$, different from the highest weight, are of all of the form $\psi = 2\lambda_1 - \sum n_i\alpha_i - \sum_i \beta_i$ where at least one of the coordinates n_i relative to the indices i, for which $(\alpha_i,\lambda_i) \neq 0$, is non negative.

If we consider the projection of the subspace $A^\ell = R_1 = \overline{T_1\tilde{h}_{2\lambda}}$, this can be analyzed as follows. We have the orbit $T_1 \cdot \bar{h}_{2\lambda_1}$ and its closure R_1' and R_1 maps to R_1'. In coordinates we know that the T_1 weights appearing in $L_{2\lambda_1}$ are of type $2\lambda_1 - \sum n_i 2\alpha_i$ and then the corresponding mapping expresses such coordinates as $\Pi\, x_i^{n_i}$, but we know that some $n_i > 0$ for one the indices $i = i_1, i_2, \ldots, i_k$. Thus we deduce that $\Pi_1(A^{\ell} \cap W_{i_1,\ldots,i_k})$ is just the point $\overline{v_{2\lambda_1} \otimes v_{2\lambda_1}}$. This proves the lemma.

5.2. We have thus established a G equivariant mapping

$$\Pi_1 : W_{i_1,\ldots,i_k} \to G \cdot \overline{v_{2\lambda_1} \otimes v_{2\lambda_1}}.$$

This last variety is of the form $G/P_{i_1,\ldots,i_k}$ for the parabolic fixing $\overline{v_{2\lambda_1} \otimes v_{2\lambda_1}}$.

Since the map is G equivariant, it is a fibration. We want to study a typical fiber. Let us study $\Pi_1^{-1}\overline{(v_{2\lambda_1} \otimes v_{2\lambda_1})} = \bar{X}_1$.

Since Π_1 is a smooth morphism $\bar{X}_1$ is smooth and is the closure of the fiber of Π_1 restricted to the open orbit in $W_{i_1,\ldots i_k}$; this is irreducible since P is connected. We start to study $\bar{X}_1$ locally always in the open set V. A point (γ, a) in $U_\Gamma \times A_{i_1,\ldots i_k}$ is in the fiber $\bar{X}_1$ if and only if $\gamma \cdot \overline{v_{2\lambda_1} \otimes v_{2\lambda_1}} = \overline{v_{2\lambda_1} \otimes v_{2\lambda_1}}$, i.e. if and only if $\gamma \in P_{i_1,\ldots,i_k}$. Now $U \cap P_{i_1,\ldots,i_k}$ is exactly the unipotent subgroup generate by the root subgroup of the roots $-\alpha_i$ where $\alpha_i \in \Gamma_1$ and also α_i is a root of the Levi subgroup of $P_{i_1,\ldots,i_k}$. The semisimple part of the Levi subgroup of $P_{i_1,\ldots i_k}$ is relative to the root system generated by the roots β_j and the roots α_k's for which $(\alpha_k, \lambda_1) = 0$. Clearly such a subgroup $L_{i_1,\ldots,i_k}$ is σ stable. Moreover, if we consider $A_{i_1,\ldots,i_k} \subseteq \mathbb{P}(V_{2\lambda_2})$, we can analyze it as follows:

$h_{2\lambda_2} = v_{2\lambda_2} \otimes v_{2\lambda_2} + \sum z_i'$ where z_i' has T_1 weight $2\lambda_2 - \sum m_j 2\alpha_j$. We can split $h_{2\lambda_2}$ as $h_{2\lambda_2} = h'_{2\lambda_2} + a'$ where a' is the sum of all terms of weight $2\lambda_2 - \sum m_j 2\alpha_j$ with $m_j \neq 0$ for some $j \in \{i_1, i_2, \ldots, i_k\}$. Consider any element $t \in T_1$ such that t commutes with the Levi subgroup $L_{i_1,\ldots,i_k}$. Consider $H_{i_1,\ldots,i_k} = L_{i_1,\ldots,i_k} \cap H$, we have if $g \in H_{i_1,\ldots,i_k}$, $t^{-1} \cdot gt = g$ and so $t\, h_{2\lambda_2} = g \cdot t \cdot h_{2\lambda_2}$. Hence,

$$h'_{2\lambda_2} + t \cdot a' = g \cdot h'_{2\lambda_2} + g \cdot t \cdot a'.$$

We deduce that $h'_{2\lambda_2} = g \cdot h'_{2\lambda_2}$ so $h'_{2\lambda_2}$ is $H_{i_1,\ldots,i_k}$ invariant. Moreover, we see that $A_{i_1,\ldots,i_k}$ can be considered as the closure of the action of the Torus $(T_1)_{i_1\ldots i_k}$ on $\tilde{h}'_{2\lambda_2}$.

Thus, we deduce that the fibre we are studing is in fact the closure of the orbit of the semisimple part of the Levi subgroup acting on $\tilde{h}'_{2\lambda_2}$. Since it is easily verified that $(T_1)_{i_1\ldots i_k}$ is a maximal anisotropic in $L_{i_1 \ldots i_k}$ and λ_2 restricted to $T \cap L_{i_1 \ldots i_k}$ is a regular special weight we can apply the general remarks and lemma 5.1, and see that X_1 is isomorphic to the minimal compactification of the corresponding symmetric algebraic variety $\bar{L}_{i_1 \ldots i_k}/\bar{H}_{i_1 \ldots i_k}$.

Thus we have proved:

THOREM. Let $\{i_1, \ldots, i_k\}$ be a subset of the indices $\{1,2,\ldots,\ell\}$ and let $S_{i_1,\ldots,i_k}$ be the corresponding stable subvariety of $\bar{X}$. Let $P_{i_1 \ldots i_k}$ be the parabolic subgroup associated to the weight $\lambda_1 = \lambda_{i_1} + \lambda_{i_2} + \ldots + \lambda_{i_k}$, then there is a G-equivariant fibration $\Pi_1 : S_{i_1,\ldots,i_k} \to G/P_{i_1,\ldots,i_k}$ with fibres isomorphic to the minimal compactification of $\bar{L}_{i_1 \ldots i_k}/\bar{H}_{i_1 \ldots i_k}$.

We should remark that in the case of the "compactification of a group $\bar{G}$", the set $\{1,\ldots,\ell\}$ can also be thought as the set of simple roots of G, for each subset the parabolic of $G \times G$ is $P \times P$ and the fiber of the $G \times G$ equivariant fibration is the "compactification of the adjoint group associated to the Levi factor of P".

5.3.

DEFINITION. $\bar{X}$ will be called simple if $\underline{g}$ = Lie G contains no proper σ-stable ideal.

It is clear that in this case either G is simple or we are in the case of a "compactification of a simple group".It also clear that in general $\bar{X}$ is the direct product of simple compactifications.

6. THE VARIETY OF LIE SUBALGEBRAS

6.1. We wish to compare our method with the one developed by Demazure in [5] and show that, in fact, his construction falls under our analysis.

The method is the following: consider the Lie algebras $\underline{g}$ and $\underline{h}$ of G, H respectively. Say $\dim \underline{g} = n$, $\dim \underline{h} = m$. Take for every $g \in G$ the subgroup gHg^{-1} and its Lie algebra $ad(g)\underline{h}$. The stabilizer in G of the subalgebra $\underline{h}$ under the adjoint action is exactly the subgroup $\tilde{H}$ considered in 2.1, so we can identify $G/\tilde{H}$ with the orbit of $\underline{h}$ in the Grassmann variety $G_{m,n}$ of m-dimensional subspaces in the n-dimensional space $\underline{g}$.

We define a compactification $\tilde{X}$ of $G/\tilde{H}$ by putting $\tilde{X} = \overline{G\underline{h}} \subseteq G_{n,m}$.

We want to show that $\tilde{X}$ coincides with our $\bar{X}$. If we use the Plücker embedding, we see that we can identify $\tilde{X}$ with the closure of the G-orbit of the point $\mathbb{P}(\overset{m}{\Lambda}\, \underline{h})$ in $\mathbb{P}(\overset{m}{\Lambda}\, \underline{g})$. If h is a vector spanning the line $\overset{m}{\Lambda}\,\underline{h}$, h is H invariant and we want to study its weight structure.

From Proposition 1.3 we know that

$$\underline{h} = \underline{t}_0 \oplus \sum_{\alpha\in\Phi_0} g_\alpha \oplus \sum_{\alpha\in\Phi_1^+} \mathbb{C}(x_\alpha + \sigma(x_\alpha))$$

so if

$$\{\beta_1,\dots,\beta_r\} = \Phi_0^+, \qquad \{\alpha_1,\dots,\alpha_t\} = \Phi_1^+$$

We have

$$\overset{m}{\Lambda}\,\underline{h} = \overset{k}{\Lambda}\,\underline{t}_0 \wedge x_{\beta_1} \wedge \dots \wedge x_{\beta_r} \wedge x_{-\beta_1} \wedge \dots \wedge x_{-\beta_r} \wedge (x_{\alpha_1} + \sigma(x_{\alpha_1}))$$

$$\wedge \dots \wedge (x_{\alpha_t} + \sigma(x_{\alpha_t})).$$

If we develop h and write it as a sum of weight vectors, we see that this sum contains a unique vector of weight $\mu = \alpha_1+\alpha_2+\dots+\alpha_t$ i.e. $\overset{\ell}{\Lambda}\underline{t}_0 \wedge x_{\beta_1} \wedge \dots \wedge x_{\beta_r} \wedge x_{-\beta_1} \wedge \dots\dots\dots \wedge x_{-\beta_r} \wedge x_{\alpha_1} \wedge \dots \wedge x_{\alpha_t}$ and the others have T_1 weitht of the form $\mu - 2\sum m_i\alpha_j$, $\alpha_j \in \Gamma_1$ and m_j non negative integers.

LEMMA. μ is a regular special weight.

PROOF. The fact that μ is special follows since $\mu^\sigma = -\mu$. To see that μ is regular recall that $2\rho = \beta_1 + \dots + \beta_r + \alpha_1 + \dots + \alpha_t$ and $(2\rho, \check{\alpha}_j) = 2$ while $(\beta_i, \check{\alpha}_j) \leq 0$ for each $\alpha_j \in \Gamma_1$ and $\beta_j \in \Phi_0^+$. Hence, clearly $(\mu, \check{\alpha}_j) \geq 2$ for each $\alpha_j \in \Gamma_1$.

We are now ready to deduce:

PROPOSITION. The compactification $\underline{\tilde{X}} = \overline{G \cdot \underline{h}} \subseteq G_{m,n}$ is isomorphic to $\underline{\bar{X}}$ of 2.1.

PROOF. Let $W \subset \overset{m}{\Lambda}\,\underline{g}$ be the minimum G-stable submodule containing $\mathbb{C}h = \overset{m}{\Lambda}\,\underline{h}$. Clearly for every irreducible component $V_i \subset W$ and G-equivariant projection Π_i: $W \to V_i$ we have $\Pi_i(h) \neq 0$.

In particular it follows from 1.5 that V_i has as its highest weight a special weight $\leq \mu$. Also, μ is a highest weight for W, we can now apply 4.1 and conclude the proof.

6.2. We can now easily see that the boundary points of $\tilde{X}$ are the Lie subalgebras (of groups related to the ones discussed in 6.2) as in Demazure's analysis.

In fact, to pass to the limit, up to conjugation, it is enough to do it under the action of T_1. If $t \in T_1$, we have:

$$t(\overset{m}{\Lambda}\, \underline{h}) = \overset{k}{\Lambda}\, \underline{t}_0 \wedge x_{\beta_1} \wedge \ldots\ldots \wedge x_{-\beta_r}$$

$$\wedge (x_{\alpha_1} + t^{-2\alpha_1} \sigma(x_{\alpha_1}) \wedge \ldots \wedge (x_{\alpha_t} + t^{-2\alpha_t} \sigma(x_{\alpha_t}))$$

Going to the limit $t^{-2\alpha_i} \to 0$ if $i = i_1,\ldots,i_k$ and $t^{-2\alpha_i} \to 1$ otherwise, we obtain the subalgebra spanned by

$$\underline{t}_0,\ x_{\beta_1},\ldots,x_{\beta_r},x_{-\beta_1},\ldots,x_{-\beta_r},x_{\alpha_k},\ldots,x_{\alpha_j} + \sigma(x_{\alpha_j})$$

where k runs over all the indices for which α_k is a root of the unipotent radical $U_{i_1,\ldots,i_k}$ of the parabolic $P_{i_1,\ldots,i_k}$ and j runs over the remaining indeces.

This is the Lie algebra of the following subgroup. Consider the automorphism σ induced on $P_{i_1,\ldots,i_k}/U_{i_1,\ldots,i_k}$. Consider the fixed points of σ in $P_{i_1\ldots i_k}/U_{i_1\ldots i_k}$ and the subgroup of $P_{i_1,\ldots,i_k}$ mapping onto this group of fixed points.

The Lie algebra is the one required by the previous analysis.

Remark that the projection from a G-orbit in $\bar{\underline{X}}$ to the corresponding variety of parabolics is the one obtained by associating to a Lie algebra the normalizer of its unipotent radical.

7. COHOMOLOGY AND PICARD GROUP

7.1. We want now to describe a cellular decomposition of $\bar{\underline{X}}$ which can be constructed, using the theory of Bialynicki-Birula [2],[26]. One of his main theorems is the following:

THEOREM. If $\bar{\underline{X}}$ is a smooth projective variety with an action of a Torus T and if $\bar{\underline{X}}$ has only a finite number of fixed points $\{x_1,\ldots,x_n\}$ under T, one can construct a decomposition $\bar{\underline{X}} = \cup\, C_{x_i}$ where each C_{x_i} is an affine cell (an affine space) centered in x_i.

The decomposition depends on certain choices. In particular, for a suitable choice of a one parameter group $\mu\colon G_m \to T$ such that $\bar{\underline{X}}^{G_m} = \bar{\underline{X}}^T$. Given such a choice, one decomposes the tangent space T_{x_i} of $\bar{\underline{X}}$ at x_i as $T_{x_i} = T^+_{x_i} \oplus T^-_{x_i}$ (where T^+ and T^- are generated by vectors of positive respectively negative weight). Then C_{x_i} is an affine space of (complex) dimension $\dim T^+_{x_i}$.

Furthermore, in [26], be shows that the variety $\bar{\underline{X}}$ is obtained by a sequence of attachments of the C_{x_i}'s and so the integral homology has, as basis, the fundamental classes of the closures of the C_{x_i}'s (in particular it is concentrated in even dimensions and has no torsion).

7.2. In order to apply 7.1 we need the following proposition due to D. Luna.

PROPOSITION. Let G be a reductive algebraic group acting on a variety with finitely many orbits. If T is a maximal Torus of G, the set of fixed points X^T is finite.

PROOF. We can clearly reduce to the case in which X is itself an orbit. In this case it is enough to show that, if $x \in X^T$, x is an isolated fixed point. We have X = Gx by assumption and $T \subseteq St_x$. The tangent space of X in x can be identified in a T equivariant way with Lie G/Lie St_x which is a quotient of Lie G/Lie T over which T acts without any invariant subspaces, proving the claim.

In particular we can apply this proposition to our variety $\underline{\bar{X}}$ in view of 3.1.

We should remark that in the case of a group G considered as G × G space, there are no fixed points on any non closed orbits. So the fixed points all lie in the closed orbit isomorphic to G/B × G/B and they are thus indexed by pairs of elements of the Weyl group.

7.3. Notice that, since $\underline{\bar{X}}$ has a paving by affine spaces, we have Pic $(\underline{\bar{X}}) \cong H^2(\underline{\bar{X}})$. We want now to compute $H^2(\underline{\bar{X}})$ by computing the number of 2 dimensional cells given by 7.1.

For this we fix a Borel subgroup and the positive roots as in § 1. Since the center of G acts trivially on $\underline{\bar{X}}$, we can use the action of a maximal Torus T of the adjoint group. Hence, the simple roots are a basis of $\underline{t}$*. We can construct a generic 1-parameter subgroup $\mu: G_m \to T$ which has the same fixed points on $\underline{\bar{X}}$ as T and in the following way:

We order lexicographically the simple roots as

$$\beta_1 > \beta_2 > \dots \beta_k > \alpha_1 > \dots > \alpha_\ell > \alpha_{\ell+1} > \dots > \alpha_h$$

where $\bar{\alpha}_i = \frac{1}{2}(\alpha_i - \alpha_i^\sigma)$ $i = 1,\dots,\ell$ are the restricted simple roots.

We can, since in our computations there are only finitely many weights involved (the set Λ of weights appearing in the tangent spaces of the fixed points), select μ in such a way that $\langle\lambda,\mu\rangle>0, \lambda \in \Lambda$ if and only if $\lambda > 0$ in the lexicographic ordering. If $x \in X$ is a fixed point of T, we analyze the tangent space τ_x as follows: x is in an orbit $\mathcal{O}$ which fibers $\Pi: \mathcal{O} \to G/P$ with fiber a symmetric variety $\bar{L}/\bar{L}^\sigma$, we can assume $x \in \bar{L}/\bar{L}^\sigma$ and decompose τ_x in T stable subspaces $\tau_1 \oplus \tau_2 \oplus \tau_3$ such that τ_1 is isomorphic to the tangent space of $\Pi(x)$ in G/$\underline{P}$, τ_2 is isomorphic to the tangent space of x in $\bar{L}/\bar{L}^\sigma$ and τ_3 is isomorphic to the normal space of $\mathcal{O}$ in $\underline{\bar{X}}$ at the point x. To compute dim τ^+ one needs to compute dim τ_i^+ for each i. Now dim τ_1^+ is given by the theory of Bruhat cells , we claim:

LEMMA. 2 dim τ_2^+ = dim τ_2.

PROOF. The T-structure of τ_2 is isomorphic to the structure of the tangent space at the identity of $\bar{L}/\bar{L}^\sigma$ under the conjugate Torus $\bar{T} = x^{-1}T x$. Such tangent space is isomorphic to $\underline{\ell}/\underline{\ell}^\sigma$ with ℓ = Lie $\bar{L}$, $\underline{\ell}^\sigma$ = Lie $\bar{L}^\sigma$. Since $\bar{T} \subset \bar{L}^\sigma$, we see that in the root space decomposition of $\underline{\ell}$ under $\bar{T}$ we have Lie $\bar{T} \subseteq \underline{\ell}^\sigma$. $\underline{\ell}^\sigma$ is a sum of root subspaces, and if $\underline{\ell}_\alpha \subset \underline{\ell}^\sigma$, also $\ell_{-\alpha} \subset \underline{\ell}^\sigma$. Thus, $\underline{\ell}/\underline{\ell}^\sigma$ is a sum of root spaces $\underline{\ell}_\beta \oplus \underline{\ell}_{-\beta}$. And then, if $\underline{\ell}_\beta \subset (\underline{\ell}/\underline{\ell}^\sigma)^+$, we have $\underline{\ell}_{-\beta} \subset (\ell/\underline{\ell}^\sigma)^-$ and the lemma is proved.

7.4. For the computation of the T weights in T_3 we have a simple analysis in the case in which the fixed point x lies in the closed orbit G/P.

In this case $x = w\,x_o$, w in the Weyl group and we have:

LEMMA. In $w\,x_o$ the dimension of τ_3^+ equals the number of restricted simple roots $\bar{\alpha}_i$ such that $w\bar{\alpha}_i > 0$.

PROOF. Using the notations of §.2, $x_o \in V \cong U \times A^\ell$ and is identified with the point (1,0), ($1 \in U$, $0 \in A^\ell$). $G/P \cap V = U \times 0$, so the normal space at x_o is isomorphic to the space $\mathbb{A}^\ell$ with the induced T-action.

Thus the normal space to a point $w\,x_o$ is isomorphic to $\mathbb{A}^\ell$ with the action twisted by w^{-1}. Since the T weights on $\mathbb{A}^\ell$ are the $-2\bar{\alpha}_i$ we have that the T weights in the normal space at $w\,x_o$ are the elements $-2w\bar{\alpha}_i$, hence the claim.

7.5. In the computation of $H^2(X)$ we need to compute the points x such that dim $\tau_x^+ = 1$. Thus, we need in particular to analyze:

LEMMA. If G/H is a symmetric variety of dimension 2, with a fixed point under a Torus T', then Lie G = $\mathfrak{sl}(2)$, Lie H = $\mathfrak{so}(2)$ = Lie T', (up to normal factors on which the automorphism σ acts trivially).

PROOF: Let us recall the consequence of the Iwasawa decomposition 1.3.

$$\underline{g} = \underline{h} \oplus (\underline{t}_1 + \sum_{\alpha\in\Phi_1^+} \mathbb{C}\, x_\alpha); \quad \text{Thus, } 2 = \dim \underline{t}_1 + |\Phi_1^+|.$$

Since we generally have $\underline{t}_1 \neq 0$ if $G/H \neq 1$ and also $|\Phi_1^+| \neq 0$ since G is semisimple, we must have $1 = \dim \underline{t}_1 = |\Phi_1^+|$. Moreover, since we want to factor out all normal subgroups of G on which σ acts trivially, we have G simple. We wish to show that Φ_0 is empty. In fact, if there is a simple root $\beta \in \Phi_0$, since G is simple we may assume that $\beta + \alpha$ is also a root. But then either β or $\beta+\alpha \in \Phi_1^+$ and we have a contradiction. Then we see that G is of rank 1 and the remaining statements easily follow.

7.6. We are now ready for the computation of Pic (X).

THEOREM. Pic (X) $\cong Z^{\ell+r}$ where r is the number of simple roots α_i, $i = 1,\ldots,\ell$ such that: there exist two distinct simple roots α, β with $\bar{\alpha}_i = \frac{1}{2}(\alpha - \alpha^\sigma) = \frac{1}{2}(\beta - \beta^\sigma)$ and either $-\alpha^\sigma \neq \beta$ or, if $-\alpha^\sigma = \beta$, $(\alpha,\beta) \neq 0$.

PROOF. Let $\bar{\alpha}_i = \frac{1}{2}(\alpha_i - \alpha_i^\sigma)$, $i = 1,\ldots,\ell$ be the simple restricted roots (cf. 2.2). Suppose $x \in \bar{X}$ is a fixed point with dim $\tau_x^+ = 1$, first of all we analyze the case in which $x \in G/P$, the unique closed orbit. In this case $\tau_x = \tau_1 + \tau_3$ and we must have either dim $\tau_1^+ = 0$, dim $\tau_3^+ = 1$ or dim $\tau_1^+ = 1$, dim $\tau_3^+ = 0$. Now x is a center of a Bruhat cell in G/P of dimension equal to dim τ_1^+ so it is either the point x_o corresponding to the 0 cell or a point $s_\alpha x_o$ with α a simple root in Φ_1^+. Thus, by Lemma 7.4 dim τ_3^+ at wx_o is the number of i such that $w\bar{\alpha}_i$ is negative. In particular we see that we can get 2 dimensional cells only centered at the points $s_\alpha x_o$ and we need to count how many $\alpha \in \Phi_1^+$ are such that $s_\alpha\bar{\alpha}_i > 0$ for all i's. Now if $\alpha \neq \alpha_i$, $-\alpha_i^\sigma$, we have $s_\alpha(\bar{\alpha}_i) > 0$ (since $s_\alpha(\beta) > 0$ if β is positive $\alpha \neq \beta$). Now given $\alpha \in \Phi_1^+$ if $\alpha = \alpha_i$, we have $s_\alpha(\bar{\alpha}_j) > 0$ if $j \neq i$. As for $s_\alpha(\bar{\alpha}_i)$ it depends on $-\alpha_i^\sigma$.
We have various cases:

i.) $-\alpha_i^\sigma = \alpha_i$,

ii.) $-\alpha_i^\sigma = \alpha_i + \beta$, $\beta \neq 0$ a positive combination of roots in Φ_0.

iii.) $-\alpha_i^\sigma = \alpha_j + \beta$, $j \neq i$.

In case i.) $s_\alpha(\bar{\alpha}) = -\bar{\alpha} < 0$,

In case ii.) $s_\alpha(\bar{\alpha}) = -\alpha + \frac{1}{2}(\beta - \frac{2(\alpha.\beta)}{(\alpha,\alpha)}\alpha) > 0$,

In case iii.) the same reasoning as in ii.) holds if $\beta \neq 0$, $s_\alpha(\alpha + \alpha_j + \beta) = \beta + \alpha_j + m\alpha > 0$ (some m).

If $\beta = 0$, we have

$$s_\alpha(\alpha + \alpha_j) = -\alpha + \alpha_j - \frac{2(\alpha,\alpha_j)}{(\alpha,\alpha)}\alpha$$

Now since $\alpha_j = -\alpha^\sigma$, we must have $(\alpha,\alpha) = (\alpha_j,\alpha_j)$. Hence, the Dynkin diagram formed by α, α_j is either disconnected and $(\alpha,\alpha_j) = 0$ or is A_2 and then $\frac{2(\alpha,\alpha_j)}{(\alpha,\alpha)} = -1$ so $s_\alpha(\alpha + \alpha_j) = \alpha_j > 0$. If $(\alpha,\alpha_j) = 0$, we have $s_\alpha(\alpha + \alpha_j) = -\alpha + \alpha_j < 0$ since $\alpha = \alpha_i$ $i \leq \ell$ and $j > \ell$.

Now we have to consider the case $\alpha = -\alpha_i^\sigma \neq \alpha_i$, since α is a simple root this occurs only in the case $-\alpha_i^\sigma = \alpha_j$, $j > \ell$. The same analysis as before shows that if $(\alpha,\alpha_i) = -\frac{1}{2}$ we have $s_\alpha(\alpha + \alpha_i) = \alpha > 0$

if $(\alpha,\alpha_i) = 0$ $s_\alpha(\alpha + \alpha_i) = \alpha_i - \alpha > 0$.

It remains to analyze the case of x lying in a non closed orbit $\mathcal{O}$. By Lemmas 7.3 and 7.5 this can occur only when $\mathcal{O}$ fibers on a variety G/P' with fiber the minimal compactification of a symmetric variety isomorphic to $SL(2)/S\mathcal{O}(2)$. This is the variety of distinct unordered pairs of points in $\mathbb{P}^1$ and its minimal compactification is the space $\mathbb{P}^2$ considered as the symmetric square of $\mathbb{P}^1$. In this case we only have 2 SL(2) orbits in $\mathbb{P}^2$ and so only 2 G orbits in $\bar{\mathcal{O}}$.
Thus by 3.1 we have $\dim \bar{\mathcal{O}} = \dim G/P + 1$ and a $\mathbb{P}^1$-fibration $G/P \to G/P'$. Thus, we can identify P' with the parabolic group generated by P and the subgroup $X_{-\alpha}$ relative to a simple root $\alpha \in \Phi_1^+$ and we have $\alpha^\sigma = -\alpha$, and $\alpha = \alpha_i$ for some $1 \leq i \leq \ell$. As in Lemma 7.3 write $\tau_x = \tau_1 \oplus \tau_2 \oplus \tau_3$. Since T acts on τ_2 by a negative and a positive weight as we have noted above in order to have that the set of T weights appearing on τ_x contains only a positive weight, we must have that the T weights in τ_1 and τ_3 consist of negative weights. This implies that $p(x) \in G/P'$ is the unique B fix point in G/P', otherwise at least one of the weights appearing in τ_1 would be positive. Furthermore, notice that the fact that p(x) is the unique B fix point in G/P' determines x uniquely since in $p^{-1}(p(x)) = \mathbb{P}^2$ there are exactly three T fix points of which two are x_o and $s_\alpha(x_o)$ both belonging to the closed orbit. But for such x we have that the set of weights appearing on τ_3 is
$\{-\frac{(\alpha_j - \alpha_j^\sigma) + s_\alpha(\alpha_j - \alpha_j^\sigma)}{2}\}$ for $1 \leq j \leq \ell$, $j \neq i$ which are all negative. This is easily seen as follows: first of all the normal bundle to $\bar{\mathcal{O}}$ in X is just the sum of the restrictions of the normal line bundles to the closures of the codimension one orbits S_j, $1 \leq j \leq \ell$, $j \neq i$, containing $\bar{\mathcal{O}}$. Thus we have to compute the weight of T for each such line bundle N_j. Let us fix $1 \leq j \leq \ell$, $j \neq i$, then the T weight of N_j in x_o is just $-(\alpha_j,-\alpha_j^\sigma)$. Now if we let $T_\alpha \subset T$ denote ker α, we have that T_α acts trivially on $\mathbb{P}^2$ hence the T_α weight in x and x_o are the same. Thus the given formula is correct for T_α. It remains to verify the formula on a "complement of T_α in T. This amounts to perform the computation in the maximal torus of PSL(2) which can be carried out directly.
So it follows that the action of T on τ_x has exactly one negative weight and the cell associated to x has dimension 2. Summarizing our result we have

1) If $\bar{\alpha}_i$ is such that there exists only one simple root α with $\frac{1}{2}(\alpha - \alpha^\sigma) = \bar{\alpha}_i$ and $\alpha^\sigma \neq -\alpha$ then we get one 2 cell whose center lies in the unique closed orbit G/P.
2) If α_i is as in one but $\alpha^\sigma = -\alpha$ then again we get one 2 cell but its center lies in the orbit $\mathcal{O}$ whose closure $\bar{\mathcal{O}}$ fibers with $\mathbb{P}^2$ fibers onto G/P', P' being the parabolic generated by P and $X_{-\alpha}$.
3) If $\bar{\alpha}_i$ is such that there exists two distinct simple roots α,β such that $\bar{\alpha}_i = \frac{1}{2}(\alpha - \alpha^\sigma) = \frac{1}{2}(\beta - \beta^\sigma)$, $-\alpha^\sigma = \beta$ and $(\alpha,\beta) = 0$ then we get exactly one 2 cell whose center lies in G/P.
4) If $\bar{\alpha}_i$ is such that $\bar{\alpha}_i = \frac{1}{2}(\alpha - \alpha^\sigma) = \frac{1}{2}(\beta - \beta^\sigma)$ and either $-\alpha^\sigma \neq \beta$ or $-\alpha^\sigma = \beta$ but $(\alpha,\beta) \neq 0$, then we get two 2 cells, whose both centers lie in G/P.

This is our theorem.

DEFINITION. $\underline{\bar{X}}$ will be called exceptional when rk Pic$(\underline{\bar{X}}) > \ell$.

7.7. REMARK. It is clear from the previous analysis that the main difficulty in computing explicitely the dimensions of the cells lies in the computation of τ_3^+. In the special case in which all fixed points lie in the closed orbit this is accomplished by Lemma 7.4.

In particular for the case of a group $\bar{G}$ considered as a symmetric variety over $\bar{G} \times \bar{G}$ we have the following computation for the Poincarè polynomial: $\Sigma b_i q^j$, $b_i = \dim H_i(\underline{\bar{X}},\mathbb{Z})$:

$$\Big(\sum_{w\in W} q^{2\ell(w)}\Big)\Big(\sum_{w\in W} q^{2(\ell(w)+L(w))}\Big)\ (*)$$

($\ell(w)$ the length of w, L(w) the number of simple reflections s_α with $\ell(s_\alpha w) < \ell(w)$).

8. LINE BUNDLES ON $\underline{\bar{X}}$

8.1. Let $\underline{\bar{X}}$ be as usual and let $\underline{Y} = G/P \subset \underline{\bar{X}}$ be the unique closed orbit in $\underline{\bar{X}}$.

PROPOSITION. Let i^*: Pic $(\underline{\bar{X}}) \to$ Pic (Y) be the homomorphism induced by the inclusion. Then i^* is injective.

PROOF. First assume that for any simple root $\alpha \in \Phi_1^+$ we have $\alpha^\sigma = -\alpha$. Then we know that Pic $(\underline{\bar{X}}) \cong \mathbb{Z}^\ell$, where ℓ is the number of simple roots in Φ_1^+. Furthermore, let $\omega_1,\ldots,\omega_\ell$ be the fundamental weights correspond-

(*) We wish to thank G. Lusztig for suggesting this formula.

ing to such α's. Then we have shown how to imbed $\bar{X} \subset \prod_{i=1}^{s} \mathbb{P}(V_{2\omega_i})$. So we get a map $h^*: \mathrm{Pic}(\prod_{i=1}^{s} \mathbb{P}(V_{2\omega_i})) \to \mathrm{Pic}(\bar{X})$. But it is clear that i^*h^* is injective since the restriction of the tautological bundle L_i on $\mathbb{P}(V_{2\omega_i})$ to G/P gives the line bundle associated to $2\omega_i$. Since $\mathrm{rk}(\mathrm{Pic}(\prod_{i=1}^{s} \mathbb{P}(V_{2\omega_i}))) = \mathrm{rk}(\mathrm{Pic}(X))$ our assertion follows.

Let us now suppose that there exists a simple root α such that $\alpha^\sigma \neq -\alpha$. Let S be the unique orbit closure associated to $\alpha - \alpha^\sigma$. Then it follows from the description of the dimension two cells given in 7, that each dimension 2 cell in $\bar{X}$ is already contained in S, so we prove that the map $\mathrm{Pic}(\bar{X}) \to \mathrm{Pic}(S)$ induced by inclusion is injective.

Let us now consider the map $\mathrm{Pic}(S) \to \mathrm{Pic}(Y)$ and recall that for a suitable parabolic P_o we get a fibration $S \to G/P_o$ whose fiber is the variety $\bar{X}_{\bar{L}}$ which is the minimal compactification of $\bar{L}/\bar{L}^\sigma$ where $\bar{L}$ is the adjoint group of the semisimple Levi factor of P_o and $\bar{L}^\sigma$ the fix points group of the involution induced by σ on $\bar{L}$. We thus get the diagram

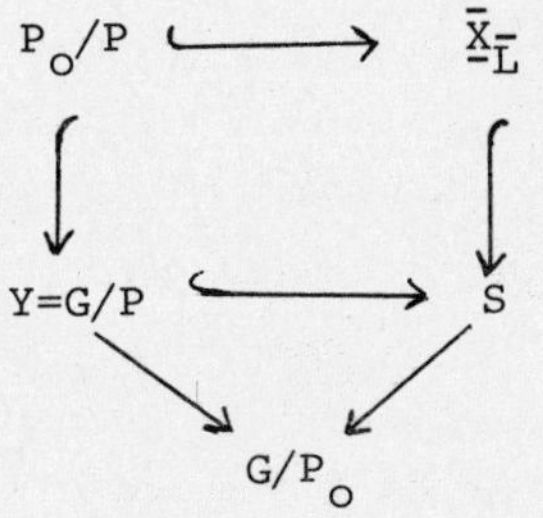

and we can identify P/P_o with the unique closed orbit in $\bar{X}_{\bar{L}}$. But notice that $\mathrm{Pic}(G/P) \cong \mathrm{Pic}(G/P_o) \oplus \mathrm{Pic}(P/P_o)$ and $\mathrm{Pic}(S) \cong \mathrm{Pic}(G/P_o) \oplus \mathrm{Pic}(\bar{X}_{\bar{L}})$. Also, by induction on the rank we can assume that the map $\mathrm{Pic}(\bar{X}_L) \to \mathrm{Pic}(P/P_o)$ induced by inclusion is injective. This clearly implies that the map $\mathrm{Pic}(S) \to \mathrm{Pic}(Y)$ is also injective.

REMARK. Notice that since we can identify Pic (Y) with the lattice spanned by the fundamental weights relative to the simple roots in Φ_1^+, our proposition implies that we can also identify Pic (X) with a sublattice of such a lattice, call it Γ. Notice also that since for each dominant special weight λ with the property that $\frac{2(\lambda,\alpha-\alpha^\sigma)}{(\alpha-\alpha^\sigma,\alpha-\alpha^\sigma)} \in \mathbb{Z}^+$ for every simple root $\alpha \in \Phi_1^+$ we have constructed a map $\Pi: X \to \mathbb{P}(V_\lambda)$ we clearly have that Γ contains the lattice spanned by such weights. In particular, this lattice contains the double of the lattice of special weights $\alpha - \alpha^\sigma \in \Gamma$ for each simple root $\alpha \in \Phi_1^+$.

We wish to collect some of the information gotten up to now for future use.

We have the weights μ_i introduced in 1.7 and a natural embedding

$$\bar{X} \to \Pi \mathbb{P}(V_{\mu_i})$$

The mapping of the closed orbit $Y \to \Pi \mathbb{P}(V_{\mu_i})$ so induced is the canonical one obtained by the diagonal morphism. We compose this with the natural projection $G/B \to Y$.

The ample generator of Pic $(\mathbb{P}(V_{\mu_i}))$ is mapped by the composed homomorphism to the element L_{μ_i} of $\mathrm{Pic}(G/B)$ corresponding to the weight μ_i (notice that under this convention $H^o(G/B, L_{\mu_i}) \cong V^*_{\mu_i}$ as a G-module).

If J is a subset of $\{1,\ldots,\ell\}$ and S_J denotes the corresponding orbit closure, the composition $S_J \to \bar{X} \to \Pi \mathbb{P}(V_{\mu_i}) \xrightarrow{P_J} \prod_{i\in J} \mathbb{P}(V_{\mu_i})$ factors through the canonical fibration $S_J \to G/P_J$ and the canonical inclusion $G/P_J \hookrightarrow \prod_{i\in J} \mathbb{P}(V_{\mu_i})$. Therefore in particular the line bundle corresponding to μ_i restricted to S_J comes from the corresponding line bundle in G/P_J.

Finally since $\mathrm{Pic}(\bar{X})$ is discrete and G is simply connected any $L \in \mathrm{Pic}(\bar{X})$ has a G linearization ([27]). Suppose now $L \in \mathrm{Pic}(\bar{X})$ is a G linearized line bundle. If we restrict this to the closed orbit Y we have the induced bundle already linearized. Now for a linearized line bundle L_λ on Y the corresponding weight λ is the character by which the maximal torus acts on the fiber over the unique B fix point, x_o, in Y.

Recall that the cell $U \times \mathbb{A}^\ell$ in $\bar{X}$ is a B^- stable affine subspace and (1,0) is the fixed point x_o in Y previously introduced. If s is a section trivializing L_λ on $U \times \mathbb{A}^\ell$ so is b^*s for any $b \in B^-$. Since the only invertible functions on $U \times \mathbb{A}^\ell$ are the constants we have $b^*s = \alpha s$, α a scalar. Restricting to the point x_o we have $\alpha = b^{-\lambda}$.

8.2. Notice that since any $L \in \mathrm{Pic}(X)$ can be G linearized we have that G acts linearly on each $H^i(\bar{X}, L)$.

LEMMA. Let $L \in \mathrm{Pic}(\bar{X})$ and consider $H^o(\bar{X}, L)$ as a G module. Then $\dim \mathrm{Hom}_G(V, H^o(\bar{X}, L)) \leq 1$ for each irreducible G-module V.

PROOF. Suppose $\mathrm{Hom}_G(V, H^o(\bar{X}, L)) \neq 0$. Let μ be the highest weight of V. Let $s_1, s_2 \in H^o(\bar{X}, L)$ be two non zero U invariant sections whose weight is μ. Then $\frac{s_1}{s_2}$ is a B invariant rational function on $\bar{X}$. Since B has a dense orbit in $\bar{X}$, it follows the $\frac{s_1}{s_2}$ is constant. Hence, s_1 is a multiple of s_2 and our claim follows.

Now let $V \subset \bar{X}$ be the open set described in 2 and identify V with

$U \times \mathbf{A}^{\ell}$. Let $\{x_i\}$ be the coordinate functions on $\mathbf{A}^{\ell}$. For any $t \in T$,

$tx_i = t^{-(\alpha_i - \alpha_i^{\sigma})} x_i$ for the corresponding simple root $\alpha_i \in \phi_1^+$, $1 \le i \le \ell$.

PROPOSITION. Let V_{μ} be the irreducible G-module whose highest weight is μ. Let $\lambda \in \Gamma$ and $L_{\lambda} \in \mathrm{Pic}\,(\bar{\underline{X}})$ be the corresponding line bundle, then if

$$\mathrm{Hom}\,(V_{\mu}^*, H^0(\bar{\underline{X}}, L_{\lambda})) \ne 0 \qquad \mu = \lambda - \sum t_i\,(\alpha_i - \alpha_i^{\sigma}), \quad t_i \in z^+$$

PROOF. Let $s \in H^0(\bar{\underline{X}}, L_{\lambda})$ be a section generating a B^- stable line. Then if we restrict s to V and we let s_0 be a section trivializing $L_{\lambda}|V$ we can write $s = s_0 f$ where f is a regular function on $V \simeq U \times \mathbf{A}^{\ell}$. Since s is U stable f is also U stable and $f = x_1^{t_1}, \ldots, x_{\ell}^{t_{\ell}}$ so our proposition follows.

COROLLARY. There exists a unique up to a scalar G-invariant section $r_i \in H^0(\bar{\underline{X}}, L_{\alpha_i - \alpha_i^{\sigma}})$ whose divisor is S_i.

PROOF. Let $r_i \in H^0(\bar{\underline{X}}, \mathcal{O}(S_i))$ be the unique, up to constant, section whose divisor is S_i. Since S_i is G-stable and G is semisimple, r_i is a G-invariant section. Also since $x_i = 0$ is a local equation of S_i on V we have $r_i|V = s_0 x_i$ where s_0 is a section trivializing $\mathcal{O}(S_i)|V$. The weight of x_i is $\alpha_i - \alpha_i^{\sigma}$ so the G-invariance of r_i implies that s_0 has weight $-(\alpha_i - \alpha_i^{\sigma})$. Hence $\mathcal{O}(S_i) \simeq L_{\alpha_i - \alpha_i^{\sigma}}$.

8.3. Now let $S_{\{i_1, \ldots, i_t\}} = S_{i_1} \cap \ldots \cap S_{i_t}$ for any subset $\{i_1, \ldots, i_t\} \subset \{1, \ldots, \ell\}$ be the corresponding G-stable subvariety. Let $\gamma \in \Gamma$ put $L_{\gamma}(i_1, \ldots, i_t) = L_{\gamma}|S_{i_1, \ldots, i_t}$. Let $\{j_1, \ldots, j_{\ell-t}\}$ denote the complement in $\{1, \ldots, \ell\}$ of $\{i_1, \ldots, i_t\}$.

PROPOSITION. Let $\gamma \in \Gamma$ be a dominant weight. Let $\{h_1, \ldots, h_s\} \subset \{j_1, \ldots, j_{\ell-t}\}$. Then

$$H^i(S_{\{i_1, \ldots, i_t\}}, L_{\gamma - \Sigma\,(\alpha_{h_i} - \alpha_{h_i}^{\sigma})}\,(i_1, \ldots, i_t)) = 0 \quad \text{for} \quad i > 0.$$

PROOF. We perform a double decreasing induction on $\{i_1, \ldots, i_t\}$ and on $\{h_1, \ldots, h_s\}$.

If $\{i_1, \ldots, i_t\} = \{1, \ldots, \ell\}$ then $\{1, \ldots, \ell\} = G/P$ is the unique closed orbit and our proposition is part of Bott's theorem [4].

Now let $\{i_1, \ldots, i_t\}$ be arbitrary and $\{j_1, \ldots, j_{\ell-t}\} = \{h_1, \ldots, h_s\}$. Then notice that by our local description of $\bar{\underline{X}}$ it follows easily that if $K(i_1, \ldots, i_t)$ denotes the canonical bundle on $S_{\{i_1, \ldots, i_t\}}$,

$$K(i_1, \ldots, i_t) = L_{-\mu - \Sigma_{m=1}^{\ell-t}\,(\alpha_{j_m} - \alpha_{i_m}^{\sigma})}\,(i_1, \ldots, i_t) \quad \text{where} \quad \mu = \sum_{\alpha \in \Phi_1^+} \alpha.$$

(Notice that $\mu \in \Gamma$ (cf. 6.1)).
Thus if we put $L = L_{\gamma-\Sigma(\alpha_{j_m}-\alpha^{\sigma}_{j_m})}(i_1\dots,i_t)$ and $K = K(i_1,\dots,i_t)$ we have that $(K \otimes L^{-1})^{-1} = L_{\gamma+\mu}(i_1,\dots,i_t)$ can be verified to be very ample. We postpone the proof of this assertion to the end of this section. It follows from Kodaira vanishing theorem that

$$H^i(S_{\{i_1,\dots,i_t\}}, K \otimes L^{-1}) = 0 \quad \text{for} \quad i < \dim S_{i_1,\dots,i_t}.$$

This implies by Serre's duality

$$H^i(S_{\{i_1,\dots,i_t\}}, L) = 0 \quad \text{for } i > 0.$$

Now by induction we have the result proved for any $S_{\{i_1,\dots,i_{t+1}\}}$ and for any $\{h_1,\dots,h_{s+1}\} \subset \{j_1,\dots,j_{\ell-t}\}$.
Corollary 8.2 implies that we have a non zero section

$$r_{i_{t+1}} \in H^o(S_{\{i_1,\dots,i_t\}}, L_{\alpha_{i_{t+1}}-\alpha^{\sigma}_{i_{t+1}}}(i_1,\dots,i_t))$$

and multiplication by $r_{i_{t+1}}$ yields an exact sequence.

$$0 \to L_{\gamma - \sum_{i=1}^{s}(\alpha_{h_i} - \alpha^{\sigma}_{h_i} - (\alpha_{i_{t+1}} - \alpha^{\sigma}_{i_{t+1}}))}(i_1,\dots,i_t) \to$$

$$L_{\gamma - \sum_{i=1}^{s}(\alpha_{h_i} - \alpha^{\sigma}_{h_i})}(i_1,\dots,i_t) \to L_{\gamma - \sum_{i=1}^{s}(\alpha_{h_i} - \alpha^{\sigma}_{h_i})}(i_1,\dots,i_{t+1}) \to 0$$

Then we get a long exact sequence that together with an inductive hypothesis immediately proves the proposition.

THEOREM. Let $\lambda \in \Gamma$ then:
1) $H^o(\underline{\bar{X}}, L_\lambda) \neq 0$ if and only if $\lambda = \gamma + \Sigma t_i(\alpha_i - \alpha^{\sigma}_i)$ for some dominant γ, $t_i \in \mathbf{Z}^+$. Assuming $H^o(X, L_\lambda) \neq 0$, if V_γ is the irreducible G-module of highest weight γ, $H^o(X, L_\lambda) = \oplus V^*_\gamma$ for all dominant γ of the form $\gamma = \lambda - \Sigma t_i(\alpha_i - \alpha^{\sigma}_i)$, $t_i \in \mathbf{Z}^+$.
2) For λ dominant $H^i(\underline{\bar{X}}, L_\lambda) = 0$, $i > 0$.

PROOF.
1) The only if part is just Proposition 8.2.
To prove the if part assume λ is dominant. Then we know that $H^o(G/P, L_\lambda|_{G/P})$ is the irreducible G-module V_λ whose highest weight is λ. Now consider the varieties

$$\underline{\bar{X}} = S_\phi \supset S_{\{1\}} \supset S_{\{1,2\}} \supset S_{\{1,2,3\}} \supset \dots\dots S_{\{1,2,\dots,\ell\}} = G/P$$

We claim that for each $\ell \geq i \geq 1$ the restriction map

$$H^0(S_{\{1,2,\dots,i-1\}}, L_\lambda|S_{\{1,2,\dots,i-1\}}) \to H^0(S_{\{1,2,\dots,i\}}, L_\lambda|S_{\{1,2,\dots,i\}})$$

is onto.

This follows at once from the cohomology exact sequence associated to the sequence

$$0 \to L_{\lambda-(\alpha_i-\alpha_i^\sigma)}(i,2,\dots,i-1) \to L_\lambda(1,2,\dots,i-1) \to L_\lambda(1,2,\dots,i) \to 0$$

considered above and the vanishing of

$$H^1(S_{\{1,\dots,i-1\}}, L_{\lambda-(\alpha_i-\alpha_i^\sigma)}(1,2,\dots,i-1))$$

proved in Proposition 8.3.

In particular, the restriction map

$$H^0(\underline{\bar{X}}, L_\lambda) \to H^0(G/P, L_\lambda|_{G/P}) \quad \text{is onto.}$$

Hence, $\mathrm{Hom}_G(V_\lambda^*, H^0(\underline{\bar{X}}, L_\lambda)) \neq 0$ and we can find a non zero lowest weight vector $\underline{v}_\lambda \in H^0(\underline{\bar{X}}, L_\lambda)$ whose weight is $-\lambda$.

Now let $\lambda = \gamma + \sum_{i=1}^{\ell} t_i(\alpha_i - \alpha_i^\sigma)$, $t_i \in \mathbf{Z}^+$, γ dominant in Γ.

Consider the section $r_1^{t_1}\cdots\cdots r_\ell^{t_\ell} \in H^0(\underline{\bar{X}}, L_{\sum_{i=1}^{\ell} t_i(\alpha_i - \alpha_i^\sigma)})$ and the section $v_{-\gamma} \in H^0(\underline{\bar{X}}, L\ )$.

Then the section $v_{-\gamma} r_1^{t_1}\cdots r_\ell^{t_\ell}$ is clearly non zero U-invariant and its weight is $-\gamma$. So $\mathrm{Hom}_G(V_\gamma, H^0(\underline{\bar{X}}, L_\lambda)) \neq 0$. This proves 1); 2) is contained in Proposition 8.3.

REMARK.

1) By a completely analogous argument we can prove that if $\lambda \in \Gamma$ then

$$\mathrm{Hom}\ (V_\gamma^*, H^0(S_{\{i_1,\dots,i_t\}}, L_\lambda|S_{\{i_1,\dots,i_t\}})) \neq 0$$

if and only if

$$\lambda = \gamma + \sum_{m=1}^{\ell-t} t_m(\alpha_{j_m} - \alpha_{j_m})$$

2) Clearly we can define a filtration of $H^0(\underline{\bar{X}}, L_\lambda)$ by putting for each ℓ-tuple of non negative integers $(t_1,\dots,t_\ell)$, $W(t_1,\dots,t_\ell)$ to be the subspace of sections $s \in H^0(\underline{\bar{X}}, L_\lambda)$ vanishing on S_1 of order $\geq t_1, \dots,$ on S_ℓ of order $\geq t_\ell$. Then we can restate our theorem as follows:

$$W_\lambda(t_1,\dots,t_\ell) \Big/ \sum_{(\bar t_1,\dots,\bar t_\ell) > (t_1,\dots,t_\ell)} W_\lambda(\bar t_1,\dots,\bar t_\ell) \cong \begin{cases} 0 & \text{if } \lambda-\sum t_i(\alpha_i-\sigma(\alpha_i)) \text{ is not dominant} \\ V^*_{\lambda-\sum t_i(\alpha_i-\sigma(\alpha_i))} & \text{otherwise} \end{cases}$$

(Here $(\bar t_1,\dots,\bar t_\ell) \geq (t_1,\dots,t_\ell)$ means $\bar t_1 \geq t_1,\dots,\bar t_\ell \geq t_\ell$).

8.4. In order to complete the proof of 8.3 we have to discuss the ampleness of $L_{\gamma+\mu}$ $(i_1 \dots i_t)$ which has been used there.

We start with a general easy fact. Let ω,ω' denote two distinct fundamental weigths V_ω, $V_{\omega'}$, $V_{\omega+\omega'}$ the irreducible representations of highest weight $\omega,\omega',\omega+\omega'$.

We have a canonical G equivariant projection $p: V_\omega \otimes V_{\omega'} \to V_{\omega+\omega'}$ and we denote by $\bar p$ the induced projection $\mathbb{P}(V_\omega \otimes V_{\omega'}) \to \mathbb{P}(V_{\omega+\omega'})$ of projective spaces: Remark that $\mathbb{P}(V_\omega) \times \mathbb{P}(V_{\omega'})$ is embedded in $\mathbb{P}(V_\omega \otimes V_{\omega'})$ via the Segre map.

LEMMA. The map $\bar p$ restricted to $P(V_\omega) \times P(V_{\omega'})$ is a regular embedding.

PROOF. We consider the irreducible representations of G as sections of line bundles an G/B so that the map p corresponds to the usual multiplication. Since G/B is irreducible the product of 2 non zero sections is always non zero. Now if $s,s' \in V_\omega$, $t,t' \in V_{\omega'}$ and $st = s't'$ we claim that $s' = cs$, $t' = c^{-1}t$, c a scalar. In fact since ω,ω' are fundamental the divisors of s,s',t,t' are all irreducible since ω,ω' are independent in Pic (G/B) the divisor of s cannot equal the divisor of t' and so we have divs = divs' and the claim.

This proves that $\bar p$ is injective when restricted to $P(V_\omega) \times P(V_{\omega'})$. To see that the map is also smooth one can use the same fact in local affine coordinates.

We are now ready to prove:

PROPOSITION. For any $\gamma \in \Gamma$ dominant the line bundle $L_{\gamma+\mu}$ is ample on $\underline{X}$ hence also on $S_{\{i_1,\dots,i_t\}}$ for any choice of $i_1,\dots,i_t$.

PROOF. We distinguish 2 cases. If γ is special, since μ is a regular special weight so is $\mu+\gamma$ hence by 3.1 and 4.1 we have that $L_{2(\mu+\gamma)}$ is very ample on X.

Assume γ not special. This can happen only if we are in the exeptional case i.e. if the $\mathrm{rkPic}(\underline{X}) > \ell$ since if a multiple of a weight γ is special so is γ and $\mathrm{Pic}(\underline{X})$ contains the double of the lattice of special weights.

First of all we can clearly reduce to the case is which X is simple (cf. 5.3).

In the group case $\underline{\bar{X}} = \overline{G \times G/G}$ we have rk Pic($\underline{\bar{X}}$)= rkG = ℓ by remark 7.7 otherwise $\underline{\bar{X}} = \overline{G/\tilde{H}}$ with G simple.
We know by 7.6 that rk Pic($\underline{\bar{X}}$) > ℓ if and only if there exists a simple root α such that:

$$\alpha^\sigma = -\alpha' - \beta \qquad \text{with} \qquad \alpha' = \alpha \quad \text{and either } \beta \neq 0 \text{ or } (\alpha^\sigma,\alpha') \neq 0.$$

Now we can inspect the tables of Satake diagrams in the classification of symmetric spaces (cf. [10], p. 532-534) and we see using the notations of such tables that the only cases to be considered are the ones denoted by AIII (first diagram) A IV, D III (second diagram), EIII. One remarks by inspecting the table V (p. 518) that these cases belong to table III (p. 515).

In all cases one can verify that there is a unique pair of simple roots α,α' with the above properties and hence rk Pic($\underline{\bar{X}}$)= ℓ+1.

Case AIII and AIV can be explicitely described as follows.

We consider in $\underline{s\ell}_n$ the automorphism σ defined as conjugation by the block matrix

$$\begin{pmatrix} I_k & 0 \\ e & -I_{n-k} \end{pmatrix} \qquad \text{with } k \neq n-k.$$

Case DIII can be described as
so(4n+2) relative to the symmetric form

$$\begin{pmatrix} 0 & I_{2n+1} \\ I_{2n+1} & 0 \end{pmatrix}$$

and conjugation relative to

$$\begin{pmatrix} I_{2n+1} & 0 \\ 0 & -I_{2n+1} \end{pmatrix}$$

For E III consider the Dynkin diagram of E_6 indexed as

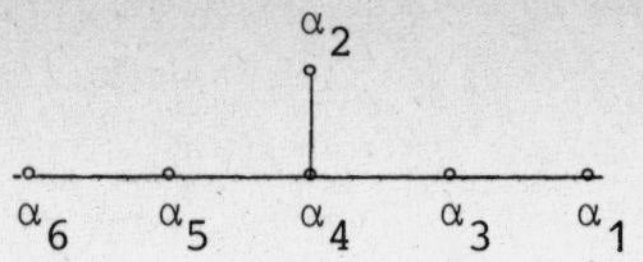

relative to a Cartan subalgebra $\underline{t}$.

Denote by x_α the generator of the corresponding root subspace and define σ as the identity on $\underline{t}$,

$$\sigma(x_{\alpha_i}) = x_{\alpha_i}, \quad i \neq 1, \quad \sigma(x_{\alpha_1}) = -x_{\alpha_1}.$$

One can now verify in each case that the fixed group H is the intersection of a suitable maximal parabolic subgroup Q of type α with its opposite Q' which in all cases is of type α'.

Let us denote by ω and ω' the dual fundamental weights to α, α' and V_ω, $V^*_{\omega'}$ the corresponding irreducible representations. We remark that $V_\omega \simeq V_{\omega'}$ and by 1.3 that $\omega^\sigma = -\omega'$, so that $\omega + \omega'$ is a special weight. If $v \in V_\omega$ (resp. $v' \in V_{\omega'}$) generate the line fixed by Q (resp. by Q') we have that v, v' are seminvariants under H and $v \otimes v'$ is an H invariant, thus if we project $v \otimes v'$ on $V_{\omega+\omega'}$ we obtain a non zero H invariant. By the analysis of section 4 we have a regular morphism π of $\bar{\underline{X}}$ onto the orbit closure Y of the class of $v \otimes v'$ in $\mathbb{P}(V_{\omega+\omega'})$.

We show now that Y is isomorphic to $G/Q \times G/Q'$. This follows from Lemma 8.4 in the following way. In $\mathbb{P}(V_\omega \otimes V_{\omega'})$ the $G \times G$ orbit of $v \otimes v'$ is clearly $G/Q \times G/Q'$ and this orbit projects isomorphically to its image in $\mathbb{P}(V_{\omega+\omega'})$ under $\bar{p}$. On the other hand an easy computation of dimensions shows that the G orbit of $v \otimes v'$ is open in $G/Q \times G/Q'$ hence its closure is $G/Q \times G/Q'$. Since $\bar{p}$ is G-equivariant everything is proved. Comparing the map $\bar{\underline{X}} \to Y \simeq G/Q \times G/Q'$ with the two projections and the respective Plücker embeddings we have two regular projective morphisms associated to the non special weights ω, ω'. We go back now to γ and claim that a suitable positive multiple of γ is of the form $\zeta + a\omega$ or $\zeta + a\omega'$ with $a > 0$ and ζ a dominant special weight.

This can be shown remarking that the subgroup Γ' of Γ generated by the special weights and ω has the same rank as Γ thus a positive multiple of γ lie in Γ'. Now if a dominant weight is in Γ', using the notations of 1.3 it is of the form

$$m\gamma = \sum n_i\omega_i + a\omega \quad \text{with} \quad n_i = n_{\tilde{\sigma}(i)},$$

and ω (resp. ω') is one of the ω_i's, for istance $\omega = \omega_1$ (resp. $\omega' = \omega_2$).

Also $m\gamma$ being dominant $n_1 + a \geq 0$ and $n_i \geq 0$ for $i \neq 1$. If $a \geq 0$ we are done otherwise

$$m\gamma = (n_1 + a)(\omega + \omega') + \sum_{i>2} n_i\omega_i - a\omega'.$$

From this it is clear that for any dominant $\gamma \in \Gamma$ the complete linear system associated to a suitable positive multiple of γ is without base points, since μ is very ample this implies that $\mu + \gamma$ is ample.

9. COMPUTATION OF THE CHARACTERISTIC NUMBERS

9.1. In section 7 we have computed Pic $(\underline{\bar{X}}) \simeq H^2(\underline{\bar{X}},\mathbb{Z})$. We want now to give an explicit algorithm to compute the characteristic numbers. This means that, given n elements $x_1,\ldots,x_n \in H^2(\underline{\bar{X}},\mathbb{Z})$, $n = \dim \underline{\bar{X}}$, we wish to evaluate the product $x_1\ldots x_n \in H^{2n}(\underline{\bar{X}},\mathbb{Z})$ against the class of a point.

Given n reduced hypersurfaces $D_1,\ldots,D_n$ in $\bar{G}/\bar{H}$ such that their closures in $\underline{\bar{X}}$, $\bar{D}_i$ do not contain the unique closed orbit, if $x_i = \mathcal{O}(\bar{D}_i) \in$ Pic $(\underline{\bar{X}}) \simeq H^2(\underline{\bar{X}},\mathbb{Z})$ the corresponding characteristic number counts exactly the number of points common to generic translates g_iD_i, $g_i \in G$, of the D_i's (this is an easy consequence of [12] since $\underline{\bar{X}}$ has a finite number of orbits).

We may work in $H^2(\underline{\bar{X}},\mathbb{Q})$ and use suitable bases for this space. We may also assume that $\underline{\bar{X}}$ is simple (cf. 5.3).

It follows from the analysis performed in section 8 that Pic $(\underline{\bar{X}}) \otimes \mathbb{Q}$ can be identified with the vector space generated by the special weights if $\underline{\bar{X}}$ is not exceptional, otherwise one has to add to the special weights a fundamental weight ω.

Let us denote with Σ the vector space spanned by the special weights and, in the exceptional case $\Gamma_{\mathbb{Q}} = \Sigma + \mathbb{Q}\omega$.

We also know that the divisors S_i correspond to twice the restricted simple roots and form a basis of Σ. Denote by $[S_i]$ these elements in Σ. We have another basis of Σ given by the elements λ_j (cf. 4.1). We notice that $(\lambda_j,[S_i]) = 0$ if $i \neq j$ (for the Killing form).

LEMMA. If $i_1,\ldots,i_k$, $j_1,\ldots,j_{\ell-k}$ is a shuffle of the indices $1,2,\ldots,\ell$, the elements $\lambda_{i_1},\ldots\ldots,\lambda_{i_k},[S_{j_1}],\ldots\ldots,[S_{j_{\ell-k}}]$ form a basis of Σ.

PROOF. Clear by the orthogonality relations.

9.2. Given an oriented compact manifold X and an oriented submanifold Y denote by [Y] the Poincarè dual of the fundamental class of Y. We shall use the following basic facts:

1) If Y_1, Y_2 are oriented submanifolds of X with transversal intersection we have:

$$[Y_1 \cap Y_2] = [Y_1] \cup [Y_2]$$

2) If $Y \subset X$ is a d-dimensional oriented submanifold and $c \in H^d(X)$ we have that the evaluation of $c \cup [Y]$ on the class of a point in X equals the evaluation of $c|_Y$ on the class of a point in Y.

The main proposition is the next one.

PROPOSITION. Let $S_{\{i_1 \dots i_k\}} = S_{i_1} \cap \dots \cap S_{i_k}$. If $S_{\{i_1 \dots i_k\}}$ is not the closed orbit in $\bar{X}$ then:

1) Every monomial $\lambda_{i_1}^{h_1} \lambda_{i_2}^{h_2} \cdot \dots \cdot \lambda_{i_k}^{h_k}$ with $\Sigma h_i = \dim S_{\{i_1 \dots i_k\}}$ vanishes on $S_{\{i_1 \dots i_k\}}$.
2) In the exceptional case every monomial $\omega^{h_o} \lambda_{i_1}^{h_1}, \lambda_{i_2}^{h_2} \cdot \dots \cdot \lambda_{i_k}^{h_k}$ with $\Sigma h_i = \dim S_{\{i_1 \dots i_k\}}$ vanishes on $S_{\{i_1 \dots i_k\}}$.

PROOF. 1) Recall that we have a projection $\pi \colon S_{\{i_1 \dots i_k\}} \to G/P_{\{i_1 \dots i_k\}}$ and the classes $\lambda_{i_1}, \dots, \lambda_{i_k}$ come via π^* from the cohomology of $G/P_{\{i_1 \dots i_k\}}$. Since $S_{\{i_1 \dots i_k\}}$ is not the closed orbit we have $\dim S_{\{i_1 \dots i_k\}} > \dim G/P_{\{i_1 \dots i_k\}}$ and everything follows.

2) We have seen in 8.4 that L_ω induces a morphism $p \colon \bar{X} \to G/Q$ for a suitable maximal parabolic Q and ω is the pullback of the ample generator of Pic (G/Q) by p^*. We wish to consider the induced map $\pi \times p \colon S_{\{i_1 \dots i_k\}} \to G/P_{\{i_1 \dots i_k\}} \times G/Q$ and denote by $\tilde{S}_{\{i_1 \dots i_k\}}$ its image. We know that $\omega + \omega^\sigma$ is one of the fundamental special weights λ_i. If the index i is one of the indeces of the set $\{i_1, \dots, i_k\}$ then the parabolic Q contains $P_{\{i_1 \dots i_k\}}$ and the projection $p \colon S_{\{i_1 \dots i_k\}} \to G/Q$ factors through $G/P_{\{i_1 \dots i_k\}}$. This case therefore follows as in 1). Otherwise $G/P_{\{i_1 \dots i_k\}} \times G/Q$ contains a unique closed orbit under G isomorphic to $G/P_{\{i_1 \dots i_k\}} \cap Q$. We claim that $\tilde{S}_{\{i_1 \dots i_k\}}$ equals this orbit. In fact first of all the fiber of the projection $G/P_{\{i_1 \dots i_k\}} \cap Q \to G/P_{i_1 \dots i_k}$ equals the variety $L_{\{i_1 \dots i_k\}}/L_{\{i_1 \dots i_k\}} \cap Q$ which is a complete homogeneous space over the semisimple part of $L_{\{i_1 \dots i_k\}}$.

If we restrict to a fiber $X_{\{i_1\ldots i_k\}}$ of π the line bundle L_ω we obtain a line bundle of the same type (relative to the minimal compactification $X_{\{i_1\ldots i_k\}}$ of $\bar{L}_{\{i_1\ldots i_k\}}/\bar{H}_{\{i_1\ldots i_k\}}$ (cf. 5.2)).
Since we know that $H^o(X_{\{i_1\ldots i_k\}}, L_\omega|X_{\{i_1\ldots i_k\}})$ is an irreducible $L_{\{i_1\ldots i_k\}}$ module we get that the restriction homomorphism

$$H^o(X,L_\omega) \to H^o(X_{\{i_1\ldots i_k\}}, L_\omega|X_{\{i_1\ldots i_k\}})$$

is onto. Hence the induced morphism on $X_{\{i_1\ldots i_k\}}$ coincides with the restriction to $X_{\{i_1\ldots i_k\}}$ of p and maps it onto $\bar{L}_{\{i_1\ldots i_k\}}/L_{\{i_1\ldots i_k\}} \cap Q$. This proves the claim. Since $S_{i_1\ldots i_k}$ is not the closed orbit $\dim \tilde{S}_{\{i_1\ldots i_k\}} < \dim S_{\{i_1\ldots i_k\}}$ and everything follows as in 1.

9.3. We are now ready to illustrate the algorithm. We treat the exceptional case, the non exceptional is the same without the appearence of ω.

Consider monomials of degree n of type $M=[S_{i_1}]\ldots[S_{i_k}]\omega^{h_o}\lambda_{j_1}\ldots\lambda_{j_s}$ with $i_1,\ldots,i_k$ distinct (in particular the ones with $k = 0$ are the monomials we wish to evaluate). We call k the index of M. We count the number of indices j_h appearing in M and different from $i_1,i_2,\ldots,i_k$ and call this the content of M.

If $j_1 \neq i_1,i_2,\ldots,i_k$ we have an explicit formula expressing λ_{j_1} in terms of $\lambda_{i_1},\lambda_{i_2},\ldots,\lambda_{i_k}$ and the $[S_j]$'s relative to the remaining indeces (Lemma 9.1).

Substituting we obtain M expressed as a linear combination of monomials of higher index and of lower content.

Iterating we obtain M as a combination of monomials of index ℓ or of content 0.

By Proposition 9.2 all monomials of contenent 0 vanish, the computation of the remaining ones can be performed:

LEMMA. The evaluation of $[S_1][S_2]\ldots[S_\ell]\omega^{h_o}\lambda_{i_1}\ \ldots\ \lambda_{i_k}$ on the class of a point in X equals the evaluation of $\omega^{h_o}\lambda_{i_1}\ldots\lambda_{i_k}$ restricted to the closed orbit on the class of a point in it.

PROOF. Clear since the closed orbit is the transversal intersection of the hypersurfaces S_i.

We summarize

THEOREM. By an explicit algorithm the computation of the characteristic numbers is reduced to the one relative to the closed orbit (for which it is known since the cohomology ring of a complete homogeneous space is known [3]).

10. AN EXAMPLE

10.1. In his fundamental work [14] H. Schubert has computed the number of space quadrics tangent to 9 quadrics in general position to be 666.841.088. We want here to perform again this computation.

The variety of non degenerate quadrics in $\mathbb{P}^n$ is symmetric, it is $X_o = SL(n+1)/\tilde{SO}(n+1)$ (the involution being $\sigma(A) = {}^tA^{-1}$).

The variety X is classically called the variety of complete quadrics ([1],[15],[17],[19],[21],[22]).

One can easily verify (by the invariant theory of the orthogonal group) that the irreducible representations of SL(n+1) containing an invariant for SO(n+1) are exactly the ones of highest weight $\sum_{i=1}^{n} n_i 2\omega_i$ (ω_i the fundamental weights). From this it follows that we can identify Pic $(\bar{X})$ with 2Λ where Λ is the lattice of weights for SL(n+1) and that the closed orbit in $\bar{X}$ is the full flag variety F. The usual maximal Torus of diagonal matrices is anisotropic and so the restricted simple roots coincide with the usual simple roots. Hence:

$$[S_1] = 2(2\omega_1) - 2\omega_2$$

$$[S_i] = 2(2\omega_i) - 2\omega_{i-1} - 2\omega_{i+1} \qquad 1 < i < n$$

$$[S_n] = 2(2\omega_n) - 2\omega_{n-1}.$$

Let us fix for each $i = 0,\dots,n-1$ a linear subspace π_i of dimension i in $\mathbb{P}^n$. Denote by D_i the hypersurfaces in X_o of quadric tangent to π_i. We also fix a non degenerate quadric Q and denote by D the hypersurface in X_o of quadrics tangent to Q. We denote as usual by $\bar{D}_i$, $\bar{D}$ their closures in $\bar{X}$.

PROPOSITION.

1) $[\bar{D}_i] = \mathcal{O}(\bar{D}_i) = L_{2\omega_i}$.

2) $[\bar{D}] = 2\sum_{i=0}^{n-1} [\bar{D}_i]$

3) $\bar{D}_i$ and $\bar{D}$ do not contain the closed orbit.

PROOF. 1) X_o is the affine variety of symmetric $(n+1)\times(n+1)$ matrices of determinant 1. The map from X_o to $\mathbb{P}(V^*_{2\omega_i})$ is easily seen to be induced by the map associating to each matrix the matrix of determinants of $i\times i$ minors, which gives a quadric in $\mathbb{P}(V_{\omega_i})$ whose intersection with the Grassmann variety $G_{i-1,n}$ of i-1 dimensional subspaces is exactly the set of tangent subspaces to the original quadric.

Given an i-1 dimensional subspace π_{i-1} in $\mathbb{P}^n$ we consider it as a point in $G_{i-1,n}$, hence, by taking the embedding of $G_{i-1,n}$ in $\mathbb{P}(V_{2\omega_i})$ as a point in $\mathbb{P}(V_{2\omega_i})$. Then it is clear that the intersection of X_o with the hyperplane in $\mathbb{P}(V^*_{2\omega_i})$ associated to this point is at least set theoretically D_{i-1}. So we have found an $s \in H^o(\bar{X}, L_{2\omega_i})$ whose divisor has support equal to D_{i-1}. But it is clear from our computation of Pic (X) that the divisor of s is reduced so it equals $\bar{D}_{i-1}$ proving 1).

2) Consider the variety $F_{0,n-1}$ of flags $p \in \pi \subset \mathbb{P}^n$ where p is a point and π is an hyperplane. Define a flag (p,π) to be tangent to a quadric $Q \in \bar{X}_o$ if $p \in Q$ and π is the hyperplane tangent to Q in p. Let $Y \subset \bar{X} \times F_{0,n-1}$ be the closure of the correspondence $\tilde{Y} = \{(Q,(p,\pi) \mid (p,\pi)$ is tangent to Q, $Q \in \bar{X}_o\}$. Clearly $\dim Y = \dim \bar{X} + n - 1 = \frac{(n+1)(n+2)}{2} + n - 2$ and we get two projections

$$\begin{array}{ccc} & Y & \\ {}^{g}\swarrow & & \searrow^{f} \\ \bar{X} & & F_{0,n-1} \end{array}$$

A simple dimension count shows that we have an homomorphism

$$g_* f^* : H^n(F_{0,n-1}, \mathbb{Z}) \to H^2(\bar{X}, \mathbb{Z})$$

Consider our complete flag $\pi_o \subset \pi_1 \subset \quad \subset \pi_{n-1} \subset \mathbb{P}^n$.
It is well known that a basis of $H^n(F_{0,n-1}, \mathbb{Z})$ is given by the classes dual to the following Schubert subvarieties:

$$Y_i = \{(p,\pi) \mid p \subset \pi_i \subset \pi\}.$$

On the other hand it follows easily from our definition of Y that $g_* f^*([Y_i]) = [\bar{D}_i]$ so that $g_* f^*$ is an isomorphism.

Furthermore if we fix a quadric $Q \in \bar{X}_o$ and we embed it in $F_{0,n-1}$ by associating to each point in Q its tangent flag we get that $g_* f^*([Q]) = [\bar{D}]$ so that in order to prove our claim it is sufficient to show that

$$[Q] = \sum_{i=0}^{n-1} 2[Y_i] \quad \text{in} \quad H^2(F_{0,n-1}, \mathbb{Z})$$

Denote by $Y'_o, \ldots, Y'_{n-1}$ the Schubert cycles dual to $Y_o, \ldots, Y_{n-1}$; i.e. $Y'_i = \{(p,\pi) \mid p \subset \pi_{n-i},\ \pi \supset \pi_{n-i-1}\}$. We are reduced to show that the evaluation on the class of a point in $F_{0,n}$ of $[Q]\cdot[Y_i]$ is 2 for each $0 \leq i \leq n-1$. This is clear by elementary considerations on the geometry of quadrics.

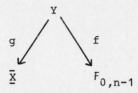

3) We first show that $\bar{D}_i \not\supset F$ for each $0 \leq i \leq n-1$. Assume the contrary and let $s \in H^o(\underline{\bar{X}}, L_{2\omega_i})$ be a section whose divisor is $\bar{D}_i$. The restriction of s to F is zero. On the other hand it follows from our results of section 8 that the restriction homomorphism

$$j^*: H^o(\underline{\bar{X}}, L_{2\omega_i}) \to H^o(F, L_{2\omega_i}|F)$$

is an isomorphism.

We now show our result for $\bar{D}$. For this, given a non singular quadric $Q \in X_o$, define a flag $f \in F$ to be tangent to Q if the point of f lies in Q and the hyperplane of f is the hyperplane tangent to Q in this point. Consider the variety $Z \subset \underline{\bar{X}} \times F$ which is the closure of the correspondence $\tilde{Z} = \{(Q,f) | Q \in X_o,$ f is tangent to $Q\}$. Consider the fibration $p: \underline{\bar{X}} \times F \to \underline{\bar{X}} \times F_{0,n-i}$ induced by the natural fibration $q: F \to F_{0,n-i}$. Then we claim $Z = p^{-1}(Y)$. This is clear since $\tilde{Z} = p^{-1}(\tilde{Y})$. This allows us to determine the fiber of the projection $g: Z \to \underline{\bar{X}}$ over a point f_o in the closed orbit.

In fact think of f_o as a flag $f_o = \{\pi_o \subset \pi_1 \subset \ldots \subset \pi_{n-1} \subset \mathbb{P}^n\}$ and for each $f \in g^{-1}(f_o)$ put $q(f) = (p,\pi)$. We claim that $g^{-1}(f_o) = \cup Z_i$, where $Z_i = \{f | p \subset \pi_i \subset \pi\}$.

To see this notice that the image of f_o in $\mathbb{P}(V^*_{2\omega_i})$ under the morphism $\underline{\bar{X}} \to \mathbb{P}(V^*_{2\omega_i})$ represents a degenerate quadric in $\mathbb{P}(V_{\omega_i})$ whose intersection with the Grassmannian of i-1 dimensional subspaces is just the set of such subspaces intersecting π_{n-1}.

Thus if $f \in g^{-1}(f_o)$ its (i-1) dimensional subspace has to meet π_{n-i}. In particular $p \in \pi_{n-1}$.

Assume $p \in \pi_i - \pi_{i-1}$. We claim $\pi \supset \pi_i$. In fact if $i \geq 1$ each (n - i) dimensional subspace τ with $p \in \tau \subset \pi$ has to meet π_{i-1} by the above remarks, and if i = 0 there is nothing to prove. So $f \in Z_i$. Having shown this it is easily seen that given f_o in the closed orbit of $\underline{\bar{X}}$ such that π_i is not tangent to Q for all $0 \leq i \leq n-1$, $f_o \notin \bar{D}$ proving 3).

COROLLARY. The evaluation at the class of a point of any monomial of the form

$$(2\omega_1)^{h_1} \ldots\ldots (2\omega_n)^{h_n} (2 \sum_{i=1}^{n} 2\omega_i)^{h_{n+1}}$$

with $\sum_{i=1}^{n+1} h_i = \frac{(n+1)(n+2)}{2} - 1 = \dim \underline{\bar{X}}$ gives the number of quadrics which are simultaneously tangent to h_1 points, h_2 lines,....., h_n hyperplanes, h_{n+1} quadrics lying in general position.

REMARK. Our proof of the fact that $\bar{D} \not\supset F$ works also in the case in which $\bar{D}$ is the closure in $\bar{X}$ of the hypersurface of X_o of quadrics tangent to any fixed subvariety in $\mathbb{P}^n$. Thus since $[\bar{D}]$ can be written as a linear combination of the $[\bar{D}_i]$'s the problem of enumerating the number of quadrics simultaneously tangent to $\frac{(n+1)(n+2)}{2} - 1$ subvarieties in general position is reduced to the same problem for linear spaces. This fact has been recently shown in a much greater generality by Fulton, Kleiman, Mac Pherson.

In the case of $\mathbb{P}^3$ working out the computations with the algorithm given in 9.2 one finds the following table which can also be found in Schubert's book (p. 105):

$x_1^9 = x_3^9 = 1$	$x_1^6x_2^2x_3 = x_3^6x_2^2x_1 = 12$
$x_1^8x_2 = x_3^8x_2 = 2$	$x_1^5x_2^3x_3 = x_3^5x_2^3x_1 = 24$
$x_1^7x_2^2 = x_3^7x_2^2 = 4$	$x_1^4x_2^4x_3 = x_3^4x_2^4x_1 = 48$
$x_1^6x_2^3 = x_3^6x_2^3 = 8$	$x_1^6x_2x_3^2 = x_3^6x_2x_1^2 = 18$
$x_1^5x_2^4 = x_3^5x_2^4 = 16$	$x_1^5x_2^2x_3^2 = x_3^5x_2^2x_1^2 = 36$
$x_1^4x_2^5 = x_3^4x_2^5 = 32$	$x_1^4x_2^3x_3^2 = x_3^4x_2^3x_1^2 = 72$
$x_1^3x_2^6 = x_3^3x_2^6 = 56$	$x_1^5x_2x_3^3 = x_3^5x_2x_1^3 = 34$
$x_1^2x_2^7 = x_3^2x_2^7 = 80$	$x_1^4x_2^2x_3^3 = x_3^4x_2^2x_1^3 = 68$
$x_1x_2^8 = x_3x_2^8 = 92$	$x_1^4x_2x_3^4 = 42$
$x_2^9 = 92$	$x_1^2x_2^6x_3 = x_3^2x_2^6x_1 = 104$
$x_1^8x_3 = x_3x_1^8 = 3$	$x_1^3x_2^5x_3 = x_3^3x_2^5x_1 = 80$
$x_1^7x_3^2 = x_3^2x_1^7 = 9$	$x_1^3x_2^4x_3^2 = x_3^3x_2^4x_1^2 = 112$
$x_1^6x_3^3 = x_3^3x_1^6 = 17$	$x_1x_2^7x_3 = 104$
$x_1^4x_3^5 = x_3^5x_1^4 = 21$	$x_1^2x_2^5x_3^2 = 128$
$x_1^7x_2x_3 = x_3^7x_2x_1 = 6$	$x_1^3x_2^3x_3^3 = 104$

and so

$$(2(x_1 + x_2 + x_3))^9 = 666.841.088$$

REFERENCES

[1] A.R. ALGUNEID: Complete quadric primals in four dimensional space. Proc. Math. Phys. Soc. Egypt, 4, (1952), 93-104.

[2] BIALYNICKI-BIRULA: Some theorems on actions of algebraic groups. Ann. of Math., 98, 1973, 480-497.

[3] A. BOREL: Sur la cohomologie des espaces fibrés principaux et des espaces homogènes de groupes de Lie compacts. Ann. of Math., 57, 1953, 116-207.

[4] R. BOTT: Homogeneous vector bundles. Ann. of Math. 66, 1957, 203-248.

[5] M. DEMAZURE: Limites de groupes orthogonaux ou symplectiques. Preprint 1980.

[6] G. GHERARDELLI: Sul modello minimo delle varietà degli elementi differenziali del 2° ordine del piano proiettivo. Rend. Acad. Lincei, (7) 2, 1941, 821-828.

[7] G.H. HALPHEN: Sur la recherche des points d'une courbe algégrique plane. In "Journal de Mathématique", 2, 1876, 257.

[8] HARISH-CHANDRA: Spherical functions on a semisimple Lie group I. Amer. J. of Math., 80, 1958, 241-310.

[9] S. HELGASON: A duality for symmetric spaces with applications to group representations. Advances in Math. 5, 1-154, (1970).

[10] S. HELGASON: Differential geometry, Lie groups, and symmetric spaces. Acad. Press 1978.

[11] S. KLEIMAN: Problem 15. Rigorous foundation of Schubert enumerative calculus. Proceedings of Symp. P. Math. 28, A. M. S., Providence (1976).

[12] S. KLEIMAN: The transversality of a general translate. Comp. Math., 28, 1974, 287-297.

[13] D. LUNA, T. VUST: Plongements d'espaces homogenès. Preprint.

[14] H. SCHUBERT: Kalkül der abzählenden geometrie. Liepzig 1879 (reprinted Springer Verlag 1979).

[15] J.G. SEMPLE: On complete quadrics I. J.London Math. Soc. 23, 1948, 258-267.

[16] J.G. SEMPLE: The variety whose points represent complete collinea tions of S_r on S'_r. Rend. Mat. 10, 201-280 (1951).

[17] J.G. SEMPLE: On complete quadrics II. J. London M. S. 27, 280-287 (1952).

[18] F. SEVERI: Sui fondamenti della geometria numerativa e sulla teo ria delle caratteristiche. Atti del R. Ist. Veneto, 75, 1916, 1122-1162.

[19] F. SEVERI: I fondamenti della geometria numerativa. Ann. di Mat., (4) 19, 1940, 151-242.

[20] E. STUDY: Uber die geometrie der kegelschnitte, insbesondere dere charakteristiken problem. Math. Ann., 26, 1886, 51-58.

[21] J.A. TYRELL: Complete quadrics and collineations in S_n. Mathematika 3, 69-79 (1956).

[22] I. VAISENCHER: Schubert calculus for complete quadrics. Preprint.

[23] B.L. VAN DER WAERDEN: Z.A.G. XV, Losung des charakteristiken-problem für kegelschnitte. Math. Ann. 115, 1938, 645-655.

[24] J. VUST: Opération des groupes réductifs dans un type de cônes presque homogènes. Bull. Soc. Math. France, 102, 1974, 317-333.

[25] H.G. ZEUTHEN: Abzählende methoden der geometrie. Liepzig 1914.

[26] A. BIALYNICKI-BIRULA: Some properties of the decomposition of algebraic varieties determined by actions of a torus. Bull. Acad. Polon. Sci. Ser. Sci. Math. Astronom. Phys. 24 (1976) n. 9, 667-674.

[27] R. STEINBERG: Générateurs, relations et revêtements de groupes algébriques, p. 113-127, Collq. Theorie des Groupes Algébriques, Gauthier Villars (1962).

[28] R. STEINBERG: Endomorphisms of linear algebraic groups, Mem. of the A.M.S. n. 80 (1968).

GEOMETRIC INVARIANT THEORY AND APPLICATIONS TO MODULI PROBLEMS

D. Gieseker
University of California
Los Angeles, California 90024 USA

These notes are a brief introduction to geometric invariant theory (GIT) and also contain two applications of that theory to the construction of moduli spaces in algebraic geometry. The first two sections sketch the basics of GIT over the complex numbers. In §3 we connect GIT and the theory of stable bundles of rank two on a non-singular curve. We then consider in §4 the relation between smooth curves and GIT. Our main result here is that there are enough projective invariants of space curves to separate any two projectively distinct smooth curves of genus g and degree d provided $d \geq 2g$ and that the curves are non-degenerate. This result can be used to construct a moduli space $\mathfrak{m}_g$ for smooth curves of genus g. In sections §5 and §6, we look at the connection between stable curves in the sense of Mumford and Deligne and stable curves in the senses of GIT. The main result is essentially that the compactification $\overline{\mathfrak{m}_g}$ of $\mathfrak{m}_g$ considered by Mumford and Deligne is a projective variety. (This result was originally obtained by F. Knutsen in characteristic zero using other methods.) Finally in §7 we indicate how GIT can be used to construct compactified generalized Jacobians of stable curves. Here we consider the example of an irreducible curve with one node. The nature of the compactification of the generalized Jacobian of a general stable curve obtained by GIT has yet to be worked out. One can also extend the results of §5,6,7 to vector bundles of rank two [G - M, G_4]. Roughly, one gets a construction of a projective moduli space of stable bundles on an irreducible curve which has one node. This can then be used to study the topology of the moduli space of stable bundles on a smooth curve by degeneration methods.

The original source for the first two sections is [M_1], but [N] also provides a more leisurely treatment. A connection between GIT and the theory of stable bundles on a smooth curve was worked out by Mumford and Seshadri. [N] contains an account of this work. In these notes, we make a slightly different connection which is more suitable for higher dimensional varieties [G_1,Ma]. Mumford gave a proof of the existence of $\mathfrak{m}_g$ using GIT in [M_1] using the Chow variety of a space curves. Here we use Grothendieck's Hilbert scheme which is arguably easier. [G_2] contains an extension of these ideas to the n canonical images of surfaces of general type. The connection between GIT and stable curves was worked out jointly by Mumford and myself using the Chow variety and Hilbert scheme [M_2,G_3]. Finally an exhaustive discussion of the developments in GIT since the first edition of Mumford's book and the present can be found in the second edition of Mumford's book.

§1. Let k be an algebraically closed field and let W be a vector space of dimension n. Let G be the algebraic group $SL(W,k)$. Suppose that V is a vector space of dimension ℓ and that G acts on $V^* = Hom(V,k)$. If $x \in V^*$ and $\sigma \in G$, we will denote the action of σ on x by x^σ. G acts on $\mathbb{P}(V)$, the hyperplanes in V. We would like to be able to form a reasonable quotient of $\mathbb{P}(V)$ by G as a projective variety. Unfortunately, it is usually impossible to form such a quotient. At least, one would hope for a map $m : \mathbb{P}(V) \to \mathbb{P}^K$ which is G invariant and separates orbits. As the following example shows, no such map can exist in general.

Example:

Let $G = SL(3)$ and let $V^* = S^3(k^3)$ with the natural action of G. Choose a $\lambda \in k$, $\lambda \neq 0,1$. For each $t \in k$, define a element of V^* by

$$P_t(X_0,X_1,X_2) = X_0^2X_2 - X_1(X_1 - tX_2)(X_1 - t\lambda X_2)$$

where X_0,X_1,X_2 is a basis for k^3. Let $\overline{P}_t \in \mathbb{P}(V)$ denote the point corresponding to P_t. Note that all the $\overline{P}_t$ are in the same G orbit as $\overline{P}_1$ if $t \neq 0$, but that $\overline{P}_0$ is not in the same G orbit as $\overline{P}_1$. Indeed $\overline{P}_1$ defines a smooth elliptic curve in $\mathbb{P}^2$, but $\overline{P}_0$ defines a cubic curve with a cusp. There is a morphism $F : \mathbb{A}^1 \to \mathbb{P}(V)$ so that $F(t) = \overline{P_t}$. Suppose there is a G invariant map m of $\mathbb{P}(V)$ to $\mathbb{P}^K$ for some N which separates G orbits. Then $m \circ F$ is constant on $\mathbb{A}^1 - \{0\}$ so $m(F(1)) = m(F(o))$. So no such map can exist.

The idea behind GIT is to find a large G invariant open set U of $\mathbb{P}(V)$ so that one can form a reasonable quotient of U by G. For instance in the above example, one must either exclude $\overline{P}_0$ or $\overline{P}_1$ from U. Since an arbitrary smooth elliptic curve is projectively equivalent to $\overline{P}_1$ for some appropriate λ, one sees that one must exclude $\overline{P}_0$ from any U.

Let $x \in V^* - \{0\}$ and let $\overline{x}$ be the corresponding point in $\mathbb{P}(V)$. We define $\rho_x : G \to V^*$ by $\rho_x(\sigma) = x^\sigma$. It is a fact that ρ_x is proper if and only if $x^G = \rho_x(G)$ is closed and the stabilizer of x is finite [N, Lemma 3.17]. Here are the the fundamental definitions of GIT:

Definition 1.1:

1) $\overline{x}$ is stable if ρ_x is proper .
2) $\overline{x}$ is semi-stable if $0 \notin \overline{x^G}$.
3) $\overline{x}$ is weakly stable if x^G is closed.

Here $\overline{x^G}$ is the closure of orbit of x in V^*. Note that we are interested in the orbit of x in V^*, not in the orbit of $\overline{x}$ in $\mathbb{P}(V)$. Let $U_{S.}$ (resp. $U_{S.S.}$) be the set of stable (resp. semi-stable) points. Note that there is much confusion in the literature over these definitions. In particular, weakly stable is not standard terminology.

We wish to define a map $m : U_{S.S.} \to \mathbb{P}^K$ enjoying various pleasant properties.

The method of GIT is to choose a large N and look at a basis $P_0,\ldots,P_K$ of the homogeneous polynomials of degree N which are invariant under G. We then define

$$m(\bar{x}) = (P_0(x),\ldots,P_K(x)) \in \mathbb{P}^K .$$

Of course m is not defined if all the P_i vanish at x. The main result of GIT is the following:

Theorem 1.2:

For N sufficiently large, the map m is defined exactly on $U_{S.S.}$. Further if $\bar{x}$ and $\bar{x}'$ are weakly stable and $m(\bar{x}) = m(\bar{x}')$, then $\bar{x}$ and $\bar{x}'$ are in the same G orbit. Also if $X \subseteq U_{S.S.}$ is closed and G invariant, then $m(X)$ is closed in $\mathbb{P}^K$.

The following corollary is a frequently used consequence of the last statement of Theorem 1.2.

Corollary 1.2.1:

Let S be a smooth curve, $P \in S$, and suppose f is a map of $S - P$ to $U_{S.S.}$. Then there is a curve S', a map $\pi : S' \to S$ and a point $Q \in S'$ so that $\pi(Q) = P$, and a map f' of $S' \to U_{S.S.}$ and a map $h : S' - Q \to G$ so that for $x \in S' - Q$,

$$f(\pi(x)) = (f'(x))^{h(x)} .$$

We may further choose f' so that $f'(Q)$ is stable or has stabilizer of positive dimension.

This corollary is often referred to as the semi-stable replacement property of GIT.

One can say more about the quotient of $U_{S.S.}$. The reader is referred to [N]. The proof of Theorem 1.2 is due to Mumford in characteristic zero. In characteristic p, Nagata first reduced the problem to a conjecture of Mumford. This conjecture was then established by Haboush and independently by Formenck and Procesi.

In these notes we will content ourselves with sketching the proof of Theorem 1.2 when $k = \mathbb{C}$. To start the proof, we put a positive definite hermetian inner product on W, and let $U \subseteq G = SL(W)$ be the special unitary group. The first observation is that U is Zariski dense in G. Indeed, let G' be the Zariski closure of U. The tangent space of G' at the identity contains the complexified tangent space of U. But the tangent space of U consists of traceless matrices with $A = -\overline{A^t}$, where — denotes conjugation. The $\mathbb{C}$ span of the tangent space of U is the tangent space to $SL(n,\mathbb{C})$. Thus $\dim G' = \dim G$, since all algebraic groups are non-singular over $\mathbb{C}$. Thus $G' = G$.

Suppose G acts linearly on a vector space V_1. We claim there is a U invariant positive definite hermetian form on V_1. Let V_2 be the vector space of forms on V_1, linear in the first variable and conjugate linear in the second. Let $h \in V_2$ be a positive definite Hermetian form. Let $B \subseteq V_2$ be the convex hull of the orbit of h under U. B is compact and U invariant, and every element of B is positive definite Hermetian. Now the centroid of B is invariant under U and gives the desired inner product.

Next we note that a subspace $V' \subseteq V$ is G invariant if and only if it is U invariant. Indeed, the stabilizer of V' is a Zariski closed subset of G.

Next we come to the Reynolds operator. Note that V is a direct sum of irreducible representations using the U invariant hermitian inner product. Let V^{inv} be the invariant vectors in V. Then we can construct a G invariant projection from V to V^{inv}. Now let $R_m : S^m(V) \to S^m(V)^{inv}$ be this projection. Let $g \in S^n(V)$ be invariant. We have the following Reynold's identity: If $f \in S^m(V)$, (1.2.2) $R_{n+m}(fg) = gR_m(f)$. Indeed, define

$$F(f) = R_{n+m}(fg) - gR_m(f).$$

Then F is a G invariant map from $S^m(V)$ to $(S^{m+n}(V))^{inv}$ which kills $S^m(V)^{inv}$. On the other hand, if $V_1 \subseteq S^m V$ is a non-trival irreducible subspace, the map of V_1 to $S^{n+m}(V)^{inv}$ is trivial. So F is trivial. So (1.2.2) is established.

Note every invariant homogeneous polynomial P of positive degree must vanish on all points which are not semi-stable. Indeed, P is constant on any orbit and hence on the closure of any orbit. So such a polynomial must vanish if $0 \in \overline{xG}$.

Let's check that if $x \in V^*$ is semi-stable, then there is a k and a $P \in S^k(V)^G$ so that $P(x) \neq 0$. First, note that since $\{0\}$ and $\overline{xG}$ are closed disjoint sets, there is a polynomial P which vanishes at 0 and which is identically one on $\overline{xG}$. (The ideals I_1 and I_2 of $\{0\}$ and $\overline{xG}$ in $\mathbb{C}[V]$ generate $\mathbb{C}[V]$ by the Nullstellensatz.) Next suppose I_1 and I_2 are invariant ideals in $\mathbb{C}[V]$ and let $\bar{f}_i \in \mathbb{C}[V]/I_i$ be invariant. The if $I_1 + I_2 = \mathbb{C}[V]$, we can find an invariant $f \in \mathbb{C}[V]$ so that f maps to $\bar{f}_i \in \mathbb{C}[V]/I_i$. Indeed, let $\mathbb{C}[V]_n$ be the polynomials of degree $\leq n$. We have a map

$$\varphi_n : \mathbb{C}[V]_n \to \mathbb{C}[V]_n/(I_1 \cap \mathbb{C}[V]_n) \oplus \mathbb{C}[V]_n/(I_2 \cap \mathbb{C}[V]_n).$$

We choose n so that $\bar{f}_i \in \mathbb{C}[V]_n/I_i \cap \mathbb{C}[V]_n$ and so that $(\bar{f}_1, \bar{f}_2)$ is in the image of φ_n. But then there is an invariant $f \in \mathbb{C}[V]_n$ so that $\varphi_n(f) = (\bar{f}_1, \bar{f}_2)$. Hence there is an invariant polynomial f which vanishes at 0 but which is identically one on $\overline{xG}$. Now write $f = \sum_{i=0}^{n} f_i$ as a sum of its homogeneous parts. Thus $f_0 = 0$, but some $f_i(x) \neq 0$. Similarly, one shows that any two weakly stable orbits can be separated by an invariant polynomial.

Next we claim that $C[V]^G$ is finitely generated. Let $R = \oplus R_m$ be the Reynolds operator. Let I be the ideal in $C[V]$ generated by homogeneous invariant polynomials of positive degree. Let $P_1,\ldots,P_m \in I$ be generators of I as a $C[V]$ module, where the P_i are homogeneous and invariant. We claim that $\{1,P_1,\ldots,P_m\}$ generate $C[V]^G$ as a C algebra. We assume inductively that $1,\ldots,P_m$ generate $C[V]^G$ in degree $< n$. Given P homogeneous and invariant of degree n, we can write

$$P = \sum Q_i M_i$$

where the Q_i are monomials of degree $< n$ and the M_i are monomials in $P_1,\ldots,P_m$. Now

$$P = R(P) = \sum R(Q_i)M_i$$

From the induction hypothesis, $R(Q_i)$ are polynomials in $1,\ldots,P_m$. So we can choose an N and $P_0,\ldots,P_K \in S^N(V)^G$ so that $1,P_0,\ldots,P_K$ generate the ring $\oplus S^{Nk}(V)^G = R'$. Let $C[Y_0,\ldots,Y_K]$ be a polynomial ring and consider the surjection

$$\psi : C[Y_0,\ldots,Y_K] \to R'$$

Let $X \subseteq U_{S.S.}$ be closed and G invariant. Let I_X be the ideal of functions in $C[V]^G$ vanishing on X, let $m : U_{S.S.} \to P^K$ be the map defined by $P_0,\ldots,P_K$, and let $I_1 = \psi^{-1}(I_X \cap R')$. Then $\overline{m(X)}$ is defined by the ideal I_1. If $\overline{m(X)} \neq m(X)$, there is a homogeneous ideal $I_2 \supseteq I_1$ so that I_2 maps to a nontrivial ideal I_3 in R', but so that for each point $Q \in U_{S.S.}$, some element of I_3 does not vanish at Q. Let I_4 be the ideal of functions in $C[V]$ vanishing on the non-semistable points. Then $I_4 \subseteq \sqrt{I_3 . C[V]}$. Using the Reynolds operator, we see that

$$M \subseteq \sqrt{I_3}$$

where M is the ideal generated by homogeneous functions of positive degree in R'. This contradicts the nontriviality of I_3.

§2. It is difficult to check directly that a point $x \in V^*$ is stable. However, there is a convenient test called the numerical critereon. Recall that $\mathbb{G}_m$ is just the multiplicative group k^* as an algebraic group. By definition, a one parameter subgroup of G(1 - P - S) is a non-trivial map $\lambda : \mathbb{G}_m \to G = SL(W)$. It is well known that there is then a basis $e_1, \dots, e_n$ of W and integers r_i so that

$$e_i^{\lambda(\alpha)} = \alpha^{r_i} e_i .$$

We will say that e_i has λ weight r_i. Of course, $\sum r_i = 0$, since $G = SL(W)$. (Over $\mathbb{C}$, it is not difficult to see that W decomposes into one dimensional subspaces using unitary techniques.) If $\mu : \mathbb{G}_m \to \mathbb{A}^n = k^n$ is any map, we will use the symbolism $\lim_{\alpha \to 0} \mu(\alpha)$. If μ extends to a map of $\mathbb{A}^1 \supseteq \mathbb{G}_m$ to W, we define

$$\lim_{\alpha \to 0} \mu(\alpha) = \mu(0)$$

Otherwise, we say $\lim_{\alpha \to 0} \mu(\alpha) = \infty$. This means that at least one component of μ goes to ∞ as $\alpha \to 0$. We say $x \in V^*$ is λ-stable (resp. λ semi-stable) if $\lim_{\alpha \to 0} x^{\lambda(\alpha)} = \infty$ (resp. $\lim_{\alpha \to 0} x^{\lambda(\alpha)} \neq 0$).

<u>The numerical criteron</u>: x is stable (resp semi-stable) if and only if x is λ stable (resp. λ semi-stable) for all 1 - P - S of G.

<u>Proof</u>:

Let $D = \operatorname{Spec} k[[t]]$ and let $D^* = \operatorname{Spec} k((t))$. Suppose $x \in V^*$ is not stable. Consider the orbit map $\rho_x : G \to V^*$. (One can regard D and D^* as the unit disk and punctured unit disk over $\mathbb{C}$.) By the valuative critereon for properness, we can find a map $\mu : D^* \to G$ which does not extend to a map of D to G, but so that $\rho_x \circ \mu$ does extend to a map of D to V^*. Let $K = k((t))$ and $\mathcal{O} = k[[t]]$. Then μ is equivalent to an element of $SL(n, K)$. Using the theory of elementary divisors, we can find σ_1 and σ_2 in $SL(n, \mathcal{O})$ so that

$$\sigma_1 \mu \sigma_2 = \begin{pmatrix} t^{r_1} & & 0 \\ & \ddots & \\ 0 & & t^{r_n} \end{pmatrix} = \lambda(t)$$

Using $\sigma_1(0)$ to change basis in k^n, we may further assume that $\sigma_1(0)$ is the identity. We claim that $\lim_{\alpha \to 0} x^{\lambda(t)}$ exists. Let $v_1, \dots, v_\ell$ be a basis of V^* so that

$$v_i^{\lambda(t)} = t^{s_i} v_i .$$

We may assume that $s_1 \le s_2 \le \dots$. Now write $x = \sum x_j v_j$, and assume $x_i \neq 0$ but

$x_j = 0$ for $j < i$. We can write

$$v_i^{\sigma_1^{-1}(t)} = \sum a_{ij}(t) v_j .$$

Thus $a_{ij}(0) = \delta_{ij}$. So

$$\begin{aligned} x^{\mu(t)\sigma_2(t)} &= \sum_i x_i v_i^{\mu(t)\sigma_2(t)} \\ &= \sum_i x_i v_i^{\sigma_1^{-1}(t)\lambda(t)} \\ &= \sum_j \left(\sum_i a_{ij}(t) x_i \right) v_j^{\lambda(t)} \\ &= \sum_j \left(\sum_i a_{ij}(t) x_i \right) t^{s_j} v_j \end{aligned}$$

Now $\lim_{t \to 0} x^{\mu(t)\sigma_2(t)}$ is defined, so if $s_j < 0$, we must have $\sum_i a_{ij}(0) x_i = x_j = 0$. Thus $\lim_{t \to 0} x^{\lambda(t)}$ exists.

A similar argument shows that if x is not semi-stable, then there is a $1 - P - S$ λ with $\lim_{t \to 0} x^{\lambda(t)} = 0$.

§3. We wish to connect geometric invariant theory to the theory of stable bundles on a smooth curve X. For simplicity we will consider bundles of rank two. Fix a line bundle L of degree $d \gg g$. Let S be the set of bundles E of rank two with $\overset{2}{\Lambda} E \cong L$. Recall that a bundle E of rank two is stable if for every quotient line bundle $E \to L \to 0$, we have

$$\deg L > \frac{\deg E}{2} .$$

Notice first that if $E \in S$ is stable, then E is generated by global sections and $H^1(E) = 0$. Indeed, recall that if s is a section of a bundle F on X, then there is subbundle $F' \subseteq F$ of rank one so that $s \in H^0(F') \subseteq H^0(F)$. In particular $\deg F' \geq 0$. Thus if $H^1(E) \neq 0$, Serre duality shows that $H^0(E^{-1} \otimes \Omega^1) \neq 0$, so $E^{-1} \otimes \Omega^1$ has a sub bundle of non-negative degree. So E has a quotient bundle of degree $\leq 2g - 2$. This contradicts our assumptions that $d \gg g$ and that E is stable. Similary $H^1(E(-P)) = 0$ for any point $P \in X$, so E is generated by global sections.

We next associate to each stable $E \in S$ a point in some projective space modulo the action of an algebraic group. We fix a basis $s_1, \ldots, s_k$ of $H^0(E)$, where $k = d + 2(1 - g)$. Such a basis gives an isomorphism φ of $H^0(E)$ with C^k. There is a canonical map

$$\overset{2}{\Lambda} H^0(E) \to H^0(\overset{2}{\Lambda} E) .$$

obtained by sending $s_1 \wedge s_2$ to the corresponding section in $\overset{2}{\Lambda} E$. We have assumed that $\overset{2}{\Lambda} E \cong L$, so there is an canonical isomorphism of $H^0(\overset{2}{\Lambda} E)$ with $H^0(L) = V$ determined up to scalar multiple. On the other hand, φ determines an isomorphism of $\overset{2}{\Lambda} H^0(E)$ with $\overset{2}{\Lambda} C^k$. Putting these together, we obtain a map depending on E and φ:

$$T_{E,\varphi} \in \mathrm{Hom}(\overset{2}{\Lambda} C^k, V) .$$

We claim that if $T_{E,\varphi} = T_{F,\varphi'}$ and E is generated by global sections, then $E \cong F$. Indeed, since E is generated by global sections, there is a map from X to G', the grassmannian of all two dimensional quotients of C^k. Via the Plücker coordinates, G' is embedded in $\mathbf{P}(\overset{2}{\Lambda} C^k)$. In fact, the induced map of X to $\mathbf{P}(\overset{2}{\Lambda} C^k)$ is given by the linear system generated by $s_i \wedge s_j$ as sections of $H^0(\overset{2}{\Lambda} E) = H^0(L)$. So $T_{E,\varphi}$ determines E. Thus isomorphism classes of bundles E generated by global sections with $H^1(E) = 0$ map injectively to the orbits of $\mathbf{P}(\mathrm{Hom}(\overset{2}{\Lambda} C^K, V))$ under $SL(k,C)$.

We claim that if E is a stable bundle, then $T_{E,\varphi}$ is stable in the sense of GIT. Suppose that $T_{E,\varphi}$ is not stable. There is a 1 - P - S λ so that $T_{E,\varphi}$ is not λ-stable. Let v_i be a basis of $H^0(E) = C^k$ so that $v_i^{\lambda(\alpha)} = \alpha^{r_i} v_i$. Thus $v_i \wedge v_j = 0$ in $H^0(\overset{2}{\Lambda} E)$ if $r_i + r_j < 0$. Order the r_i so that $r_1 \le r_2 \le \cdots$. Let M be the subline bundle of E so that v_1 is a section of M. Note that $v_1 \wedge v_j = 0$ if $r_1 + r_j < 0$. Thus $v_j \in H^0(M)$ if $r_1 + r_j < 0$. $N = \#\{j \mid r_j < |r_1|\}$. Then $h^0(M) \ge N$. If $h^1(M) \ne 0$, then $h^0(M) \le g$, while if $h^1(M) = 0$, $h^0(M) = \deg M + 1 - g < \frac{1}{2}(\deg E + 2(1-g)) = k/2$, since we have assumed E is stable. Thus $N < k/2$. For more than $\frac{k}{2}$ of the j, we have $r_j > |r_1|$. This contradicts $\sum r_i = 0$.

One can continue this line of reasoning to produce a quasi-projective moduli space for stable bundles, even over smooth varieties of higher dimension [G_1,Ma].

§4. Our object in this section is to construct a moduli space for smooth curves of genus g. This was first done by GIT by Mumford $[M_1, M_2]$ using Chow points. Here we will work with the Hilbert point of a curve. Full details may be found in $[G_3]$.

Let $X \subseteq \mathbb{P}(W)$ be a closed subscheme. We let $P_X(n) = \chi(\mathcal{O}_X(n))$. Recall that P_X is a polynomial in n, called the Hilbert polynomial of X. Let P be a fixed polynomial and consider the set U of subschemes of $\mathbb{P}(W)$ with Hilbert polynomial P. We will assign to each $X \in U$ a point $H_m(X)$ in some projective space $\mathbb{P}^M$ which will classify X as a subscheme of $\mathbb{P}(W)$. Further $G = SL(W,k)$ will operate on $\mathbb{P}^M$ and two subschemes X and X' will be projectively equivalent if and only if $H_m(X)$ and $H_m(X')$ are in the same orbit. We will then investigate the stability of $H_m(X)$. We will say X is m-stable (or m-Hilbert stable) if $H_m(X)$ is stable. The main result of invariant theory can be phrased as follows: There are enough projective invariants of subschemes of $\mathbb{P}(W)$ to separate any two m-stable subschemes which are not projectively equivalent.

One case which is of particular interest is that of n canonically embedded curves. Let $\dim W = (2n-1)(g-1)$ and let X be a smooth projective curve of genus g. Choosing an isomorphism of W with $H^0(X,\mathcal{O}(nK))$ determines an embedding of X into $\mathbb{P}(W)$ if $n \geq 2$. One of the main results of these notes will be that this embedding is m-stable for $m >> 0$. Notice that the n canonical images of two curves are projectively equivalent if and only if the curves are isomorphic. Thus the projective invariants of an n canonical curve form moduli for the curve.

We turn to Grothendieck's construction of the mth Hilbert point $H_m(X)$. Let $X \in U$. Note $H^0(\mathbb{P}(W), \mathcal{O}(m)) = S^m(W)$. There is a natural map

$$\psi : S^m(W) \to H^0(X, \mathcal{O}_X(m))$$

The elements of $S^m(W)$ are just homogeneous polynomials of degree m on $\mathbb{P}(W)$, and ψ is just the restriction map. The kernel of ψ consists in the homogeneous polynomials which vanish on X. According to fundamental results of Grothendieck, there is an M depending only on P so that for $m \geq M$, ψ is surjective, the kernel of ψ determines X, and $\dim H^0(X,\mathcal{O}(m)) = P(m)$. Consider

$$\psi' = \overset{P(m)}{\Lambda}\psi : \overset{P(m)}{\Lambda} S^m(W) \to \overset{P(m)}{\Lambda} H^0(X,\mathcal{O}(m)) \cong k.$$

Then ψ' is onto and hence determines a point $H_m(X)$ in $\mathbb{P}(\overset{P(m)}{\Lambda} S^m(W))$. Note that $H_m(X)$ determines $\overset{P(m)}{\Lambda}\psi$ up to a scalar. Further, $\overset{P(m)}{\Lambda}\psi$ are the Plücker coordinates of $\ker \psi$. Note finally that $\ker \psi$ determines X. Hence $H_m(X)$ determines X.

We will now discuss the stability of $H_m(X)$ under G. Let λ be a 1 - P - S of G and let $X_o, \ldots, X_{n-1}$ be a basis of W so that

$$X_i^{\lambda(\alpha)} = \alpha^{r_i} X_i .$$

We will say that X_i has λ weight r_i. A monomial $M = X_0^{i_0} \cdots X_{n-1}^{i_{n-1}}$ will be said to have λ weight $\sum i_j r_j$. The λ weight of M is denoted $w_\lambda(M)$.

Proposition 4.1.

$H_m(X)$ is λ-stable if and only if there are monomials $M_1,\ldots,M_{P(m)}$ in $S^m(W)$ so that the following two conditions hold:

i) $\psi(M_1),\ldots,\psi(M_{P(m)})$ are a basis for $H^0(X,\mathcal{O}(m))$.

ii) $\sum w_\lambda(M_i) < 0$.

Proof:

$$\left(\overset{P(m)}{\Lambda}\psi\right)^{\lambda(\alpha)}(M_1 \wedge \cdots \wedge M_{P(m)}) = \alpha^{\sum w_\lambda(M_i)} \psi(M_1) \wedge \cdots \wedge \psi(M_{P(m)}).$$

For $H_m(X)$ to be λ stable, there must exist monomials $M_1,\ldots,M_{P(m)}$ sc that $\sum w_\lambda(M_i) < 0$ and $\psi(M_1) \wedge \cdots \wedge \psi(M_{P(m)}) \neq 0$.

We can obtain a critereon for semi-stability by substituting (ii bis) $\sum \omega_\lambda(M_i) \leq 0$ for condition (ii).

Next, we suppose X is a smooth curve of degree d and genus g. We suppose $X \subseteq \mathbb{P}(W)$ embedded by a complete linear system and that the genus of X is at least one.

Our aim is to sketch a proof of the following result:

Theorem 4.2.

For $m \gg 0$, $H_m(X)$ is stable if

i) X is not contained in any hyperplane

ii) $d \geq 2g$.

Theorem 4.2 is essentially due to Mumford [M_1] except that Mumford considers the stability of the Chow point of X rather than the mth Hilbert point of X. More precisely, we should show that there is an M depending only on d and g so that if $m \geq M$, and conditions (i) and (ii) are satisfied, then $H_m(X)$ is stable.

Let λ be a 1 - P - S and let X_i be a basis of $H^0(X,\mathcal{O}_X(1)) = W$ so that $X_i^{\lambda(\alpha)} = \alpha^{r_i} X_i$ with $r_1 \leq r_2 \leq \cdots \leq r_\ell$. Our aim is to examine the λ-stability of $H_m(X)$. To this end, we let F_i be the subsheaf of $\mathcal{O}_X(1)$ generated by $X_1,\ldots,X_i$. Thus F_i is the smallest subsheaf of $\mathcal{O}_X(1)$ so that $X_j \in H^0(X,F_i) \subseteq H^0(X,\mathcal{O}_X(1))$ for $j \leq i$. Now the F_i are line bundles on X. Let $e_i = \deg F_i$.

Thus $e_1 = 0$, since $F_1 = \mathcal{O}_X$ and $e_\ell = d$, since $\mathcal{O}_X(1)$ is generated by $X_1,\ldots,X_\ell$.

Proposition 4.3.

Suppose there are integers i_j with $1 = i_1 < \ldots < i_k = \ell$ so that

$$\sum_f \frac{e_{i_f} + e_{i_{f+1}}}{2}\left(r_{i_{f+1}} - r_{i_f}\right) > r_\ell d .$$

Then $H_m(X)$ is λ stable if $m = (L+1)N$ with $N \gg L \gg 0$.

Proof.

If L_1 and L_2 are two line bundles on X and V_i is a subspace of $H^0(X,L_i)$, we let $V_1.V_2$ denote the subspace of $H^0(X,L_1 \otimes L_2)$ spanned by elements of the form $v_1 \otimes v_2$, $v_i \in V_i$. With this notation, for any L and $0 \leq p \leq L$ consider

$$V'_{i,j} = V_i^{L-p}.V_j^p \subseteq H^0(X,F_i^{L-p} \otimes F_j^p).$$

Notice that the sections in $V'_{i,j}$ generate the sheaf $F_i^{L-p} \otimes F_j^p$. V_ℓ is a very ample linear system. Hence $V'_{i,j}.V_\ell = V_{i,j}$ is a very ample linear system.

$$V_{i,j} \subseteq H^0(X,\ F_i^{(L-p)} \otimes F_j^p \otimes \mathcal{O}_X(1))$$

Fix i,j,L and p momentarily and let

$$M = F_i^{(L-p)} \otimes F_j^p \otimes \mathcal{O}_X(1).$$

and let $V = V_{i,j}$. We claim that for N sufficiently large, $V^N = H^0(X,M^{\otimes N})$. For V determines a very ample linear system and hence an embedding $X \subseteq \mathbf{P}^m$, where $m+1 = \dim V$. There is a short exact sequence

$$0 \to \mathcal{I}_X(N) \to \mathcal{O}_{\mathbf{P}^m}(N) \to \mathcal{O}_X(N) \to 0$$

For $N \gg 0$, $H^1(\mathcal{I}_X(N)) = 0$, so

$$S^N(V) = H^0(\mathbf{P}^m, \mathcal{O}(N)) \twoheadrightarrow H^0(X,M^{\otimes N}).$$

Thus our claim follows. Releasing i,j,L and p, we have

$$V_{i,j}^N = H^0(X,(F_i^{\otimes(L-p)} \otimes F_j^p \otimes \mathcal{O}_X(1))^{\otimes N}) \tag{4.3.1}$$

By Riemann-Roch, the right hand side of (4.3.1) has dimension $N((L-p)e_i + pe_j + d) + 1 - g$. Note that $V_{i,j}^N$ is generated by monomials whose λ-weight is less than or equal to $((L-p)r_i + pr_j + r_\ell)N$. Consider the following filtration:

$$(V_{i_1}^L \cdot V_\ell)^N \subseteq (V_{i_1}^{L-1} \cdot V_{i_2} \cdot V_\ell)^N \subseteq \dots \subseteq (V_{i_1} \cdot V_{i_2}^{L-1} \cdot V_\ell)^N \subseteq$$
$$(V_{i_2}^L \cdot V_\ell)^N \subseteq (V_{i_2}^{L-1} \cdot V_{i_3} \cdot V_\ell)^N \subseteq \dots \subseteq (V_{i_2} \cdot V_{i_3}^{L-1} \cdot V_\ell)^N \subseteq$$
$$\vdots$$
$$(V_{i_{k-1}}^L \cdot V_\ell)^N \subseteq \dots \qquad \subseteq (V_{i_{k-1}} \cdot V_{i_k}^{L-1} \cdot V_\ell)^N \subseteq (V_{i_k}^L \cdot V_\ell)^N$$

We pick a basis of $(V_{i_k}^L \cdot V_\ell)^N = H^0(X, \mathcal{O}((L+1)N))$ by picking a basis of $(V_{i_1}^L \cdot V_\ell)^N$, then extending this to a basis of $(V_{i_1}^{L-1} \cdot V_{i_2} \cdot V_\ell)^N$, etc. We estimate the total weight T of such a basis

$$T \le N(Lr_{i_1} + r_\ell)Ne_\ell +$$
$$+ \sum_{p=1}^{L} N((L-p)r_{i_1} + pr_{i_2} + r_\ell)(e_{i_2} - e_{i_1})N$$
$$\vdots$$
$$+ \sum_{p=1}^{L} N((L-p)r_{i_{k-1}} + pr_{i_k} + r_\ell)(e_{i_\ell} - e_{i_{\ell-1}})N.$$

Thus

$$T \le N^2L^2\left\{\sum_f \frac{e_{i_{f+1}} - e_{i_f}}{2}(r_{i_{f+1}} + r_{i_f})\right\} + o(N^2,L^2)$$

where $o(N^2,L^2)$ denotes terms which are much smaller that N^2L^2 if $N \gg L \gg 0$. On the other hand, one sees that the term in the braces is

$$\left\{r_\ell e_\ell - \sum_f \frac{e_{i_f} + e_{i_{f+1}}}{2}\left(r_{i_{f+1}} - r_{i_f}\right)\right\}$$

Thus Proposition 4.3 is established.

To deduce Theorem 4.2 from Proposition 4.3 we use the Riemann Roch Theorem and Clifford's Theorem to estimate e_k. If $e_k \ge 2g-1$, $e_k + 1 - g \ge k$, since $H^1(F_k) = 0$ and $H^0(F_k)$ has at least k sections. If $e_k \le 2g-2$, Clifford's Theorem says that $e_k \ge 2(k-1)$. If $d > 2g$, we see that in either case

$$e_k \ge \frac{d}{d-g}(k-1)$$

with strict inequality except for $k = 1$ and $k = \ell$. Given a sequence $1 = i_1 < \dots < i_k = \ell$, to show

$$\sum \frac{e_{i_f} + e_{i_{f+1}}}{2} \left(r_{i_{f+1}} - r_{i_f} \right) > r_\ell d,$$

it suffices to show that

$$(4.3.2) \qquad \frac{d}{d-g} \sum_f \frac{i_f + i_{f+1} - 2}{2} \left(r_{i_{f+1}} - r_{i_f} \right) \geq r_\ell d$$

(The r_i's are assumed non-constant.) Consider the Newton polygon of the points (k, r_k). Let the i_f be the break points of this polygon, and let ρ_k be the point on the polygon above k. Then $\rho_k \leq r_k$, so $\sum \rho_k \leq 0$. To show (4.3.2) holds, it suffices to show that

$$(4.3.3) \qquad \frac{d}{d-g} \sum \frac{i_f + i_{f+1} - 2}{2} \left(\rho_{i_{f+1}} - \rho_{i_f} \right)$$

is $\geq \rho_\ell d$. Since the ρ_k are linear functions of k between i_f and i_{f+1}, (4.3.3) can be replaced by

$$(4.3.4) \qquad \frac{d}{d-g} \sum_{k=1}^{\ell-1} \frac{2k-1}{2} (\rho_{k+1} - \rho_k)$$

But (4.3.4) is just

$$\frac{d}{d-g} \left((\ell - 1)\rho_\ell + \frac{1}{2} (\rho_1 + \rho_\ell) - \sum \rho_k \right)$$

Note $\sum \rho_k \leq 0$ and since the ρ's are convex functions of k, we have $\frac{\ell}{2} (\rho_1 + \rho_\ell) \geq \sum \rho_i$. Finally, $\ell - 1 = d - g$, so (4.3.2) is valid.

§5. Let us fix d and g and let $P(n) = nd + 1 - g$ and let U be the set of all subschemes of $\mathbf{P}(W)$ with Hilbert polynomial P. We defined the Hilbert point mapping

$$H_m : U \to \mathbf{P}(\overset{P(m)}{\Lambda} S^m(W))$$

which is injective. According to Grothendieck, $H_m(U)$ is the set of (closed) points of a closed subscheme $\mathcal{H} \subseteq \mathbf{P}(\overset{P(m)}{\Lambda} S^m(W))$. We are going to investigate the properties of a curve C so that $H_m(C)$ is semi-stable for $m >> 0$ and C is connected.

We assume $g > 1$ and $d \geq 1000g^2$. Our aim is to establish the following:

Proposition 5.1.

There is an M so that if $m \geq M$ and C is any curve of degree d and genus g which is m-Hilbert stable in $\mathbf{P}(W)$, then C is reduced, has only ordinary nodes as singularities and ω_C has non-negative degree on each component of C. Further, any chain of rational curves on which ω_C is trivial has degree 1, i.e. consists of a straight line.

By being more careful, one can show the above holds in $d \geq 10(g-1)$. [G_3,M_2].

The method of proof of Proposition 5.1 is to exhibit a 1-P-S λ of $SL(W)$ for each C which does not satisfy the criteria of the Proposition. We begin with some general definitions. Let $\mathfrak{F}$ be a coherent sheaf on a scheme and let $W \subseteq H^0(X,\mathfrak{F})$ be a subspace so that $\mathfrak{F}$ is generated at each point by the sections in W.

Definition 5.2

A weighted filtration on $\mathfrak{F}$

$$B = \begin{pmatrix} \mathfrak{F}_1 & \cdots & \mathfrak{F}_k \\ & & \\ r_1 & \cdots & r_k \end{pmatrix}$$

is a sequence of subsheaves $\mathfrak{F}_1 \subseteq \mathfrak{F}_2 \subseteq \cdots \subseteq \mathfrak{F}_k = \mathfrak{F}$ and rational numbers r_i, $r_1 \leq r_2 \leq \cdots \leq r_k$. Let $B' = \begin{pmatrix} \mathfrak{F}'_i \\ r'_i \end{pmatrix}$ be another weighted filtration of $\mathfrak{F}$. If $\mathfrak{F}_i \subseteq \mathfrak{F}'_j$ whenever $r_i \leq r'_j$, we say B' dominates B.

Let $\pi : Y \to X$. Given a weighted filtration $B = \begin{pmatrix} \mathcal{G}_i \\ r_i \end{pmatrix}$ on $\pi^*(\mathfrak{F})$, there is an induced filtration $B' = \begin{pmatrix} W_i \\ r_i \end{pmatrix}$ on W, where

$$W_i = \{s \in W \mid \pi^*(s) \in H^0(Y,\mathcal{G}_i)\}.$$

Conversely, given a weighted filtration $\begin{pmatrix} W_i \\ r_i \end{pmatrix}$ on W, there is an induced filtration on $\mathfrak{F}$. (These two processes of induced filtration do not commute).

Let L be a line bundle on a curve C and suppose $V \subseteq H^0(C,L)$ is a very ample linear system. Let $\begin{pmatrix} V_i \\ r_i \end{pmatrix} = B$ be a weighted filtration on V. We choose a basis (X_j, ρ_j) of V compatible with $\begin{pmatrix} V_i \\ r_i \end{pmatrix}$. (Thus $X_j \in V_i$ if $\rho_j \leq r_i$). We let $w_B(m,C)$ be the minimum weight of a basis of $H^0(C,L^{\otimes m})$ consisting in monomials in the X_i's . (One can show $w_B(m,C)$ is a polynomial of degree 2 in m if $m \gg 0$) If $f(m)$ is a function of m, we write

$$w_B(m,C) \geq f(m) + 0(m)$$

to mean that there is a constant K depending only on d and g so that

$$w_B(m,C) \geq f(m) + Km.$$

We now give two lemmas which will enable us to estimate $w_B(m,C)$ from below.

Lemma 5.3.

Suppose $C_i \subseteq C$ are subcurves. Suppose the natural map:

$$\varphi : \mathcal{O}_C \to \oplus\, \mathcal{O}_{C_i}$$

has kernel and cokernel of finite length. Then

$$w_B(m,C) \geq \sum_i w_B(m,C_i) + 0(m).$$

Proof.

Let q be the maximum of the lengths of ker φ and coker φ. Then for $m \gg 0$, the kernel and cokernel of

$$\varphi_m : H^0(C,L^{\otimes m}) \to \oplus\, H^0(C_i,L^{\otimes m})$$

have dimension $\leq q$. Given a basis $P_1,\ldots,P_r$ of $H^0(C,L^{\otimes m})$, we can by suitably reordering the P_i and partitioning $P_1,\ldots,P_{r-q}$ into sets $Q_i \subseteq \{P_1,\ldots,P_{r-q}\}$, assume that Q_i gives an independent set in $H^0(C_i,L)$. Thus

$$w_B(m,C) \geq \sum_i w_B(m,C_i) + m(r_1 - r_\ell) \cdot q.$$

For the next lemma, assume that C is irreducible. Let $\tilde{C}$ be the normalization of C_{red} and let $\mathcal{I}$ be the ideal of C_{red} in C. Let ℓ be the length of the local ring of the generic point of C. Suppose R is an effective divisor on $\tilde{C}$. Let p be an integer and suppose the $r_1,\dots,r_k$ are integers.

Proposition 5.4.

Suppose V_j maps to zero in $H^0(\tilde{C},\tilde{L})$ for $j < p$ and that V_i maps to $H^0(\tilde{C},\tilde{L}(r_i - r_k)R))$. Then if $\deg L \geq (r_k - r_p)\deg R$, we have

$$w_B(m,C) \geq \frac{1}{2}((r_k - r_p)^2 \deg R + 2\ell r_p \deg \tilde{L})m^2 + O(m)$$

Proof.

First replace C by the subscheme defined by $\mathcal{I}^\ell$. Since $\mathcal{I}^\ell$ is supported at a finite number of points, neither the hypothesis nor the conclusion of the theorem are changed.

Let B' be the weighted filtration

$$\begin{pmatrix} V_p & \cdots & V_k \\ r_p & \cdots & r_k \end{pmatrix},$$

that is, we change the weights of the V_i for $i \leq p$ from r_i to r_p. Let (X_i,ρ_i) be a basis of V compatible with B. Let M be a monomial in the X_i's which is nonzero in $H^0(C,L^{\otimes m})$. Then M can involve at most ℓ of the X_i's with $X_i \in V_{p-1}$, since $\mathcal{I}^\ell = 0$. Thus

$$w_B(m,C) \geq w_{B'}(m,C) + O(m),$$

since the B and B' weight of a monomial can differ at most by $\ell(r_p - r_1)$.

Thus we may assume $B = B'$. Notice that

$$h^0(C,L^{\otimes m}) = m\ell \deg_{\tilde{C}}(\tilde{L}) + O(1).$$

Consider the new weighted filtration

$$B' = \begin{pmatrix} V_i \\ r_i - r_p \end{pmatrix}, \quad i \geq p.$$

We have

$$w_B(m,C) = w_{B'}(m,C) + mr_p h^0(C,L^{\otimes m}) = w_{B'}(m,C) + m^2\ell r_p \deg \tilde{L} + O(m).$$

Hence it suffices to establish our Proposition for B'. So we may assume $r_p = 0$.

Since $r_i \geq 0$,

$$w_B(m,C) \geq w_B(m,C_{red}),$$

so we may assume C is reduced. Now let M be a monomial in $V^{\otimes m}$ of weight Q The image of M is in $H^0(\tilde{C},\tilde{L}^{\otimes m}((Q-r_k m)R))$. Thus there is a constant C_1 so that the image of any monomial of weight Q or less lies in a subspace of codimension at least $(r_k m - Q)\deg R + C_1$ in $H^0(C,L^{\otimes m})$. Adding up the possible contributions for each weight Q, we see any basis must have weight at least

$$\sum_{Q=0}^{mr_k} (Q \deg R + O(1)) = r_k^2(\deg R)\frac{m^2}{2} + O(m)$$

Next suppose C is m-Hilbert semi-stable in $\mathbb{P}(W)$ and let $\begin{pmatrix} W_i \\ r_i \end{pmatrix}$ be a weighted filtration of W and assume $r_i \in \mathbb{Z}$. We let (X_i,ρ_i) be a compatible basis of W and let $w_B = \sum \rho_i$ be the weight of the basis. We claim

$$w_B(m,C) \leq (m^2 d)\frac{w_B}{n} \qquad (n = \dim W)$$

if $h^1(L^{\otimes m}) = 0$. Indeed, assign to weight ρ_i' to X_i by

$$\rho_i' = n\rho_i - w_B$$

Letting B' be the weighted filtration dervived from the ρ_i', we see that

$$n\, w_B(m,C) = w_{B'}(m,C) + h^0(L^{\otimes m})w_B m.$$

$$h^0(L^{\otimes m}) = md + 1 - g < md$$

$$\sum \rho_i' = 0$$

We can define a 1 - P.S. λ by $X_i^{\lambda(\alpha)} = \alpha^{\rho_i'}X_i$. Proposition 4.1 shows that $H_m(C)$ is unstable if $w_B(m,C) \geq \frac{1}{n}(m^2 d w_B)$. We will let

$$\frac{d}{n} = 1 + \varepsilon.$$

Since $n = d + 1 - g$, we see $\varepsilon = \frac{g-1}{n} \leq \frac{1}{1000g}$. <u>From now on, we assume there are no destabilizing</u> 1 - PS.

<u>Lemma 5.5.</u>

The components of C are generically reduced.

Proof.

Let $C_1 \subseteq C$ be a reduced and irreducible subcurve of C and let $\mathcal{J}$ be the ideal sheaf of C_1. Consider the filtration B induced on W by

$$\begin{pmatrix} L \otimes \mathcal{J} & L \\ 0 & 1 \end{pmatrix}$$

Let x be a generic point of C_1 and let ℓ be the length of $\mathcal{O}_{C,x}$. We can find two subcurves C_2 and C_3 in C so that $(C_2)_{red} = C_1$ and the natural map

$$\mathcal{O}_C \to \mathcal{O}_{C_2} \oplus \mathcal{O}_{C_3}$$

has kernel and cokernel of finite length.

Note that the filtration B is nontrivial. Let $d' = \deg_{C_1}(L)$. If B were trivial, W would map injectively to $H^0(C_1,L)$. But if $\ell \geq 2$, $d \geq 2d'$. Further $h^0(C_1,L) \geq n$ and $h^0(C_1,L) \leq d' + 1$, which is absurd. Note that $w_B \leq d' + 1$, so

$$w_B(m,C) \leq m^2 w_B(1 + \varepsilon) \leq (1 + \varepsilon)m^2(d' + 1)$$

On the other hand

$$w_B(m,C) \geq w_B(m,C_2) + w_B(m,C_3) + O(m)$$

Further,

$$w_B(m,C_2) \geq \ell d'm^2 + O(m)$$

from Proposition 5.4. Thus $\ell d' \leq d' + 1$ for $m >> 0$, since $w_B(m,C_3) \geq 0$. Since $\ell \geq 2$, we see that $\ell = 2$, and $d' = 1$. In particular, $C_2 \cap C_3 \neq \emptyset$, since otherwise $C_3 = \emptyset$ and $d = 2$. Let C_4 be a component of C_3 meeting C_2. Let R be a point of $\tilde{C}_4$ mapping to C_1. Applying Proposition 5.4 we see that

$$w_B(C_4,m) \geq \frac{1}{2} m^2 + O(m)$$

Hence

$$w_B(m,C) \geq (2d' + \frac{1}{2}) m^2 + O(m)$$

So

$$2d' + 1/2 \leq d' + 1.$$

Lemma 5.6.

If $C' \subseteq C$ is a subcurve and $h^0(C'_{red}, L) \leq d'$, then $d' \geq 20\, g$.

Proof.

Look at the filtration on W induced by

$$\begin{pmatrix} L \otimes \mathcal{I}_{C'_{red}} & L \\ 0 & 1 \end{pmatrix}.$$

Then $w_B \leq d'$ by hypothesis. Further, if $d' < 20g$, $C' \neq C$. Let $C'' = \overline{C - C'}$. As in Lemma 5.5,

$$w_B(m, C'') \geq \frac{1}{2} m^2 + 0(m)$$

$$w_B(m, C') \geq m^2 d' + 0(m)$$

So

$$(d' + \frac{1}{2}) m^2 \leq d'(1 + \varepsilon) m^2 + 0(m).$$

Since $\varepsilon \leq \frac{1}{1000\, g}$, our claim follows

Lemma 5.7.

If C' is a reduced irreducible subcurve of C, then the map $\pi : \tilde{C}' \to C$ is unramified.

Proof.

Note that if $\tilde{C}' \to C'$ is ramified, C' is a singular curve, so Lemma 5.6 is applicable. Let $R \in \tilde{C}'$ be a point of ramification and $\tilde{L} = \pi^*(L)$. Consider the filtration B induced on W by

$$\begin{pmatrix} \tilde{L}(-3R) & \tilde{L}(-2R) & \tilde{L} \\ 0 & 1 & 3 \end{pmatrix}$$

$$B = \begin{pmatrix} W_1 & W_2 & W_3 \\ 0 & 1 & 3 \end{pmatrix}$$

Since π is ramified at R, $\dim W_3/W_2 \leq 1$ and of course $\dim W_2/W_1 \leq 1$. So $w_B \leq 4$. From Proposition 5.4

$$w_B(m,C') \geq \frac{9}{2} m^2 + O(m) .$$

We reach a contradiction.

Lemma 5.8.

C_{red} has no triple points.

Proof.

Suppose first that three distinct components C_1, C_2 and C_3 meet in a point P . Let B be the weighted filtration induced on W by

$$\begin{pmatrix} L(-P) & L \\ 0 & 1 \end{pmatrix}$$

Now $w_B = 1$. As in the preceding lemmas,

$$w_B(m,C_i) \geq \frac{1}{2} m^2 + O(m)$$

So

$$w_B(m,C) \geq \frac{3}{2} m^2 + O(m)$$

But

$$w_B(m,C) \leq (1 + \varepsilon)m^2 .$$

Next, suppose that C_1 and C_2 meet in a singular point of C_1. Then C_1 is singular and so $\deg_{C_1}(L) \geq 20g$. Applying Proposition 5.4, we see that

$$w_B(m,C_1) \geq m^2 + O(m) .$$

So we again obtain a contradiction.

Similarly, C_1 cannot have a triple point.

Lemma 5.8.

C_{red} has no tacnodes.

Proof.

Suppose C_1 and C_2 meet at P and that their tangent lines are identical. Let d_i be the degree of L on C_i. We may suppose $d_1 \leq d_2$. Consider the filtration B_i on W induced by

$$\begin{pmatrix} \tilde{L}_i(-2P) & \tilde{L}_i(-P) & \tilde{L}_i \\ 0 & 1 & 2 \end{pmatrix}$$

Our assumption implies that $B_1 = B_2$ and so $w_{B_1} \leq 3$.

If $d_i \geq 2$, we see that

$$w_{B_1}(m, C_i) \geq 2m^2 + 0(m)$$

If $d_i = 1$,

$$w_{B_1}(m, C_i) \geq \frac{1}{2}(1 + 2)m^2 = \frac{3}{2}m^2 + 0(m).$$

Since we cannot have $d_1 = d_2 = 1$, we see that

$$w_{B_1}(m, C) \geq \frac{7}{2}m^2 + 0(m).$$

As before, we reach a contradiction.

Putting together our results so far, we see that C_{red} has only nodes as singularities. Next, we want to show that $H^1(C_{red}, L) = 0$. We begin with a version of Clifford's Theorem:

Lemma 5.9.

Let C be a reduced curve with only nodes and let L be a line bundle on C generated by global sections. If $H^1(C,L) \neq 0$, there is a curve $C' \subseteq C$ so that

$$h^0(C', L) \leq \frac{\deg_{C'} L}{2} + 1.$$

This is Lemma 9.1 of [G - M].

Suppose $H^1(C_{red}, L) \neq 0$ and let C' be the curve of the above lemma and let $d' = \deg_{C'} L$. Consider the filtration induced on W by

$$\begin{pmatrix} L \otimes \mathcal{I}_{C'_{red}} & L \\ 0 & 1 \end{pmatrix}$$

Then $w_B \leq \frac{d'}{2} + 1$. On the other hand, Proposition 5.4 shows that

$$w_B(m,C) \geq d'm^2 + O(m) .$$

As usual, we reach a contradiction unless $d' \leq 2$. Thus C' is either $\mathbb{P}^1$ or two copies of $\mathbb{P}^1$ meeting at one point. But then we do not have

$$h^0(C',L) \leq \frac{d'}{2} + 1.$$

Proposition 5.10

C is reduced and $W = H^0(C,L)$.

Proof.

Consider $\mathcal{I}$, the ideal defining C_{red} in C. $\mathcal{I}$ is supported at a finite number of points. We claim

(5.10.1) $$W \cap H^0(C,\mathcal{I}.L) \neq 0$$

if $\mathcal{I} \neq 0$. Suppose not. Let g' be the genus of C_{red} and let ℓ be the length of $\mathcal{I}$. We have

$$g' = g + \ell.$$

Thus if $\ell > 0$,

$$h^0(C_{red},L) < \deg L + (1 - g) = \dim W.$$

since $H^1(C_{red},L) = 0$. Thus (5.10.1) is established.

Consider the filtration B on W induced by

$$\begin{pmatrix} L \otimes \mathcal{I} & L \\ 0 & 1 \end{pmatrix}.$$

Then $w_B < \dim W$, but

$$w_B(m,C) \geq m^2 \dim W + O(m).$$

Once again we reach a contradition.

Next let $C' \subseteq C$ be a subcurve and let ω_C be the dualizing sheaf. Let e be $\deg_{C'} L$ and suppose k is the number of points in $C' \cap \overline{C - C'}$.

Proposition 5.11.

$$(1 + \varepsilon)h^0(C',L) \geq e + k/2 .$$

Proof.

Suppose not. Let $\mathcal{I}_{C'}$ be the ideal of C' and consider the filtration B induced on W by

$$\begin{pmatrix} \mathcal{I}_{C'}.L & L \\ 0 & 1 \end{pmatrix}$$

Then $w_B \leq h^0(C',L)$.

Now

$$w_B(m,C') \geq em^2 + 0(m)$$

For the moment, assume that for every irreducible component C_j of $\overline{C - C'}$, we have

(5.11.1) $$2 \deg_{C_j}(L) \geq \#(C_j \cap C')$$

On the one hand, if $\deg_{C_j}(L) \geq \#(C_j \cap C')$, we have

$$w_B(m,C_j) \geq \frac{1}{2}(\#(C_j \cap C'))m^2 + 0(m)$$

On the other hand, if $\deg_{C_j}(L) < \#(C_j \cap C')$, then every section of $\mathcal{I}_{C'}.L$ vanishes on C_j, so

$$w_B(m,C_j) \geq \deg_{C_j}(L)m^2 + 0(m)$$

$$\geq \frac{1}{2}(\#(C_j \cap C')) + 0(m)$$

So in either case, we have

$$w_B(m,C_j) \geq \frac{1}{2}(\#(C_j \cap C')) .$$

Hence in the presence of our hypothesis (5.11.1) we see that

$$w_B(m,C) \geq (e + \frac{1}{2}\sum_j(\#(C_j \cap C)))m^2 + 0(m) \geq (e + \frac{1}{2}k)m^2 + 0(m) .$$

So the Proposition is established if (5.11.1) holds.

Suppose C' is irreducible and that $e < k/2$ and let C_j be a component of $\overline{C - C'}$. We claim

$$(5.11.2) \qquad \deg_{C_j}(L) \geq \frac{1}{2}(C_j \cap C')$$

This is certainly true if $\#(C_j \cap C') = 1$. So assume $\#(C_j \cap C') \geq 2$. Assume $\deg_{C_j}(L) < \frac{1}{2}(C_j \cap C')$. Then the genus of $C_j \cup C'$ is positive, so $C + \deg_{C_j}(L) \geq 20g$ by Lemma 5.6. $\#(C_j \cap C')$ is not greater that the genus of $C_j \cup C'$, so

$$\#(C_j \cap C') \leq g.$$

We have reached a contradiction, so (5.11.2) is established.

We can now conclude from the first part of the proof that

$$(1 + \varepsilon)(e + 1) \geq (1 + \varepsilon)h^0(C',L) \geq e + \frac{k}{2}$$

This contradicts $e < k/2$, since $k \leq g$. So $e > k/2$. So Proposition 5.11 is established.

Note that C is semi-stable as a curve, meaning that it has no nonsingular rational component C'' meeting the rest of the curve in exactly one point. Indeed, one just applies Proposition 5.11 to the curve $\overline{C - C''}$. One also sees that any chain of rational curves meeting the rest of C in two points must have length one and degree 1 by applying Proposition 5.11 to $\overline{C - C''}$. Also note that since $L_{C'}$ is a quotient of L, we must have $H^1(L_{C'}) = 0$, so Proposition 5.11 may be restated as

$$(1 + \varepsilon)(e + 1 - g') \geq e + k/2$$

where g' is the genus of C'.

<u>Definition 5.12</u>.

A line bundle L on a semi-stable curve X of genus g is said to be potentially stable if it has positive degree on all components and if $X' \subseteq X$ is any subcurve of genus g' and L degree e, we have

$$(1 + \varepsilon)(e + 1 - g') \geq e + k/2.$$

We have established that all m-Hilbert stable curves are potentially stable.

§6.

Proposition 6.1.

Suppose C is a stable curve and that $\mathcal{O}(1) \cong \omega_C^{\otimes\nu}$, for some $\nu \geq 1000g^2$. Then C is m-Hilbert stable for $m \gg 0$.

Proof.

We can find a smooth curve S, a point $P \in S$, and a flat family of curves of genus g $\pi : X \to S$ so that $X_P = C$ and so that X_Q is smooth for $Q \neq P$. We choose a basis for $\pi_*(\omega_{X/S}^{\otimes n})$ over $U - P$, where U is some neighborhood of P. Such a basis determines a map Φ of $U - P$ into m-Hilbert stable points of the Hilbert scheme $\mathcal{H}_{S.S.}$. Let $m : \mathcal{H}_{S.S.} \to \mathbb{P}^k$ be the invariant map of Theorem 1.2. Then using Corollary 1.2.1, we can assume by replacing $C - P$ by some ramified cover that Φ extends to a morphism of U to $\mathcal{H}_{S.S.}$ and that $\Phi(P)$ has infinite stabilizer if $\Phi(P)$ is not stable. Thus there is a family $p : Y \to S$ and a line bundle L on Y so that (Y,L) is isomorphic to $(X, \omega^{\otimes n})$ over $S - P$, and so that $H^1(L_P) = 0$ and the embedding of Y_P determined by the sections of L_P is very ample and is m-Hilbert stable for some fixed large m. The singularities of Y are given locally by $\{xy = t^k\}$ as they are deformations of simple nodes. By resolving these singularities we can obtain a new smooth surface $p' : Y' \to S$ so that Y'_P is semi-stable as a curve, since the resolution of $\{xy = t^n\}$ is just a chain of $(n-1)$ rational curves.

Denote the pullback of L to Y' by L again. There is an isomorphism of $\omega^{\otimes n}$ with L defined on Y' over $S - P$. It follows that we can write $\omega^{\otimes n} = L(-n_i E_i)$ on Y', where the E_i are components of Y'_P. We may assume $\min\{n_i\} = 0$. Let C' be a connected subcurve so that $n_i = 0$ if $E_i \subseteq C'$, but $n_i > 0$ if $E_i \not\subseteq C'$ and $E_i \cap C' \neq \emptyset$. Both $\omega^{\otimes n}$ and L are trivial on the new rational chains introduced by resolution of the singularities of Y, so C' either contains such a chain or meets it at one point. So C' is the proper transform of a subcurve of Y_P. Let e be the degree L on C', let g' be the genus of C' and let $k = \#\overline{C' \cap (Y'_P - C')}$. We have

$$e \leq \deg_{C'} \omega^{\otimes\nu} - k \tag{6.1.1}$$

$$(1 + \varepsilon)(e + 1 - g') \geq e + k/2 \tag{6.1.2}$$

$$\deg_{C'} \omega = \deg_{C'} \omega' + k = (2g' - 2) + k. \tag{6.1.3}$$

$$1 + \varepsilon = \frac{d}{n} = \frac{2\nu}{2\nu - 1} \tag{6.1.4}$$

Substituting (6.1.3) and (6.1.4) into (6.1.2), we obtain

$$\frac{2\nu}{2\nu - 1}\left(e - \frac{1}{2}(\deg_{C'}(\omega) - k)\right) \geq e + k/2$$

Simplifying, we get

$$e - \nu \deg_{C'} \omega \geq k/2 .$$

Thus we contradict (6.1.1.) unless $C' = Y'_P$, i.e., $\omega^{\otimes \nu} \cong L$.

One can use Proposition 6.1 to produce a projective moduli space for stable curves. See $[G_3, M_1]$ for details.

§7. Let X_o be an automorphism free irreducible stable curve of genus $g > 1$ with exactly one node N and let d be a large integer. Using techniques similar to those used in the proof of Theorem 6.1, one can show that if L is a line bundle of degree d, then the image of X_o under the linear system $W = H^0(X_o, L)$ is m-Hilbert stable for $m >> 0$.

Now let U_{X_o} be the subscheme of the Hilbert scheme $\mathcal{H}$ of P^{n-1} consisting of m-Hilbert stable curves of degree d which are isomorphic to X_o. $SL(n)$ operates on $\mathcal{H}$ and hence on U_{X_o}. Since we have assumed that X_o is automorphism free, the set of points of the quotient of U_{X_o} by $SL(W)$ can be identified with the generalized Jacobian $J^d_{X_o}$ of X_o. Indeed, every curve isomorphic to X_o in P^{n-1} determines a line bundle on X_o, namely $\mathcal{O}_{X_o}(1)$. So there is a map from U_{X_o} to J_{X_o}. One can show this map induces an isomorphism of U_{X_o}/G with $J^d_{X_o}$. Let V_{X_o} be $\overline{U}_{X_o} \cap \mathcal{H}_{S.S.}$. So a point of V_{X_o} corresponds to an m-Hilbert semi-stable curve which is a limit of curves in U_{X_o}. V_{X_o} is closed in $\mathcal{H}_{S.S.}$, so V_{X_o}/G is projective. Thus V_{X_o}/G is a projective compactification of $J^d_{X_o}$. Our aim is explicate the nature of this compactification.

First of all, we introduce the semi-stable models of X_o. Let X be the normalization of X_o and let $P_1, P_2 \in X$ be the two points mapping to the node of X_o. Define a series of semi-stable curves X_k for $k \geq 0$ by letting X_k be the curves whose components are X and non-singular rational curves $R_1, \ldots, R_k$ arranged as follows:

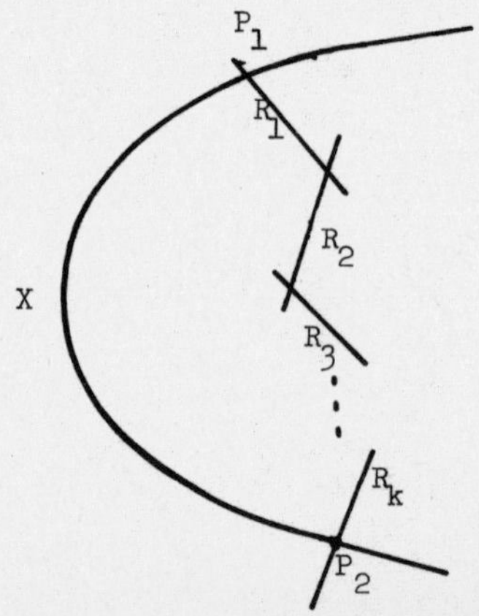

Fig. 1

Suppose S is a smooth curve and $p_i : Y \to S$ is a proper flat family of curves parameterized by S. Suppose that Y_s is isomorphic to X_o for $s \neq s_o$. One can show that Y_{s_o} must be isomorphic to X_k for some k.

Thus the underlying curve C of an point of V_{X_o} must be X_k for some k. We have already noted that $k \leq 1$ and if $k = 1$, that the degree of the one rational curve is one. We claim that no non-trivial element of $SL(n)$ fixes C as a set. (Note: X_1 has automorphism group k^* as an abstract curve, but none of these automorphisms are induced by projective linear transformations.) Indeed any projective transformation σ which fixes $C \cong X_1$ must fix X and R_1 and hence P_1 and P_2. Since we assume X_o automorphism free, σ is the identity on X. But X does not lie in a hyperplane, since $X \cup R_1$ does not lie in a hyperplane and R_1 is a straight line joining two points of X. Since σ fixes X, σ fixes all of projective space. It follows that all the curves in V_{X_o} are actually stable, since any invariant set of strictly semi-staple curves has semi-stable curves with infinite stablilzer in its closure.

There is a simple construction which illustrates how the points of V_{X_o} are limits of points in U_{X_o}. Let S be a smooth curve and let $R \in S$ be fixed. Consider the surface $S \times X$. $S \times X$ has two natural sections, $S \times P_1$ and $S \times P_2$. If E is a bundle on $S \times X$, we let E_{P_i} be the result of restricting E to $S \times P_i$. Usually, we regard E_{P_i} as a bundle on S. Given an isomorphism $\varphi : E_{P_1} \to E_{P_2}$ over S, we can define a bundle on $X_0 \times S$ by using φ as descent data. Suppose that E is a line bundle L of degree d on $X \times S$ but that $\varphi : L_{P_1} \to L_{P_2}$ is only an isomorphism over $U = S - R$. Then φ determines a family of line bundles on X_o parameterized by U and hence a map ψ of U to U_{X_o}/G. We know that one can extend ψ to a map of S to V_{X_o}/G. Our aim is to see explictly how such an extension can be obtained for certain φ.

We assume first that φ is an isomorphism of L_{P_1} with $L_{P_2}(-R)$. In this case, we blow up the point $P_2 \times R$ in $X \times S$ to obtain a new surface $\mathfrak{X}_1$:

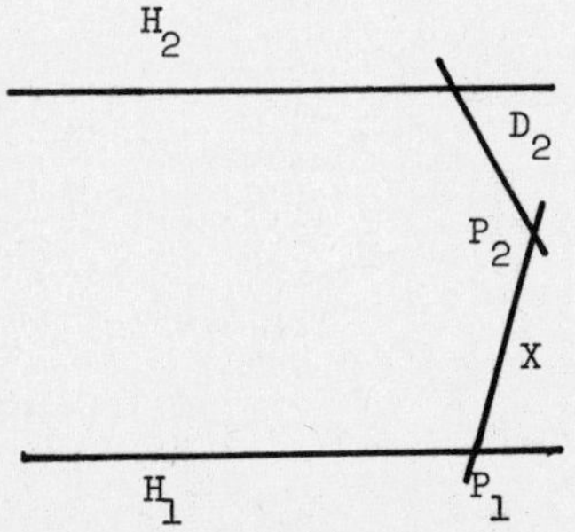

Fig. 2

D_2 denotes the new exceptional divisor, H_1 and H_2 the proper transforms of $P_i \times S$, and X the proper transform of $X \times R$. Let L again denote the pullback of L to $\mathfrak{X}_1$ and consider $L' = L \otimes \mathcal{I}_{D_2}$ where $\mathcal{I}_{D_2}$ is the ideal sheaf of D_2. Note $\deg_{D_2} L' = 1$, since $D_2^2 = -1$. Further, $L'_{H_2} = L_{P_2}(-R)$ and $L'_{H_1} = L_{P_1}$. Hence φ actually gives an isomorphism of L'_{H_1} with L'_{H_2}. We form a new family of curves $p : Z \to S$ by gluing H_1 to H_2 and then use φ to glue L'_{H_1} to L'_{H_2} to obtain a line bundle $\mathcal{L}$ on Z. We call $(Z, \mathcal{L})$ the geometric realization of φ. Note that Z_R is just X_1 and $\mathcal{L}_R$ has degree one on $R_1 \subseteq X_1$ and degree $d - 1$ on X. One can show that the embedding of X_1 by the linear system $H^0(X_1, \mathcal{L}_R)$ is m-Hilbert stable for $m \gg 0$ by using an argument similar to that used to establish Proposition 6.1. Note the $(\mathcal{L}_R)_X$ is just $L_R(-P_2)$.

Define a map ψ from $V_{X_o} - U_{X_o}$ to J_X^{d-1} by sending a m-Hilbert stable curve $X_1 \subseteq P^{n-1}$ to the line bundle $\mathcal{O}(1)_X$. ψ is onto, since given a line bundle M of degree $d - 1$ on X, we let L on $X \times S$ be the constant bundle $M(+P_2)$ and choose an isomorphism φ of L_{P_1} with $L_{P_2}(-R)$. Then the geometric realization of φ yields a line bundle on X_1 which is isomorphic to M on X. On the other hand, the line bundle $\mathcal{O}(1)_X$ determines $X_1 \subseteq P^{n-1}$ up to projective equivelence. Indeed, $\mathcal{O}_X(1)$ determines the map of X to P^{n-1} up to projective equivelence. But given $X \subseteq P^{n-1}$, X_1 is determined since X_1 is just the union of X and the straight line joining P_1 and P_2. So we see that $V_{X_o} - U_{X_o}/SL(W)$ is just J_X^{d-1}, at least set theoretically.

We next give a more global version of geometric realization. This time start with $S_o = J_X^{d-1}$. On $S_o \times X$ we have the Poincaré bundle $\mathcal{L}$. Consider $S_1 = P(\underset{\sim}{Hom}(\mathcal{L}_{P_2}, \mathcal{L}_{P_1}) \oplus \mathcal{O})$ and let M_1 be the tautological bundle for $\pi_1 : S_1 \to S_0$

$$\mathcal{O} \oplus \underset{\sim}{Hom}(\mathcal{L}_{P_2}, \mathcal{L}_{P_1}) \to M_1 \to 0 .$$

There are disjoint divisors H_1 and H_2 of S_1 so that the map from $\mathcal{O}$ to M_1 vanishes on H_1 and the map from $\underset{\sim}{Hom}(\mathcal{L}_{P_2}, \mathcal{L}_{P_1})$ to M_2 vanishes on H_2. H_1 and H_2 are both sections of π_1. Thus we obtain isomorphisms $M_1 \cong \mathcal{O}(H_1)$ and $M_1 \cong \underset{\sim}{Hom}(L_{P_2}, L_{P_1})(H_2)$. So we obtain an isomorphism $\mathcal{O} \to \underset{\sim}{Hom}(L_{P_2}, L_{P_1})(H_2 - H_1)$. This can be interpreted as an isomorphism

$$\Theta : L_{P_1}(-H_1) \to L_{P_2}(-H_2) .$$

Now on $S_1 \times X$, blow up $H_1 \times P_1$ and $H_2 \times P_2$ to obtain new exceptional divisors E_1 and E_2. Consider $L' = L(-E_1 - E_2)$. We have $L'_{P_1'} \cong L_{P_1}(-H_1)$ and $L'_{P_2'} \cong L_{P_2}(-H_2)$, where the P_i' are the proper transfons of $P_1 \times S_1$ and $P_2 \times S_2$.

We obtain a new family of semi-stable curves Y' by gluing P_1' to P_2' and using φ to descend L' to obtain a line bundle L' on Y'. Let $\pi' : Y' \to S_1$ be the projection. $\pi'_*(L')$ is locally free on S_1. Choosing a basis of $\pi'_*(L')$ over an open U determines a map of U to V_{X_o} and hence to $V_{X_o}/SL(n)$. This map is independent of the choice of the basis of $\pi'_*(L')$. We obtain a map ψ of S_1 to V_{X_o}/G. One can check that H_1 and H_2 both map isomorphically to $(V_{X_o} - U_{X_o})/SL(W)$. Both H_1 and H_2 are isomorphic to J_{d-1}. Let $\varphi : H_1 \to H_2$ be the map induced on J_{d-1} by sending a line bundle L to $L(P_2 - P_1)$. One can check that $\psi \circ \varphi = \psi$ on H_1. Although it is not obvious, one can show that $V_{X_o}/SL(W)$ is isomorphic to the normal crossing variety obtained by gluing H_1 to H_2 by φ. One first shows that V_{X_o} is a normal crossing variety by studying the deformations of X_1 and the uses the fact that $SL(W)$ is operating freely on V_{X_o} to show that $V_{X_o}/SL(W)$ has normal crossings. Thus $S_1 \to V_{X_o}/SL(W)$ is identified with the normalizion of $V_{X_o}/SL(W)$. Cf [G_4].

References

[G_1] Gieseker, D., On the moduli of vector bundles on an algebraic surface. Ann. of Math 106, 45 (1977).

[G_2] ________, Global moduli for surfaces of general type. Inv. Math. 43, 233 (1977).

[G_3] ________, Lectures on Stable Curves, Tata Lecture Notes.

[G_4] ________, A degeneration of the moduli space of stable bundles. (To appear)

[G-M] ________ and I. Morrison, Hilbert stability of rank two bundles on curves.

[Ma] Maruyama, M., Moduli of stable sheaves I, II J. Math. Soc. Kyoto 17, 91 (1977) and 18, 557, (1978).

[M_1] Mumford, D. and J. Fogarty. Geometric Invariant Theory. Second Enlarged Edition. Springer Verlag 1982.

[M_2] ________, Stability of projective varieties. L'Ens Math. 24 (1977).

[N] Newstead, P. E. Introduction to Moduli Problems and Orbit Spaces. Tata Inst. Lecture Notes, Springer Verlag, 1978.

ROOT SYSTEMS, REPRESENTATIONS OF QUIVERS AND INVARIANT THEORY

Victor G. Kac

In these notes I will discuss two approaches to the study of the orbits, invariants, etc, of a linear reductive group G operating on a finite dimensional vector space V. The two techniques are the "quiver method" and the "slice method", which are discussed in Chapters I and II respectively.

Undoubtedly, the slice method, based on Luna's slice theorem [1], is one of the most powerful methods in geometric invariant theory. Even in the case of binary forms the slice method gives results which were out of reach of mathematucs of 19th century (cf. [15]). For example, I show that for the action of $SL_2(\mathbb{C})$ on the space of binary forms of odd degree $d > 3$ the minimal number of generators of the algebra of invariant polynomials is greater than $p(d-2)$, where $p(n)$ is the classical partition function.

On the other hand, the quiver method can be applied to a (very special) class of representations for which the slice method often fails.

Most of the results of Chapter I are contained in [4] and [5]; on the most part I just give simpler versions of the proofs. Chapter II contains some new results (as, it seems, the one mentioned above).

I am mostly greatful to the organizers of the summer school in Montecatini Terme (Italy) for inviting me to give these lectures, especially to F. Gherardelli who convinced me to write the notes. My thanks go to J. Dixmier for sharing his knowledge and enthusiasm about invariant theory of binary forms, and to H. Kraft and R. Stanley for several important observations.

Chapter I. Representations of quivers.

The whole range of problems of linear algebra can be formulated in a uniform way in the context of representations of quivers introduced by Gabriel [2]. In this chapter I discuss the links of this with invariant theory and theory of generalized root systems ([3], [4]).

§1.1. Given a connected graph Γ with n vertices $\{1,\dots,n\}$ we introduce the associated root system $\Delta(\Gamma)$ as a subset in $\mathbb{Z}^n$ as follows. Let b_{ij} denote the number of edges connecting vertices i and j, if $i \neq j$, and twice the number of loops at i if $i=j$. Let $\alpha_i = (\delta_{i1},\dots,\delta_{in})$, $i=1,\dots,n$, be the standard basis of $\mathbb{Z}^n$. Introduce a bilinear form $(\ ,\ )$ on $\mathbb{Z}^n$ by:

([redacted] $\delta_{ij} - \frac{1}{2}b_{ij} \quad (i,j=1,\dots,n).$

Denote by $Q(\alpha)$ the associated quadratic form. It is clear that this is a $\mathbb{Z}$-valued form. The element α_i is called a fundamental root if there is no edges-loops at the vertex i. Denote by Π the set of fundamental roots. For a fundamental root α define the fundamental reflection $r_\alpha \in \operatorname{Aut} \mathbb{Z}^n$ by

$$r_\alpha(\lambda) = \lambda - 2(\lambda,\alpha)\alpha \quad \text{for} \quad \lambda \in \mathbb{Z}^n.$$

This is a reflection since $(\alpha,\alpha) = 1$ and hence $r_\alpha(\alpha) = -\alpha$, and also $r_\alpha(\lambda) = \lambda$ if $(\lambda,\alpha) = 0$. In particular, $(r_\alpha(\lambda), r_\alpha(\lambda)) = (\lambda,\lambda)$. The group $W(\Gamma) \subset \operatorname{Aut} \mathbb{Z}^n$ generated by all fundamental reflections is called the Weyl group of the graph Γ (for example $W = \{1\}$ if there is an edge-loop at any vertex of Γ). Note that the bilinear form $(\ ,\ )$ is $W(\Gamma)$-invariant. Define the set of real roots $\Delta^{re}(\Gamma)$ by:

$$\Delta^{re}(\Gamma) = \bigcup_{w\in W} w(\Pi).$$

For an element $\alpha = \sum_i k_i\alpha_i \in \mathbb{Z}^n$ we call the height of α (write: $\operatorname{ht}\alpha$) the number $\sum_i k_i$; we call the support of α (write: $\operatorname{supp}\alpha$) the subgraph of Γ consisting of those vertices i for which $k_i \neq 0$ and all the edges joining these vertices. Define the fundamental set $M \subset \mathbb{Z}^n$ by:

$$M = \{\alpha \in \mathbb{Z}_+^n\setminus\{0\} \mid (\alpha,\alpha_i) \leq 0 \text{ for all } \alpha_i \in \Pi, \text{ and } \operatorname{supp}\alpha \text{ is connected}\}.$$

(Note that $(\alpha,\alpha_i) \leq 0$ if $\alpha_i \notin \Pi$, automatically).

Here and further on, $\mathbb{Z}_+ = \{0,1,2,\dots\}$.

Define the set of <u>imaginary roots</u> $\Delta^{im}(\Gamma)$ by:

$$\Delta^{im}(\Gamma) = \bigcup_{w\in W} w(M\cup -M).$$

Then the <u>root system</u> $\Delta(\Gamma)$ is defined as

$$\Delta(\Gamma) = \Delta^{re}(\Gamma) \cup \Delta^{im}(\Gamma).$$

An element $\alpha \in \Delta(\Gamma) \cap \mathbb{Z}_+^n$ is called a <u>positive root</u>. Denote by $\Delta_+(\Gamma)$ (resp. $\Delta_+^{re}(\Gamma)$ or $\Delta_+^{im}(\Gamma)$) the set of all positive (resp. positive real or positive imaginary) roots. When it does not cause a confusion we will write W, Δ, etc. instead of $W(\Gamma)$, $\Delta(\Gamma)$, etc.

It is obvious that $(\alpha,\alpha) = 1$ if $\alpha \in \Delta^{re}$. On the other hand, $(\alpha,\alpha) \leq 0$ if $\alpha \in \Delta^{im}$. (Indeed, one can assume that $\alpha = \Sigma k_i\alpha_i \in M$; but then $(\alpha,\alpha) = \sum_i k_i(\alpha,\alpha_i) \leq 0$.) Hence

$$\Delta^{re} \cap \Delta^{im} = \emptyset.$$

Furthermore, one has:

$$\Delta = \Delta_+ \bigsqcup -\Delta_+.$$

This statement is less obvious but will follow from the representation theory of quivers.

We shall need two more easy facts:

$$\Delta_+^{im} = \bigcup_{w\in W} w(M) = \{\alpha \in \Delta_+ | W(\alpha) \subset \Delta_+\};$$

$\Delta_+^{re} = \{\alpha \in \Delta_+\backslash\{\alpha_1,\dots,\alpha_n\} |$ there exist $\alpha_1,\dots,\alpha_s \in \Pi$ such that $r_{\alpha_i} r_{\alpha_{i+1}} \dots r_{\alpha_s}(\alpha) \in \Delta_+$ for $1 \leq i \leq s$ and $r_{\alpha_1}\dots r_{\alpha_s}(\alpha) \in \Pi\} \cup \Pi$.

The proof of these facts can be found in [3].

§1.2. According as the bilinear form (,) is positive definite, positive semidefinite or indefinite the (connected) graph Γ is called a graph of <u>finite</u>, <u>tame</u> and <u>wild</u> <u>type</u> respectively. The complete lists of finite and tame graphs are given in Tables F and T.

<u>Table F.</u>

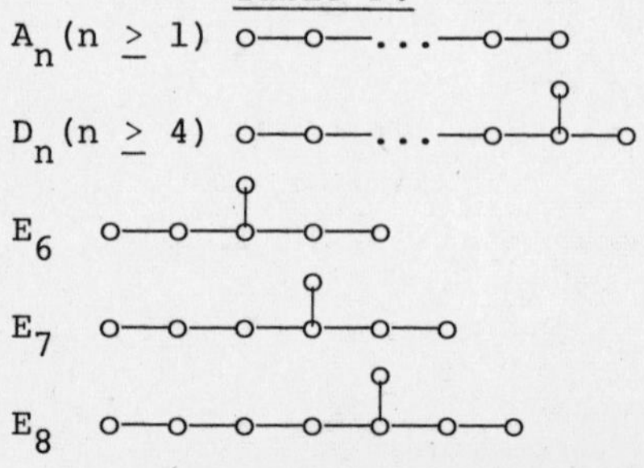

The subscript in the notation of a graph in Table F equals to the number of vertices.

Table T.

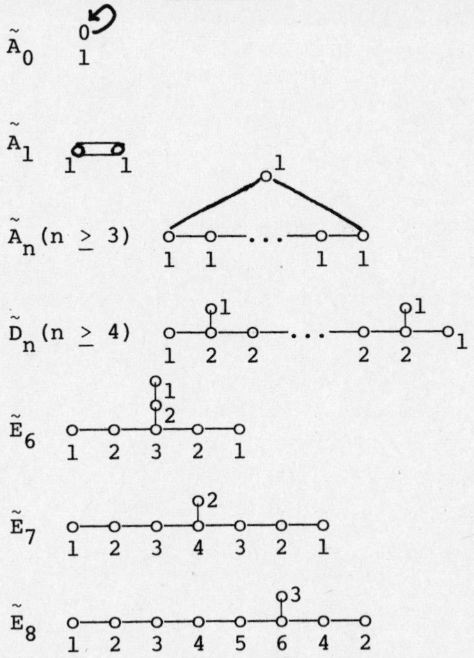

The subscript in the notation of a graph in Table T plus 1 equals to the number of vertices. The kernel of the bilinear form (,) is $\mathbb{Z}\delta$, where $\delta = \sum_i a_i\alpha_i$, a_i being the lables by the vertices. It is easy to show that the converse is also true (see e.g. [3], p.61):

Proposition. If there exists $\delta \in \mathbb{Z}_+^n$ such that $(\delta,\alpha_i) = 0$ for all i and $\delta \neq 0$ then Γ is of tame type.

Here are some characterisations of graphs of finite, tame and wild types:

Γ is finite $\iff |W(\Gamma)| < \infty \iff |\Delta(\Gamma)| < \infty \iff \Delta^{im}(\Gamma) = \emptyset$,

Γ is tame $\iff \Delta^{im}(\Gamma)$ lies on a line;

Γ is wild $\iff$ there exists $\alpha \in \Delta_+(\Gamma)$ such that $(\alpha,\alpha_i) < 0$ for all i and supp $\alpha = \Gamma$.

A graph of wild type is called hyperbolic if every one of its proper connected subgraph is of finite or tame type. In the case of a finite, affine or hyperbolic graph, there is a simple description of the root system $\Delta(\Gamma)$.

<u>Proposition</u>. <u>If a graph</u> Γ <u>is of finite, affine or hyperbolic type, then</u>

$$\Delta(\Gamma) = \{\alpha \in \mathbb{Z}^n \setminus \{0\} \mid (\alpha,\alpha) \leq 1\}.$$

<u>In particular, if</u> Γ <u>is of finite type, then</u>

$$\Delta(\Gamma) = \{\alpha \in \mathbb{Z}^n \mid (\alpha,\alpha) = 1\},$$

<u>and if</u> Γ <u>is of affine type, then</u>

$$\Delta^{re}(\Gamma) = \{\alpha \in \mathbb{Z}^n \mid (\alpha,\alpha) = 1\}; \Delta^{im}(\Gamma) = (\mathbb{Z}\setminus\{0\})\delta.$$

<u>Proof</u>. Let $\alpha \in \mathbb{Z}^n\setminus\{0\}$ be such that $(\alpha,\alpha) \leq 1$. We have to show that $\alpha \in \Delta(\Gamma)$. Note that supp α is connected as in the contrary case $\alpha = \beta + \gamma$, where supp β and supp γ are unions of subgraphs of finite type and $(\beta,\gamma) = 0$, but then $(\alpha,\alpha) \geq 2$. Next, either α or $-\alpha \in \mathbb{Z}^n_+$. Indeed, in the contrary case, $\alpha = \beta-\gamma$, where $\beta,\gamma \in \mathbb{Z}^n_+$, supp $\beta \cap$ supp $\gamma = \emptyset$, supp β is a union of subgraphs of finite type and supp γ is either a union of subgraphs of finite type or is a subgraph of affine type. But $(\alpha,\alpha) = (\beta,\beta) + (\gamma,\gamma)-2(\beta,\gamma) \leq 1$ and $(\beta,\gamma) \leq 0$. Hence the only possiblility is that $(\beta,\beta) = 1$, $(\gamma,\gamma) = 0$ and $(\beta,\gamma) = 0$. But then supp γ is a subgraph of affine type and (β,γ) must be < 0, a contradiction.

So, supp α is connected and we can assume that $\alpha \in \mathbb{Z}^n_+$. We can assume that $W(\alpha) \cap \Pi = \emptyset$, otherwise there is nothing to prove. But then, clearly, $W(\alpha) \in \mathbb{Z}^n_+$. Taking in $W(\alpha)$ an element of minimal height, we can assume that $(\alpha,\alpha_i) \leq 0$ for $\alpha_i \in \Pi$. Since, in addition, supp α is connected, we deduce that α lies in the fundamental set. □

Using the proposition, one can describe $\Delta^{re}_+(\Gamma)$ for a tame graph $\Gamma = \tilde{A}_n, \tilde{D}_n$ or $\tilde{E}_n$ via the subset $\mathring{\Delta}_+ \subset \Delta^{re}_+(\Gamma)$ of positive roots of the subgraph A_n, D_n or E_n respectively as follows:

$$\Delta^{re}_+(\Gamma) = \{\alpha + n\delta \mid \pm\alpha \in \mathring{\Delta}_+, n \geq 1\} \cup \mathring{\Delta}_+.$$

One can show, that conversely, if $\Delta(\Gamma) = \{\alpha \in \mathbb{Z}^n \mid (\alpha,\alpha) \leq 1\}$, then Γ is of finite, affine or hyperbolic type.

<u>Remark</u>. The graphs of finite type are the so called simply laced Dynkin diagrams. They correspond to simple finite-dimensional Lie algebras with equal root length. The other graphs correspond to certain infinite-dimensional Lie algebras, the so called Kac-Moody algebras.

§1.3. Examples.

a) Denote by S_m the graph with one vertex and m edges loops. The associated quadratic form on $\mathbb{Z}$ is $Q(k\alpha) = (1-m)k^2$. $\Delta(S_0) = \{\pm \alpha\}$ and $\Delta(S_m) = (\mathbb{Z}\setminus\{0\})\alpha$ if $m > 0$. $W(S_0) = \{r_\alpha, 1\}$ and $W(S_m) = \{1\}$ if $m > 0$. S_m is of finite, tame or hyperbolic type iff $m = 0$, $m=1$ or $m > 1$ respectively.

b) Denote by P_m the graph with two vertices and m edges connecting these vertices. The associated quadratic form on $\mathbb{Z}^2$ is $Q(k_1\alpha_1+k_2\alpha_2) = k_1^2 - mk_1k_2 + k_2^2$. The set of positive roots is as follows:

$\Delta_+(P_1) = \{\alpha_1,\alpha_2,\alpha_1 + \alpha_2\}$;

$\Delta_+(P_2) = \{k\alpha_1 + (k-1)\alpha_2, (k-1)\alpha_1 + k\alpha_2, k\alpha_1 + k\alpha_2;\ k \geq 1\}$

the set $\{k\alpha_1 + k\alpha_2; k \geq 1\}$ being $\Delta_+^{im}(\Gamma)$;

$m \geq 3$: $\Delta_+(P_m) = \{k_1\alpha_1 + k_2\alpha_2 \mid k_1^2-m\, k_1k_2 + k_2^2 \leq 1,\ k_1 \geq 0, k_2 \geq 0, k_1+k_2>0\}$;

more explicitly,

$\Delta_+^{re}(P_m) = \{c_j\alpha_1 + c_{j+1}\alpha_2, c_{j+1}\alpha_1 + c_j\alpha_2,\ j \in \mathbb{Z}_+\}$, where $c_j\ (j \in \mathbb{Z}_+)$ are defined by the recurrent formula:

$$c_{j+2} = mc_{j+1} - c_j,\quad c_0 = 0,\ c_1 = 1.$$

$W(P_1)$ is the dihedral group of order 6 and $W(P_m)$ is the infinite dihedral group if $m \geq 2$. P_m is of finite, tame or hyperbolic type iff $m = 1$, $m = 2$ or $m \geq 3$ respectively.

c) Denote by V_m the graph:

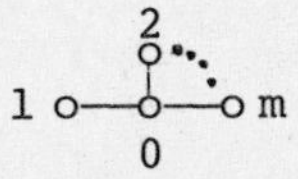

Note that $V_1 = P_1$. The associated quadratic form on $\mathbb{Z}^{m+1}$ is:

$$Q(k_0\alpha_0+\ldots+k_m\alpha_m) = \sum_{i=0}^{m} k_i^2 - k_0 \sum_{i=1}^{m} k_i.$$

$\Delta(V_2) = \{\alpha_0,\alpha_1,\alpha_2,\alpha_0 + \alpha_1,\alpha_0 + \alpha_2,\alpha_0 + \alpha_1 + \alpha_2\}$.

$\Delta(V_3) = \{\alpha_0,\alpha_1,\alpha_2, \alpha_3,\alpha_0 + \alpha_1, \alpha_0 + \alpha_2, \alpha_0 + \alpha_3, \alpha_0 + \alpha_1 + \alpha_2, \alpha_0 + \alpha_1 + \alpha_3, \alpha_0 + \alpha_2 + \alpha_3, \alpha_0 + \alpha_1 + \alpha_2 + \alpha_3, 2\alpha_0 + \alpha_1 + \alpha_2 + \alpha_3\}$.

$\Delta^{im}(V_4) = (\mathbb{Z}\setminus\{0\})\delta$, where $\delta = 2\alpha_0 + \alpha_1 + \alpha_2 + \alpha_3 + \alpha_4$;

$\Delta^{re}(V_4) = \{n\delta \pm \alpha_i,\ i = 0,1,\ldots,4;\ n\delta \pm (\alpha_0 + \alpha_{i_1} +\ldots+\alpha_{i_s})$, where $1 \leq i_1 < i_2 <\ldots<i_s$, $1 \leq s \leq 4$; $n \in \mathbb{Z}\}$.

V_m is of finite, tame or wild type iff $m \leq 4$, $m = 4$ or $m \geq 5$ respectively; V_m is hyperbolic iff $m = 5$.

d) Denote by $T_{p,q,r}$ the graph:

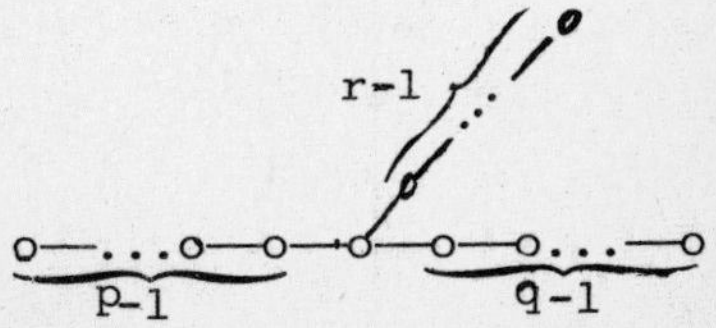

set $c = \frac{1}{p} + \frac{1}{q} + \frac{1}{r}$. Then $T_{p,q,r}$ is of finite, tame or wild type iff $c > 1$, $c = 1$ or $c < 1$ respectively. The only hyperbolic graphs among them are $T_{7,3,2}$, $T_{5,4,2}$ and $T_{4,3,3}$.

§1.4. There are at least two more equivalent difinitions of the set of positive roots $\Delta_+(\Gamma)$:

a) $\Delta_+(\Gamma)$ is a subset of $\mathbb{Z}^n_+\setminus\{0\}$ such that:

(i) $\alpha_1,\ldots,\alpha_n \in \Delta_+(\Gamma)$; $2\alpha_i \notin \Delta_+(\Gamma)$ for $\alpha_i \in \Pi$;

(ii) if $\alpha_i \notin \Pi$, $\alpha \in \Delta_+(\Gamma)$, then $\alpha + \alpha_i \in \Delta_+(\Gamma)$;

(iii) if $\alpha \in \Delta_+(\Gamma)$, then supp α is connected;

(iv) if $\alpha \in \Delta_+(\Gamma)$, $\alpha_i \in \Pi$ and $\alpha \neq \alpha_i$,

then $[\alpha, r_{\alpha_i}(\alpha)] \cap \mathbb{Z}^n \subset \Delta_+(\Gamma)$.

b) Assume that Γ has no edges-loops. Extend the action of the group $W(\Gamma)$ to the lattice $\mathbb{Z}^n \oplus \mathbb{Z}\rho$ by: $r_{\alpha_i}(\rho) = \rho - \alpha_i$. Then one can show that $s(w) := \rho - w(\rho) \in \mathbb{Z}^n_+\setminus\{0\}$ for all $w \in W$. For $w \in W$ set $\varepsilon(w) = \det(w)$; it is clear that $\varepsilon(w) = \pm 1$. Introduce the notation: $x^\alpha = x_1^{k_1}\ldots x_n^{k_n}$, where $\alpha = \sum_i k_i\alpha_i$, and take the product decomposition of the following sum:

$$\sum_{w\in W} (\det w) x^{s(w)} = \prod_{\alpha\in\mathbb{Z}^n_+} (1 - x^\alpha)^{m_\alpha}.$$

Then one can show that $m_\alpha \in \mathbb{Z}_+$, and that

$$\Delta_+(\Gamma) = \{\alpha \in \mathbb{Z}^n_+ \mid m_\alpha > 0\}.$$

The positive integer m_α is called the <u>multiplicity</u> of the root α. Note that $m_\alpha = m_{w(\alpha)}$ for $w \in W$.

I do not know how to extend this definition to the case when Γ has edges-loops.

<u>Examples</u>. The multiplicity of a real root is 1. The multiplicity of an imaginary root of a tame graph $\tilde{A}_n$, $\tilde{D}_n$ or $\tilde{E}_n$ is n. This gives the multiplicity of any root α such that $(\alpha,\alpha) = 0$ since any such root is W-equivalent to a unique imaginary root β such that supp β is a tame graph. One knows that if $(\alpha,\alpha) < 0$, then mult $k\alpha$ growth exponentially as $k \to \infty$.

§1.5. Now we turn to the representation theory of quivers. If every edge of a graph Γ is equipped by an arrow, we say that Γ is equipped by an <u>orientation</u>, say Ω; an oriented graph (Γ,Ω) is called a <u>quiver</u>.

Fix a base field $\mathbb{F}$. A <u>representation</u> of a quiver (Γ,Ω) is a collection of finite dimensional vector spaces V_j, $j = 1,\dots,n$, and linear maps $\phi_{ij}: V_i \longrightarrow V_j$ for every arrow $i \longrightarrow j$ of the quiver (Γ,Ω), everything defined over $\mathbb{F}$. The element $\alpha = \sum_i (\dim V_i)\alpha_i \in \mathbb{Z}_+^n$ is called the dimension of the representation. Morphisms and direct sums of representations of (Γ,Ω) are defined in an obvious way (the dimension of a direct sum is equal to the sum of dimensions). A representation is called indecomposable (resp. absolutely indecomposable) if it is not zero and cannot be decomposed into a direct sum of non-zero representations defined over $\mathbb{F}$ (resp. $\bar{\mathbb{F}}$, the algebraic closure of $\mathbb{F}$).

The main problem of the theory is to classify all representations of a quiver up to isomorphism. One knows that the decomposition of a representation into indecomposable ones is unique. So, for classification purposes it is sufficient to classify indecomposable representations.

Note that there exists a unique up to isomorphism representation of dimension $\alpha_i (i = 1,\dots,n)$ and it is absolutely indecomposable, namely: $V_i = \mathbb{F}$, $V_j = 0$ for $j \neq i$ and all the maps are zero.

§1.6. Examples.

a) The graph S_m has a unique orientation. The problem of classification of the representations of this quiver is equivalent to the classification of m-tuples of $k \times k$-matrices up to a simultaneous conjugation by a non-degenerate matrix. For $m = 1$ this problem is "tame" and was solved by Weierstrass and Jordan (the so called Jordan normal form). For $m \geq 2$ the problem remains open and provides a typical example of a "wild" problem.

b) Put on P_m the orientation Ω for which all arrows point into the same direction. The corresponding problem is to classify all m-types of linear maps from one vector space into another. For $m = 1$ this is a trivial "finite" problem. For $m = 2$ this is a "tame" problem, which was solved by Kronecker. For $m \geq 3$ the problem becomes "wild".

c) Put on the graph V_m the orientation Ω for which all arrows point to the vertex 0. The corresponding problem is essentially equivalent to the problem of classification of m-tuples of subspaces in a vector space V up to an automorphism of V. For $m \leq 3$ the problem is "finite". For $m = 4$ the problem is "tame" and was solved by Nazarova and Gelfand-Ponomarev, for $m \geq 5$ the problem becomes "wild".

Now I shall give precise definitions. A quiver is called *finite* if it has only a finite number of indecomposable representations (up to isomorphism). Following Nazarova [10], we call a quiver (Γ,Ω) *wild* if there is an imbedding of the category of representations of the quiver S_2 into the category of representations of (Γ,Ω); a quiver which is not finite or wild is called *tame*.

Gabriel [2] proved that the quiver (Γ,Ω) is finite iff Γ is finite (i.e., appears in Table F); this will follow from our general theorems. Nazarova [10] proved that (Γ,Ω) is tame iff Γ is tame (i.e., appears in Table T).

Let me show on examples how to prove that (Γ,Ω) is wild.

For the quiver from c) take a vector space V, put $V_1 = V_2 = V$, and take the 1'st map $V_1 \longrightarrow V_2$ to be an isomorphism. Then the category of representations of S_{m-1} is naturally imbedded in the category in question. So (P_m,Ω) is wild if $m \geq 3$.

For the quiver from c) take $V_0 = V \oplus V$, $V_1=V_2=\ldots=V_m=V$. Let $A_1,A_2,\ldots,A_{m-3}\colon V \longrightarrow V$ be some linear operators. Define the maps $\phi_i\colon V \longrightarrow V \oplus V$ $(i=1,\ldots,m)$ by:

$$\phi_i(x) = x \oplus A(x) \quad \text{for} \quad i = 1,\ldots,m-3;$$

$$\phi_{m-2}(x) = x \oplus 0,\ \phi_{m-1}(x) = 0 \oplus x,\ \phi_m(x) = x \oplus x.$$

It is easy to see that this is an imbedding of the category of representation of S_{m-3} in the category in question. So the quiver (V_m,Ω) is wild if $m \geq 5$.

§1.7. One of the main technical tools of the representation theory of quivers are the so called reflection functors. Given an orientation Ω of a graph Γ and a vertex k, define a new orientation $\tilde{r}_k(\Omega)$ of Γ by reversing the direction of arrows along all the edges containing the vertex k. A vertex k of a quiver (Γ,Ω) is called a *sink* (resp. *source*) if for all edges for which k is a vertex, the arrows point to the vertex k (resp. to the other vertex). Note that if there is a loop at k, then k is neither a sink nor a source.

Proposition [1]. *Let* (Γ,Ω) *be a quiver and* k *a sink (resp. source). Then there exists a functor* R_k^+ *(resp.* R_k^-*) from the category of representations of the quiver* (Γ,Ω) *to the category of representation of the quiver* $(\Gamma,\tilde{r}_k(\Omega))$ *such that:*

a) $R_k^{\pm}(U \oplus U') = R_k^{\pm}(U) \oplus R_k^{\pm}(U')$;

b) *If* U *is a representation of dimension* α_k, *then* $R_k^{\pm}(U) = 0$;

c) *If U is an indecomposable representation of* (Γ,Ω) *and* $\dim U \neq \alpha_k$, *then*

$$R_k^- R_k^+(U) \simeq U \quad (\textit{resp.}\ R_k^+ R_k^- (U) \simeq U),$$
$$\textit{and } \dim \ R_k^{\overset{+}{(-)}}(U) = r_{\alpha_k}(\dim U).$$

Corollary. Under the assumptions of c) *of the proposition,* $R_k^{\overset{+}{(-)}}(U)$ *is an indecomposable representation of* $(\Gamma,\tilde{r}_k(\Omega))$, *and* End U *and* End $R_k^{\overset{+}{(-)}}(U)$ *are canonically isomorphic.*

We shall explain the construction of the *reflection functors* R_k^+ and R_k^- in the next section in a more general situation.

§1.8. Now we establish a link between representation theory of quivers and invariant theory.

Fix $\alpha = \sum_i k_i\alpha_i \in \mathbb{Z}_+^n$. Then the set of all, up to isomorphism, representations of dimension α of the quiver (Γ,Ω) is in 1-1 correspondence with the orbits of the group

$$G^\alpha(\mathbb{F}) := GL_{k_1}(\mathbb{F})\times\ldots\times GL_{k_n}(\mathbb{F})$$

operating in a natural way on the vector space

$$M^\alpha(\Gamma,\Omega) := \bigoplus_{i\to j} \mathrm{Hom}_{\mathbb{F}}(\mathbb{F}^{k_i},\mathbb{F}^{k_j})$$

(here the summation is taken over all arrows of the quiver (Γ,Ω)).

Note that the subgroup $C = \{(t,\ldots,t), t\in\mathbb{F}^*\}$ operates trivially and that

$$(=) \qquad \dim G^\alpha - \dim M^\alpha(S,\Omega) = (\alpha,\alpha).$$

Furthermore, note that $U \in M^\alpha(\Gamma,\Omega)$ is an indecomposable representation of the quiver (Γ,Ω) iff End U contains no nontrivial projectors (recall that P is a projector if $P^2 = P$). U is an absolutely indecomposable representation iff End U contains no nontrivial semisimple elements, i.e., iff the stabilizer $(G^\alpha/C)_U$ of $U \in M^\alpha(S,\Omega)$ is a unipotent group.

Another observation: the group G_U^α is connected since it is the set of invertible elements in the ring End U.

Now I shall explain, what are the reflection functors. Let G be a group and π_1,π_2 some representations of G on vector spaces V_1 and V_2, $\dim V_1 = m \geq k$. Then the group $G\times GL_k$ acts naturally on the space

$$M^+ = \mathrm{Hom}(V_1,\mathbb{F}^k)\oplus V_2.$$

Set $M_0^+ = \{\phi\oplus v \in M^+ \mid \phi\in \mathrm{Hom}(V_1,\mathbb{F}^k),\ v\in V_2,\ \mathrm{rank}\,\phi = k\}$. Furthermore, the group $GL_{m-k} \times G$ acts naturally on $M^- = \mathrm{Hom}(\mathbb{F}^{m-k}, V_1) \oplus V_2$. We set $M_0^- = \{\phi \oplus v \in M^- \mid \phi \in \mathrm{Hom}(\mathbb{F}^{m-k}, V_1), v \in V_2, \text{ rank } \phi = m-k\}$. We define a map R^+ from the set of orbits on M_0^+ to the set of orbits on M_0^- as follows. If $\phi \oplus v$ lies on an orbit $\sigma \subset M_0^+$, choosing an isomorphism $\mathbb{F}^{m-k} \longrightarrow \mathrm{Ker}\,\phi$, we get a map $r^+(\phi)\colon \mathbb{F}^{m-k} \longrightarrow V_1$; denote by $R^+(\sigma)$ the orbit of $r^+(\phi) \oplus v \in M_0^-$. It is easy to see that R^+ is a well-defined map. Similarly we define the "dual" map R^- from the set of orbits on M_0^- to the set of orbits on M_0^+. One easily checks that R^-R^+ (resp. R^+R^-) is an identity map on the set of orbits in M_0^+ (resp. M_0^-).

Many people have discovered independently from each other this type of construction. For example, Sato and Kimura call it the "castling transform".

In order to get the reflection functor R_j^+ we apply the above construction to the group $G = \prod_{i \neq j} GL_{k_i}$, $k = k_j$, $V_1 = \bigoplus_{s \to j} \mathrm{Hom}_{\mathbb{F}}(\mathbb{F}^{k_s}, \mathbb{F}^{k_j})$, $V_2 = \bigoplus_{\substack{s \to i \\ i \neq j}} \mathrm{Hom}_{\mathbb{F}}(\mathbb{F}^{k_s}, \mathbb{F}^{k_j})$.

§1.9. We need a general remark about actions of a connected algebraic group G. Let G act on an irreducible algebraic variety X over field $\mathbb{F}$. Then by a theorem of Rosenlicht, there exists a dense open subset $X_0 \subset X$, an algebraic variety Z and a surjective morphism $X_0 \longrightarrow Z$, everything defined over $\mathbb{F}$, whose fibers are G-orbits. Z is called a geometric quotient of X_0.

Now, given an action of G on a constructible set X we can decompose X into a union of irreducible subsets and take a (finite) set of G-invariant algebraic subvarieties $Y_1, \ldots, Y_s \subset X$ such that $\dim X \backslash (Y_1 \cup \ldots \cup Y_s) < \dim X$ and each Y_i has a geometric quotient Z_i. Next, we apply the same procedure to $X \backslash \{Y_1 \cup \ldots \cup Y_s\}$, etc. After at most $\dim X$ steps we obtain (absolutely) irreducible varieties Z_1, $Z_2, \ldots$. We set $\mu(G,X) = \max_i \dim Z_i$. It is clear that this number is well defined. We say that the set of orbits of G on X depends on $\mu(G,X)$ parameters.

Denote by $M_{\mathrm{ind}}^{\alpha}(\Gamma,\Omega)$ the set of all absolutely indecomposable representations from $M^{\alpha}(\Gamma,\Omega)$. This is a G^{α}-invariant set, which is constructible and defined over the prime field. Indeed, there exists a finite number of projectors $P_1, \ldots, P_s$ such that $M_{\mathrm{ind}}^{\alpha}(\Gamma,\Omega) = M^{\alpha}(\Gamma,\Omega) \backslash (\bigcup_i G^{\alpha}(M^{\alpha}(\Gamma,\Omega)^{P_i}))$. Applying the above construction we obtain that the set of absolutely indecomposable representations (considered up to isomorphism) is parametrized by a finite union of algebraic varieties $Z_1, \ldots Z_2, \ldots$, defined over the prime field. We denote for short:

$$\mu_{\alpha}(\Gamma,\Omega) = \mu(G^{\alpha}, M_{\mathrm{ind}}^{\alpha}(\Gamma,\Omega)).$$

§1.10 Now we can state the main theorem.

Theorem. Suppose that the base field $\mathbb{F}$ is algebraically closed. Let (Γ,Ω) be a quiver. Then

a) There exists an indecomposable representation of dimension $\alpha \in \mathbb{Z}_+^n\setminus\{0\}$ iff $\alpha \in \Delta_+(\Gamma)$.

b) There exists a unique indecomposable representation of dimension α iff $\alpha \in \Delta_+^{re}(\Gamma)$.

c) If $\alpha \in \Delta_+^{im}(\Gamma)$, then $\mu_\alpha(\Gamma,\Omega) = 1-(\alpha,\alpha) > 0$.

The proof of the theorem is based on two lemmas. We defer their proof to the next sections.

Lemma 1. Suppose that α lies in the fundamental set M and that, moreover, $(\alpha,\alpha_i) < 0$ for some i. Then

a) The set $M_0^\alpha(\Gamma,\Omega)$ of representations in $M^\alpha(\Gamma,\Omega)$ with a trivial endomorphism ring is a dense open G^α-invariant subset. In particular, $\mu(G^\alpha, M_0^\alpha(\Gamma,\Omega)) = 1-(\alpha,\alpha)$.

b) $\mu(G^\alpha, M_{ind}^\alpha(\Gamma,\Omega)\setminus M_0^\alpha(\Gamma,\Omega)) < 1-(\alpha,\alpha)$.

Lemma 2. The number of indecomposable representation of dimension α (if it is finite) and $\mu_\alpha(\Gamma,\Omega)$ are independent of the orientation Ω.

Proof of the theorem. Note that using the reflection functors, $\mu_{r_i(\alpha)}(\Gamma,\Omega) = \mu_\alpha(\Gamma,\Omega)$ if $\alpha \neq \alpha_i$ and i is a sink or a source of the quiver (Γ,Ω) (the same is true for the number of indecomposable representations). But using Lemma 2, we can always make the vertex i a sink provided that there is no loops at i. Hence the above statement always holds if there is no loops at i.

If $\alpha \in \Delta_+^{im}(\Gamma)$, by the above remarks we can assume that $\alpha \in M$. If $(\alpha,\alpha_i) = 0$ for all i, then $\operatorname{supp}\alpha$ is a tame graph and $\alpha = k\delta$ (see §1.2), and case by case analysis in [10] gives the result. Now the part c) of the theorem follows from Lemma 1.

Similarly, part b) of the theorem follows from the (trivial) fact that there exists a unique up to isomorphism representation whose dimension is equal to a fundamental root.

To prove c) take $\alpha \in \mathbb{Z}_+^n\setminus W(\Pi)$, $\alpha \neq 0$, and suppose that there exists an indecomposable representation of dimension α. Then, as before, there exists an indecomposable representation of dimension $r_\gamma(\alpha)$ for $\gamma \in \Pi$; in particular, $r_\gamma(\alpha) \in \mathbb{Z}_+^n$. Also $\operatorname{supp}\alpha$ is connected. Hence $W(\alpha) \subset \mathbb{Z}_+^n\setminus W(\Pi)$. Taking $\beta \in W(\alpha)$ of minimal height, we have:

$(\beta,\alpha_i) \leq 0$ for all $\alpha_i \in \Pi$ and supp β is connected. Hence $\beta \in M$ and $\alpha \in \Delta_+^{im}(\Gamma)$. □

Remark. One can show (see e.g. [4]) that a generic representation of dimension $k\delta$ of a tame quiver decomposes into a direct sum of k representations of dimension δ.

§1.11. In this section we prove Lemma 1. Let first $\alpha = \Sigma k_i\alpha_i$ be an arbitrary non-zero element from $\mathbb{Z}_+^n$. Let $\alpha = \beta_1+\ldots+\beta_s$, where $\beta_1 \geq \beta_2 \geq \ldots$ (i.e., each coordinate $\geq$) be a decomposition of α into a sum of non-zero elements from $\mathbb{Z}_+^n$; let $\beta_k = \sum_i m_i^{(k)} \alpha_i$. Taking distinct elements $\lambda_1,\lambda_2,\ldots \in \mathbb{F}^*$ defines a conjugacy class of semi-simple elements in G^α consisting of the elements $g = (g_1,\ldots,g_n)$ such that λ_j is an eigenvalue of g_i with multiplicity $m_i^{(j)}$ for all $i = 1,\ldots,n$. Denote by $S_{\beta_1,\ldots,\beta_s} \subset G^\alpha$ the union of all such conjugacy classes. Then an easy computation shows that the dimensions of the centralizer G_g^α of $g \in G^\alpha$ and of the fixed point set $M^\alpha(\Gamma,\Omega)^g$ of g in $M^\alpha(\Gamma,\Omega)$ are independent of the choice of $g \in S_{\beta_1,\ldots,\beta_s}$ and, moreover, we have:

$$(!) \qquad \dim G_g^\alpha - \dim M^\alpha(\Gamma,\Omega)^g = \sum_i (\beta_i,\beta_i).$$

It follows from the theory of sheets in GL_k[6] that $S_{\beta_1,\ldots,\beta_s}$ is a locally closed irreducible subvariety in G^α; denote by $\hat{S}_{\beta_1,\ldots,\beta_s}$ the union of orbits of the same dimension in the Zarisky closure of $S_{\beta_1,\ldots,\beta_s}$. Then, as we saw, $S_{\beta_1,\ldots,\beta_s} \subset \hat{S}_{\beta_1,\ldots,\beta_s}$. Futhermore, it follows from the theory of sheets in GL_k[6] that $\hat{S}_{\beta_1,\ldots,\beta_s}$ contains a unique unipotent conjugacy class u, which corresponds to the conjugate partition of α. A similar (but slightly more delicate computation, which can be found in [4]) shows that the above properties hold for $g = u$. By a deformation argument it follows that these properties hold for arbitrary $g \in G^\alpha$ (this also can be checked by a direct computation, cf §1.13). So, we have proved the following

Lemma. For $g \in \hat{S}_{\beta_1,\ldots,\beta_s}$, dimensions of G_g^α and $M^\alpha(\Gamma,\Omega)^g$ are independent of g and formula (!) holds.

Note that $\hat{S}_{\beta_1,\ldots,\beta_s}$ is a sheet in G^α, i.e., an irreducible component of the union of the orbits of G^α of the same dimension, so that G^α is a disjoint union of the sets $\hat{S}_{\beta_1,\ldots,\beta_s}$ [6]. Note also that the trivial sheet $\hat{S}_\alpha$ coincides with C.

We need one more lemma. Its proof is based on the following identity:

$$(\#)\qquad \sum_{i,j=1}^{n} a_{ij}m_i(k_j-m_j) = \sum_{j=1}^{n} m_j(k_j-m_j)k_j^{-1}\Big(\sum_{i=1}^{n} a_{ij}k_j\Big) + \tfrac{1}{2}\sum_{i,j=1}^{n} a_{ij}\Big(\frac{m_i}{k_i}-\frac{m_j}{k_j}\Big)^2 k_ik_j,$$

provided that $a_{ij} = a_{ji}$ and $k_j \neq 0$ for all $i,j = 1,\dots,n$. This can be checked directly.

Lemma. Let $\alpha \in M$. Then

$$(!!)\qquad \dim G^{\alpha} - \dim G^{\alpha}_{g} \leq \dim M^{\alpha}(\Gamma,\Omega) - \dim M^{\alpha}(\Gamma,\Omega)^{g}.$$

The equality holds only in the following situation:

(N) $(\alpha,\alpha_i) = 0$ for all $i \in \operatorname{supp}\alpha$ or $g \in C$.

Proof. Using formula (=) from §1.8 and formula (!), we have only to show that $(\alpha,\alpha) \leq \sum_i (\beta_i,\beta_i)$ and the equality holds only in the situation (N). This is equivalent to: $\sum_i (\alpha-\beta_i,\beta_i) \leq 0$ and the equality holds only on the situation (N). We can assume that $\operatorname{supp}\alpha = \Gamma$. Applying identity (#) we deduce:

$$(\alpha-\beta_t,\beta_t) = \sum_j m_j^{(t)}(k_j-m_j^{(t)})k_j^{-1}\,(\alpha,\alpha_i) + \tfrac{1}{2}\sum_{i,j}(\alpha_i,\alpha_j)\Big(\frac{m_i^{(t)}}{k_i}-\frac{m_j^{(t)}}{k_j}\Big)^2 k_ik_j.$$

Since $(\alpha_i,\alpha_j) \leq 0$ for $i \neq j$ and $(\alpha,\alpha_i) \leq 0$, we deduce that both summands of the right-hand side are ≤ 0. This proves the inequality in question. In the case of equality, both summands are zero. Since the second summand is zero and Γ is connected, we deduce that α and β_t are proportional. Since $\alpha \neq \beta_t$, and the first summand is zero, we deduce that $(\alpha,\alpha_i) = 0$ for all i. □

Now we can easily complete the proof of Lemma 1. Indeed, if $g \in G^{\alpha}\setminus C$, then, by inequality (!!) we have:

$$\dim M^{\alpha}(\Gamma,\Omega) > \dim M^{\alpha}(\Gamma,\Omega)^{g} + (\dim G^{\alpha} - \dim G^{\alpha}_{g}).$$

It follows that $\dim M^{\alpha}(\Gamma,\Omega) > \dim(G^{\alpha}(M^{\alpha}(\Gamma,\Omega)^{g}))$ and therefore there exists a dense open set $M(g)$ in $M^{\alpha}(\Gamma,\Omega)$ such that the intersection of the conjugacy class of g with G^{α}_{U} is trivial for any $U \in M(g)$. Since there exists only a finite number of conjugacy classes of projectors in $\bigoplus_i gl_{k_i}(\mathbb{F})$, we deduce that there is a dense open set M'

in $M^\alpha(\Gamma,\Omega)$ such that $(G^\alpha/C)_U$ is a unipotent group for any $U \in M'$. Since there is only a finite number of unipotent classes in G_u^α we deduce that there is a dense open set in $M^\alpha(\Gamma,\Omega)$ which consists of representations with a trivial endomorphism ring. This proves Lemma 1a).

To prove b) note that $\mu(M_{ind}^\alpha(\Gamma,\Omega)\setminus M_0^\alpha(\Gamma,\Omega)) \leq \max_u(\dim M^\alpha(\Gamma,\Omega))^u - \dim(G^\alpha/C)_u$ where u ranges over a set of representatives of all non-trivial unipotent classes of G^α. But the right-hand side is (by (!!)) $< \dim M^\alpha(\Gamma,\Omega)-\dim G^\alpha + 1$, which is equal to $1-(\alpha,\alpha)$. □

§1.12. Unfortunately, I do not know a direct proof of Lemma 2. The only known proof requires a reduction mod p argument and counting over a finite field. In this section we recall the necessary facts.

Let X be an absolutely irreducible N-dimensional algebraic variety over a finite field $\mathbb{F}_q$ of $q = p^s$ elements (p is a prime number). Then the number of points in X over the field $\mathbb{F}_{q^t}$ is equal to $q^{Nt} + \phi(t)$, where $\phi(t)/q^{Nt} \longrightarrow 0$ as $t \longrightarrow \infty$. This is a simplest fact of the Weil philosophy. In other words knowing the number of points in X over all finite fields $\mathbb{F}_{q^t}$ we can compute the dimension of X.

Let now X be an absolutely irreducible N-dimensional algebraic variety over $\mathbb{Q}$. Then X can be represented as a union of open affine subvarieties, each of which is given by a system of polynomial equations over $\mathbb{Z}$, the transition functions being polynomials over $\mathbb{Z}$. Now we can reduce this modulo a prime p. Then for all but a finite number of primes we get an absolutely irreducible variety $X^{(p)}$ over $\mathbb{F}_p$ of dimension N.

This reduces the proof of Lemma 2 to the case when $\mathbb{F}$ is a field of prime characteristic p.

§1.13. In order to count the number of orbits of $G^\alpha(\mathbb{F}_q)$ on $M^\alpha(\Gamma,\Omega)(\mathbb{F}_q)$ we employ the Burnside lemma: for the action of a finite group G on a finite set Y the number of orbits is:

$$|Y/G| = \frac{1}{|G|}\sum_{g\in G}|Y^g|.$$

(Here Y^g denote the fixed point set of g on Y and $|Z|$ denotes the cardinality of Z). Denoting by C_g the congugacy class of $g \in G$ and using $|C_g| = |G|/|G_g|$ we can rewrite this formula:

$$|Y/G| = \sum_g |Y^g|/|G_g|,$$

where the summation is taken over a set of representatives of conjugacy classes in G.

Now we need a Jordan canonical form for the elements from $GL_k(\mathbb{F}_q)$ (this information can be found, e.g., in [9], Chapter IV).

Denote by Φ the set of all irreducible polynomials in t over $\mathbb{F}_q$ with leading coefficient 1, excluding the polynomial t. Such a polynomial of degree d has the form

$$P(t) = \prod_{i=0}^{d-1} (t - \alpha^{q^i}), \quad \text{where} \quad \alpha \in \mathbb{F}_q^*, \alpha^{q^d} = \alpha.$$

It follows that the number of polynomials from Φ of degree d is equal to

$$q - 1 \text{ if } d = 1 \text{ and } d^{-1} \sum_{j|d} \mu(j) q^{d/j} \text{ if } q > 1,$$

where μ denotes the classical Möbius function.

Let Par denote the set of all partitions, i.e., non-increasing finite sequences of non-negative integers: $\lambda = \{\lambda_1 \geq \lambda_2 \geq \dots\}$. We denote by λ' the conjugate partition and by $m_i(\lambda)$ the multiplicity of i in λ; we denote: $|\lambda| = \sum_i \lambda_i$, $\langle\lambda,\mu\rangle = \sum_i \lambda_i \mu_i$.

Conjugacy classes C_ν in $GL_k(\mathbb{F}_q)$ are parametrizes by maps ν: $\Phi \longrightarrow \text{Par}$ such that $\sum_{P \in \Phi} (\deg P)|\nu(P)| = k$ as follows.

To each $f = t^d - \sum_{i=1}^{d} a_i t^{i-1}$ we associate the "companion matrix"

$$J(f) = \begin{bmatrix} 0 & 1 & 0 & \dots & 0 \\ 0 & 0 & 1 & \dots & 0 \\ \dots & \dots & \dots & \dots & \dots \\ 0 & 0 & 0 & \dots & 1 \\ a_1 & a_2 & a_3 & \dots & a_d \end{bmatrix},$$

and for each integer $m \geq 1$ let

$$J_m(f) = \begin{bmatrix} J(f) & 1 & 0 & \dots & 0 \\ 0 & J(f) & 1 & \dots & 0 \\ \dots & \dots & \dots & \dots & \dots \\ 0 & 0 & 0 & \dots & J(f) \end{bmatrix}$$

with m diagonal blocks $J(f)$. Then the Jordan canonical form for elements of the conjugacy class C_ν is the diagonal sum of matrices $J_{\nu(f)_i}(f)$ for all $i \geq 1$ and $f \in \Phi$.

The order of the centralizer of each $g \in C_\nu$ is

$$a_\nu(q) = q^{\sum_{P\in\Phi} (\deg P) \langle \nu(P)', \nu(P)' \rangle} \prod_{P\in\Phi} b_{\nu(P)}(q^{-\deg P})$$

where for $\lambda \in \mathrm{Par}$, $b_\lambda(q) = \prod_{i\geq 1} (1-q^{-1})(1-q^{-2})\dots(1-q^{-m_i(\lambda)})$

Finally, if $g \in C_\nu$ and $h \in C_\gamma \subset GL_m$, then (see e.g. [4]):

$$\dim(\mathbb{C}^k \otimes \mathbb{C}^m)^{g\otimes h} = \sum_{P\in\Phi} (\deg P) \langle \nu(P)', \gamma(P)' \rangle .$$

Let now (Γ,Ω) be an oriented graph, and $\alpha = \Sigma k_i\alpha_i \in \mathbb{Z}_+^n$. The conjugacy classes of G^α are parametrized by the maps $\Phi \longrightarrow \mathrm{Par}^n$. An element $\lambda \in \mathrm{Par}^n$ is an n-tuple of partitions $\lambda^{(i)} = \{\lambda_1^{(i)} \geq \lambda_2^{(i)} \geq \dots\}$; set $\lambda_j = (\lambda_j^{(1)},\dots,\lambda_j^{(n)}) \in \mathbb{Z}_+^n$. For $\lambda,\mu \in \mathrm{Par}^n$ we define $(\lambda,\mu) = \sum_j (\lambda_j,\lambda_j)$, where the bilinear form $(\ ,\ )$ on $\mathbb{Z}^n$ is the one associated to Γ. This pairing depends on the graph Γ but is independent of Ω.

Using the Burnside formula we easily deduce the following formula for the number of orbits $d_\alpha(q)$ of $G^\alpha(\mathbb{F}_q)$ on $M^\alpha(\Gamma,\Omega)(\mathbb{F}_q)$:

$$d_\alpha(q) = \sum_\nu \frac{q^{-\sum_{P\in\Phi}(\deg P)\,(\nu',\nu')_q}}{\prod_{k=1}^{n} \prod_{P\in\Phi} b_{\nu(P)^{(k)}}(q^{\deg P})} ,$$

where ν ranges over all maps $\nu\colon \Phi \longrightarrow \mathrm{Par}^n$ such that $\sum_{P\in\Phi} (\deg P)|\nu(P)^{(i)}| = k_i$. This formula (derived jointly with R. Stanley) is quite intractible. However, the following two important corollaries of this formula are clear:

The number $d_\alpha(q)$ of isomorphism classes of representations of the quiver (Γ,Ω) over $\mathbb{F}_q$ is independent of the orientation Ω, and is a polynomial in q with rational coefficients.

(It is immediate that $d_\alpha(q)$ is a rational function in q over $\mathbb{Q}$, but since $d_\alpha(q) \in \mathbb{Z}$ for all $q = p^s$, p prime, $s \in \mathbb{Z}_+$, it follows, that, in fact, $d_\alpha(q)$ is a polynomial.)

We deduce by induction on $\mathrm{ht}\,\alpha$ the following

<u>Lemma</u>. <u>The number of isomorphism classes of indecomposable representations of dimension</u> α <u>of the quiver</u> (Γ,Ω) <u>over the field</u> $\mathbb{F}_q$ <u>is a polynomial in</u> q <u>with rational coefficients, independent of the orientation</u> Ω.

§1.14. It remains to pass from indecomposable representations to absolutely indecomposable ones. For that we need the following general result, the proof of which can be found e.g., in [13] (see also [3]).

Proposition. Let G be a connected algebraic group operating transitively on an algebraic variety X over a finite field $\mathbb{F}_q$. Suppose that the stablizer G_x of $x \in X$ is connected. Then the set $X(\mathbb{F}_q)$ of points defined over $\mathbb{F}_q$ is non-empty and $G(\mathbb{F}_q)$ operates transitively on it.

Since all the stabilizers of the action of G^α on $M^\alpha(\Gamma,\Omega)$ are connected, by the proposition, counting the points over $\mathbb{F}_q$ of the geometric quotients $Z_1, Z_2, \ldots$ is the same as counting the orbits of $G^\alpha(\mathbb{F}_q)$ in $M^\alpha_{\text{ind}}(\Gamma,\Omega)(\mathbb{F}_q)$.

In order to count the number of absolutely indecomposable representations over $\mathbb{F}_q$ we need the following lemma, which follows easily from the proposition (see [3],p.90).

Lemma. a) A representation U of (Γ,Ω), defined over a finite field $\mathbb{F}$, has a unique minimal field of definition $\mathbb{F}'$. If $\sigma \in \mathrm{Gal}(\mathbb{F}' : \mathbb{F}_p)$ and U is isomorphic to U^σ, then $\sigma = 1$.

b) Let $U \in M^\alpha(\Gamma,\Omega)$ be an absolutely indecomposable representation of (Γ,Ω) with a finite minimal field of definition $\mathbb{F}'$. Let $\mathbb{F}_q \subset \mathbb{F}'$ and set $G = \mathrm{Gal}(\mathbb{F}' : \mathbb{F}_q)$. Set $\tilde{U} = \bigoplus_{\sigma \in G} U^\sigma$. Then

(i) $\tilde{U} \in M^{n\alpha}(\Gamma,\Omega)$ is indecomposable over $\mathbb{F}_q$ and $\mathbb{F}_q$ is the minimal field of definition for $\tilde{U}$;

(ii) two such representations $\tilde{U}$ and $\tilde{V}$ are isomorphic over $\mathbb{F}_q$ iff U is isomorphic over $\mathbb{F}'$ to a G-conjugate of V;

(iii) every indecomposable representation for which $\mathbb{F}_q$ is the minimal field of definition can be obtained in the way described above.

Now we can easily finish the proof of Lemma 2 (and of the theorem). Denote by $m(\Gamma,\alpha;q)$ (resp. $m'(\Gamma,\alpha,q)$) the set of absolutely indecomposable (resp. indecomposable) representations over $\mathbb{F}_q$ of dimension α of the quiver (Γ,Ω). Then we deduce from the lemma that for an indivisible $\alpha \in \mathbb{Z}^n_+$ one has:

$$(\alpha) \qquad m'(\Gamma, r\alpha; q) = \sum_{d|r} \frac{1}{d} \sum_{k|d} \mu(k)\, m(\Gamma, \tfrac{r}{d}\alpha; q^{\frac{d}{k}}),$$

where μ is the classical Möbius function. From this one expresses

$m(\Gamma,r\alpha;q)$ via $m'(\Gamma,d\alpha;q^s)$ where $d|r$. Hence $m(\Gamma,r\alpha;q)$ is independent of Ω. □

§1.15. Note that we have also the following

Proposition. $m(\Gamma,\alpha;q) = q^{\mu_\alpha} + a_1 q^{\mu_\alpha - 1} + \ldots + a_{\mu_\alpha}$, where $\mu_\alpha = 1-(\alpha,\alpha)$ and $a_1, a_2, \ldots$ are integers, independent of Ω and q; moreover, $m(\Gamma,w(\alpha);q) = m(\Gamma,\alpha;q)$ for any $w \in W$.

Proof. It follows from the remarks in §1.12 and the main theorem that

$$m(\Gamma,\alpha;q^t) = q^{\mu_\alpha t} + \phi(t), \quad \text{where} \quad \phi(t)/q^{\mu_\alpha t} \longrightarrow 0 \quad \text{as} \quad t \longrightarrow \infty.$$

On the other hand, by §1.13, $m'(\Gamma,\alpha;q)$ and hence $m(\Gamma,\alpha;q)$ is a polynomial in q with rational coefficients. Since $m(\Gamma,\alpha;q) \in \mathbb{Z}$ for all $q = p^s$, it follows that the coefficients are integers. The rest of the statements were proved in the previous sections. □

Conjecture 1. a_{μ_α} = mult α (provided that Γ has no edges-loops).

Conjecture 2. $a_i \geq 0$ for all i.

I have no idea what is the meaning of the rest of a_i's.

Examples. a) If $\alpha \in \Delta_+^{re}$, then $m(\Gamma,\alpha;q) = 1$.

b) If Γ is a tame quiver with $n + 1$ vertices, and $\alpha \in \Delta_+^{im}$, then $m(\Gamma,\alpha;q) = q + n$.

c) Let (Γ,Ω) be the quiver V_k from §1.6 and let $\beta_k = 2\alpha_0 + \alpha_1 + \alpha_2 + \ldots + \alpha_k \in \Delta_+(V_k)$. Then one can show (using Peterson's reccurent formula) that the multiplicity of β_k satisfies the following reccurent relation:

$$(k-1)(\text{mult } \beta_k) = k(\text{mult } \beta_{k-1}) + 2^{k-2}(k-2); \text{ mult } \beta_3 = 1.$$

From this we deduce: mult $\beta_k = 2^{k-1} - k$.

On the other hand one has (as D. Peterson pointed out):

$$m(V_k,\beta_k) = (q+1)^{k-3} + 3m(V_{k-1},\beta_{k-1}) - 2m(V_{k-2},\beta_{k-2}),$$

which gives:

$$m(V_k,\beta_k) = q^{k-3} + \binom{k}{1}q^{k-4} + \left(\binom{k}{2} + \binom{k}{0}\right)q^{k-5} + \left(\binom{k}{3} + \binom{k}{1}\right)q^{k-6} +$$

$$+ \left(\binom{k}{4} + \binom{k}{2} + \binom{k}{0}\right)q^{k-7} + \ldots$$

and $\quad m(V_k,\beta_k;0) = 2^{k-1} - k.$

All the examples agree with the conjectures!

Remark. The constant terms of the polynomials $m(\Gamma,\alpha;q)$ and $m'(\Gamma,\alpha;q)$ are equal. Indeed, by formula (α) we have:

$$m'(\Gamma,r\alpha;0) = \sum_{d|r} \frac{1}{d} \sum_{k|d} \mu(k)m(\Gamma,\frac{r}{d}\alpha;0) = \sum_{d|r} \frac{1}{d} m(\Gamma,\frac{r}{d}\alpha;0)\left(\sum_{k|d}\mu(k)\right) =$$

$$= m(\Gamma,r\alpha;0), \quad \text{since} \quad \sum_{k|d}\mu(k) = 0 \quad \text{unless} \quad d = 1.$$

Conjecture 2 naturally suggests one more

Conjecture 3. The set of isomorphism classes of indecomposable representations of a quiver admits a cellular decomposition by locally closed subvarieties isomorphic to affine spaces, a_i's being the number of cells of dimension $\mu_\alpha - i$.

It follows from the proofs that the minimal field of definition of the (unique) representation of (Γ,Ω) of dimension $\alpha \in \Delta_+^{re}$ is $\mathbb{F}_p$ if char $\mathbb{F} = p$. Ironically enough, I do not know how to prove

Conjecture 4. If char $\mathbb{F} = 0$, the representation of (Γ,Ω) of dimension $\alpha \in \Delta_+^{re}$ is defined over $\mathbb{Q}$.

It would be interesting to give an explicit construction of this representation.

More general is the following

Conjecture 5. The main theorem holds over an arbitrary field $\mathbb{F}$.

Note that it is clear that if $\alpha \notin (\mathbb{Z}_+\Delta_+^{re}) \cup \Delta_+^{im}$, then there is no indecomposable representations of dimension α over $\mathbb{F}$. It would follow from Conjecture 4 that this is the case also for $\alpha \notin \Delta_+$.

§1.16. It is easy to see (see [3]) that the theorem, as well as Conjectures 4 and 5 would follow from the following

Conjecture 6. Let G be a linear algebraic group operating on the vector space V, all defined over $\mathbb{F}$, such that char $\mathbb{F} = 0$ or char $\mathbb{F} > \dim V$. Denote by V_0 (resp V_0^*) the sets of points with a unipotent stabilizer in V (resp. V^*). Then the number of orbits of G on V_0 is equal to that of G on V_0^* and $\mu(G,V_0) = \mu(G,V_0^*)$.

Example: $\mathbb{F} = \mathbb{R}$, $G = \{\begin{pmatrix} a & b \\ 0 & 1\end{pmatrix}$, where $a > 0\}$, $V = \mathbb{R}^2$, action on V

(resp. V^*) is the multiplication on a vector-column from the left (resp. vector row from the right). For the action on V the orbits of $\binom{1}{0}$ and $\binom{-1}{0}$ are the two (1-dimensional) orbits with a (1-dimensional) unipotent stabilizer. For the action of G on V^* the orbits of (10) and (-10) are the two open orbits with a trivial stabilizer.

The following generalization of Conjecture 6 was suggested by Dixmier.

Conjecture 7. Let $S \subset G$ be a reductive subgroup of G. Denote by V_S (resp. V_S^*) the set of points $x \in V$ (resp. $\in V^*$) such that a Levi factor of G_x is a conjugate of S. Then the numbers of orbits of G in V_S and V_S^* are equal and $\mu(G,V_S) = \mu(G,V_S^*)$.

Remark. It is easy to deduce from Conjecture 6 the following statement: Fix a miximal torus $T \subset G$ and denote by V_T(resp. V_T^*) the set of $x \in V$ (resp. $\in V^*$) such that a maximal torus of G_x is a conjugate of T. Then the numbers of orbits of G in V_T and V_T^* are equal and $\mu(G,V_T) = \mu(G,V_T^*)$.

More general is the following

Conjecture 8. Let N be the unipotent radical of G; G/N acts on the sets of orbits V/N and V^*/N. These two actions are equivalent.

§1.17. Examples. a) If (Γ,Ω) is a finite type quiver there is no imaginary roots and we recover Gabrial's theorem: $U \longmapsto \dim U$ gives a 1 - 1 correspondence between the set of isomorphism classes of indecomposable representations of a finite type quiver (Γ,Ω) and the set $\Delta_+(\Gamma)$.

We consider in more detail the finite type quiver V_3, which corresponds to the problem of classification of triples of subspaces U_1,U_2,U_3 in a given vector space U_0 up to an automorphism of U_0. There are 12 roots in $\Delta_+(\Gamma)$. Apart from the roots $\alpha_1,\alpha_2,\alpha_3$ which correspond to $U_0 = 0$ we have 9 nontrivial indecomposable triples. The corresponding dimensions $(\dim U_0;\ \dim U_1,\ \dim U_2,\ \dim U_3)$ are

(1;0,0,0), (1;1,0,0), (1;0,1,0), (1;0,0,1), (1;1,1,0), (1;1,0,1),

(1;0,1,1), (1;,1,1,1) and (2;1,1,1).

Let $a_0,a_1,\ldots,a_8$ be the number of times these representations appear as indecomposable direct summands in a given representation $(U_0;U_1,U_2,U_3)$. Then we have:

$\sum_{i=0}^{7} a_i + 2a_8 = \dim U_0$, $\sum_{i=1}^{8} a_i + 2a_8 = \dim(V_1 + V_2 + V_3)$,

$a_1 + a_4 + a_5 + a_7 + a_8 = \dim V_1$, $a_2 + a_4 + a_6 + a_7 + a_8 = \dim V_2$,

$a_3 + a_5 + a_6 + a_7 + a_8 = \dim V_3$, $a_4 + a_7 = \dim V_1 \cap V_2$,

$a_5 + a_7 = \dim V_1 \cap V_3$, $a_6 + a_7 = \dim V_2 \cap V_3$, $a_7 = \dim V_1 \cap V_2 \cap V_3$.

It is clear from this system of equation that the nine discrete parameters $\dim U_i (i = 0,\ldots,4)$, $\dim U_i \cap U_j$ $(i,j = 1,2,3,\ i = j)$, $\dim U_1 \quad U_2 \quad U_3$ and $\dim(U_1 + U_2 + U_3)$ determine the triple of subspaces U_1, U_2, U_3 is the vector space U_0 up to isomorphism.

b) The quiver P_2 from §1.6 corresponds to the problem of classification of pairs of linear maps $A,B: V_1 \longrightarrow V_2$, the problem solved by Kronecker. We assume that $\mathbb{F}$ is algebraically closed. Then the complete list of indecomposable pairs is in some bases $\{e_i\}$ and $\{f_i\}$ of V_1 and V_2 as follows $(k = 1,2,\ldots)$:

$\dim V_1 = k$, $\dim V_2 = k + 1$:

$A(e_i) = f_i; B(e_i) = f_{i+1} (i = 1,\ldots,k)$.

$\dim V_1 = k + 1$, $\dim V_2 = k$:

$A(e_i) = f_i (i = 1,\ldots,k)$, $A(e_{k+1}) = 0$;

$B(e_i) = f_{i-1} (i = 2,\ldots,k + 1)$, $B(e_1) = 0$.

$\dim V_1 = \dim V_2 = k$:

$A(e_i) - f_i (i = 1,\ldots,k)$: $B(e_i) = \lambda f_i + f_{i+1} (i = 1,\ldots,k - 1)$,

$B(e_k) = \lambda f_k$. Here $\lambda \in \mathbb{F}$ is arbitrary .

$A(e_i) = f_{i+1} (i = 1,\ldots,k - 1)$, $A(e_k) = 0$;

$B(e_i) = f_i$.

§1.18. Since the problems of classifcation of all representations of an arbitrary quiver (Γ,Ω) seems to be too difficult, we shall try to understand a simpler question: what is the structure of a generic representation of given dimension α. It is easy to see that there exists a unique decomposition

$$\alpha = \beta_1+\ldots+\beta_s \ , \quad \text{where} \quad \beta_i \in \mathbb{Z}_+^n\backslash\{0\},$$

such that the set $M_0^\alpha(\Gamma,\Omega) := \{U \in M^\alpha(\Gamma,\Omega) \mid U \simeq \bigoplus_{i=1}^{s} U_i,\ \dim U_i = \beta_i$ and all U_i are indecomposable$\}$ is a dense open subset in $M^\alpha(\Gamma,\Omega)$. This is called the <u>canonical decomposition</u> of α. Further on we assume the base field $\mathbb{F}$ to be algebraically closed.

In order to study this decomposition we need the following definition. A representation $U \in M^\alpha(\Gamma,\Omega)$ is called a <u>Schur representation</u> if $\operatorname{End} U = \mathbb{F}$ (or, equivalently, $(G^\alpha/C)_U = 1$). An element $\alpha \in \mathbb{Z}_+^n\backslash\{0\}$ is called a <u>Schur root</u> for the quiver (Γ,Ω) if $M^\alpha(\Gamma,\Omega)$ contains a Schur representation. In this case the set of Schur representations form a dense open subset $M_0^\alpha(\Gamma,\Omega)$ in $M^\alpha(\Gamma,\Omega)$. Note that $\alpha = \alpha$ is the canonical decomposition of a Schur root. Conversely, if there exists a dense open subset in $M^\alpha(\Gamma,\Omega)$ consisting of indecomposable representations (i.e., $\alpha = \alpha$ is the canonical decomposition of α), then α is a Schur root. Indeed, otherwise,

$$\mu_\alpha(\Gamma,\Omega) > \dim M^\alpha(\Gamma,\Omega) - \dim G^\alpha + 1 = 1 - (\alpha,\alpha),$$

a contradiction with the statement c) of the main theorem.

The set of Schur roots is a subset in $\Delta_+(\Gamma)$ (by the main theorem); we denote it by $\Delta_+^{\mathrm{Schur}}(\Gamma,\Omega)$. As will be clear from examples, this set (as well as the canonical decomposition) depends on the orientation Ω of the quiver.

<u>Remark</u>. One can show that even a stronger result holds [4]: If a representation $U \in M^\alpha(\Gamma,\Omega)$ is stably indecomposable, i.e., all representations from a neighbourhood of U are indecomposable, then U is a Schur representation (the converse is obvious). The quiver S_1 with relation $A^2 = 0$ shows that this property fails for quiver with relations. It might be interesting to study the rings R which have the property that every its stably indecomposable representation has a trivial endomorphism ring.

<u>Example</u>. Consider in the 3-dimensional space V_0 a quadruple of subspaces V_1,V_2,V_3,V_4 of dimensions 2,2,1,1 respectively. This quadruple is indecomposable iff $V_1 = V_2$, and $V_3 + V_4$ is a 2-dimensional

subspace different from V_1 and V_2 and $\dim V_1 \cap V_2 \cap (V_3 + V_4) = 1$ (all such quadruples are equivalent). However, the generic quadruple is decomposable and the canonical decomposition is as follows:

$$\alpha := 3\alpha_0 + 2\alpha_1 + 2\alpha_2 + \alpha_3 + \alpha_4 = (2\alpha_0 + \alpha_1 + \alpha_2 + \alpha_3 + \alpha_4) + (\alpha_0+\alpha_1+\alpha_2)$$

So, α is a (real) root but not a Schur root.

§1.19. Given a quiver (Γ,Ω), let r_{ij} denote the number of arrows with the initial vertex i and the final vertex j. We define the (in general non symmetric) bilinear form R (R in honour of Ringel) on $\mathbb{Z}^n$ by [11]:

$$R(\alpha_i,\alpha_j) = \delta_{ij} - r_{ij}.$$

Note that $(\alpha,\beta) = \frac{1}{2}(R(\alpha,\beta) + R(\beta,\alpha))$ is the associated symmetric bilinear form. The following proposition is crucial.

Proposition [11]. Let U and V be representations of a quiver (Γ,Ω) of dimensions α and β respectively. Then

$$\dim \mathrm{Hom}(U,V) - \dim \mathrm{Ext}(U,V) = R(\alpha,\beta).$$

Using formula (=) from §1.8 we deduce the following well-known formula for an arbitrary representation $U \in M^\alpha(\Gamma,\Omega)$:

$$(\equiv) \qquad \dim M^\alpha(\Gamma,\Omega) - \dim G^\alpha(U) = \dim \mathrm{Ext}(U,U).$$

We need another formula, which also can be derived from the proposition by a straight forward computation.

Lemma. Let $U_j \in M_0^{\beta_j}(\Gamma,\Omega)$ $(j = 1,\dots,s)$ and $\alpha = \sum_{j=1}^{s} \beta_j$. Let $S \in G^\alpha$ be a semisimple element, such that $\alpha = \sum_j \beta_j$ is the corresponding partition of α. Then

$$\dim M^\alpha(\Gamma,\Omega) - \dim G^\alpha\,(M^\alpha(\Gamma,\Omega)^S) = \sum_{i\neq j} \dim \mathrm{Ext}(U_i,U_j)$$

Proof. It is clear that the left-hand side of the formula is equal to:

$\dim M^{\alpha}(\Gamma,\Omega) - \dim G^{\alpha}(U) - \dim M^{\alpha}(\Gamma,\Omega)^{S} + \dim G^{\alpha}_{S} = \dim M^{\alpha}(\Gamma,\Omega) - \dim G^{\alpha} +$ $+ \dim G^{\alpha}_{U} + \dim G^{\alpha}_{S} - \dim M^{\alpha}(\Gamma,\Omega)^{S} = -(\alpha,\alpha) + \sum_{j} (\beta_j,\beta_j) + \dim G^{\alpha}_{U}$ by formula (=) from §1.8 and (!) from §1.11. Since $\dim G^{\alpha}_{U} = \dim \mathrm{Hom}(U,U)$ the lemma follows from the proposition. □

Corollary. Let $U_i \in M_0^{\beta_i}(\Gamma,\Omega)$ and $U = \bigoplus_i U_i$, $\alpha = \sum_i \beta_i$. Then $U \in M_0^{\alpha}(\Gamma,\Omega)$ iff $\mathrm{Ext}(U_i,U_j) = 0$ for all $i \neq j$. In particular, if all β_i are Schur roots, then $\alpha = \sum_i \beta_i$ is the canonical decomposition of α.

§1.20. We call an element $\alpha \in \mathbb{Z}^n_+\setminus\{0\}$ indecomposable if α cannot be decomposed into a sum $\alpha = \beta + \gamma$, where $\beta,\gamma \in \mathbb{Z}^n_+\setminus\{0\}$ and $R(\beta,\gamma)\geq 0$, $R(\gamma,\beta) \geq 0$. One deduces immediately from the remarks in §1.17 and the corollary from §1.18 the following facts:

Proposition. a) If α is an indecomposable element, then α is a Schur root.

b) Let $\alpha \in \mathbb{Z}^n_+\setminus\{0\}$ and $\alpha = \beta_1 + \ldots + \beta_s$ be the canonical decomposition of α. Then all β_i are Schur roots and $R(\beta_i,\beta_j) \geq 0$ for all $i \neq j$.

Conjecture 9. If α is a Schur root, then α is indecomposable.

Conjecture 10. Provided that (Γ,Ω) has no oriented cycles, each $\alpha \in \mathbb{Z}^n\setminus\{0\}$ admits a unique decomposition $\alpha = \sum_j \beta_j$ such that β_j are indecomposable and $R(\beta_i,\beta_j) \geq 0$ for $i \neq j$. (See [4] for a version of this conjecture without assumptions on (Γ,Ω)).

If the conjectures 9 and 10 were true, we obtain that the decomposition of α given by conjecture 10 coincides with its canonical decomposition.

In [4] conjectures 9 and 10 are checked for finite and tame quivers, and for rank 2 quivers.

Example. If (Γ,Ω) is finite, then conjecture 9 holds since then any root is indecomposable. Indeed, if $\alpha = \beta + \gamma$, where $R(\beta,\gamma) \geq 0$ and $R(\gamma,\beta) \geq 0$, then $1 = (\alpha,\alpha) = (\beta,\beta) + (\gamma,\gamma) + R(\beta,\gamma) + R(\gamma,\beta) \geq 2$. Similarly, we show that if (Γ,Ω) is a tame quiver and α is a root such that its defect $R(\delta,\alpha) \neq 0$ then α is indecomposable.

Remark. If $(\alpha,\alpha_i) \leq 0$ for all i and $(\alpha,\alpha_i) < 0$ for some i, then α is indecomposable by the identity (#), and hence is a Schur root. This gives another proof of Lemma 1a).

§1.21. The following simple facts proved in [4] show that many questions about the action of G^α on $M^\alpha(\Gamma,\Omega)$ can be answered in terms of the canonical decomposition.

Proposition. Let $\alpha \in \mathbb{Z}_+^n \setminus \{0\}$ and let $\alpha = \beta_1+\ldots+\beta_k$ be the canonical decomposition of α.

a) $\operatorname{tr\,deg} \mathbb{F}(M^\alpha(\Gamma,\Omega))^{G^\alpha} = \sum_{i=1}^{k} (1-(\beta_i,\beta_i))$.

b) $\operatorname{tr\,deg} \mathbb{F}(M^\alpha(\Gamma,\Omega)^{(G^\alpha,G^\alpha)}) = \sum_{i=1}^{k} (1-(\beta_i,\beta_i)) + |\operatorname{supp}\alpha| - s - r$,

where s and r are the number of distinct real roots and the dimension of the $\mathbb{Q}$-span of all imaginary roots in the canonical decomposition of α, respectively.

c) G^α has a dense orbit in $M^\alpha(\Gamma,\Omega)$ iff all β_i are real, the principal stabilizer being reductive iff $R(\beta_i,\beta_j) = 0$ whenever $\beta_i \neq \beta_j$.

d) If G^α has a dense orbit $\mathcal{O}$ in $M^\alpha(\Gamma,\Omega)$, then we have for the categorical quotient:

$$M^\alpha(\Gamma,\Omega)/(G^\alpha,G^\alpha) \simeq \mathbb{F}^{|\operatorname{supp}\alpha| - s},$$

where s is the same as in b) (and also is the number of distinct indecomposable summands of a representation from $\mathcal{O}$).

e) The generic (G^α,G^α)-orbit in $M^\alpha(\Gamma,\Omega)$ is closed iff $R(\beta_i,\beta_j) \neq 0$ whenever $\beta_i \neq \beta_j$.

Remarks. a) If (Γ,Ω) is a finite type quiver, then G^α always has a dense orbit in $M^\alpha(\Gamma,\Omega)$ (since it has a finite number of orbits or by part c) of the proposition), and hence formula from d) always holds. This has been found by Happel.

b) If (Γ,Ω) is a tame quiver then G^α has a dense orbit in $M^\alpha(\Gamma,\Omega)$ iff the defect $R(\delta,\alpha) \neq 0$ (by part c) of the proposition).

Chapter II. The slice method.

The slice method is based on Luna's slice theorem [7] and was for the first time applied in [5] for the classification of irreducible representations of connected simple linear groups for which the ring of invariants is a polynomial ring. In this chapter I discuss some examples of applications of this method, mainly to invariant theory of binary forms.

§2.1. Let G be a linear reductive group operating on a finite-dimensional vector space V, both defined over $\mathbb{C}$. For $p \in V$ let G_p denote the stabilizer of p and T_p the tangent space to the orbit $G(p)$ of p. Then T_p is G_p-invariant and we can consider the action of G_p on the vector space $S_p := V/T_p$. If the orbit $G(p)$ is closed, the action of G_p on the space S_p is called a slice representation. Note that G_p is a reductive group (since G/G_p is an affine variety and by Matsushima criterion G/H is affine iff H is a reductive subgroup); therefore, we can identify S_p with a G_p-invariant complementary to T_p subspace in V.

The slice method is based on the following principle:

Given a representation of a reductive group, every its slice representation is "better" than the representation itself.

§2.2. In order to make this principle more precise we have to introduce the so called categorical quotient. Let $\mathbb{C}[V]$ denote the ring of polynomials on V and $R = \mathbb{C}[V]^G$ the subring of G-invariant polynomials. Then it follows from the complete reducibility of the action of G on $\mathbb{C}[V]$ that there exists a linear map $\mathbb{C}[V] \longrightarrow \mathbb{C}[V]^G$, denoted by $f \longmapsto f^\natural$ with the following properties:

(i) if $U \subset \mathbb{C}[V]$ is G-invariant, then $U^\natural \subset U$;

(ii) if $f \in \mathbb{C}[V]^G$, $g \in \mathbb{C}[V]$, then $(fg)^\natural = fg^\natural$.

One immediately deduces the classical fact that the algebra $\mathbb{C}[V]^G$ is finitely generated. Indeed, let $I \subset \mathbb{C}[V]$ be the ideal generated by all homogeneous invariant polynomials of positive degree. By Hilbert's basis theorem, it is generated by a finite number of invariant polynomial, say $P_1,\ldots,P_N$. We prove by induction on the degree of a homogeneous polynomial $P \in \mathbb{C}[V]^G$ that P lies in the subalgebra generated by $P_1,\ldots,P_N$. We have

$$P = \sum_{i=1}^{N} Q_i P_i, \quad \text{where} \quad \deg Q_i < \deg P.$$

Applying to both sides the operator $\natural$ we get:

$$P = \sum_i Q_i^{\natural} P_i, \quad \text{where} \quad \deg Q_i^{\natural} = \deg Q_i < \deg P,$$

and applying the inductive assumption to $Q_i^{\natural}$ completes the proof.

Denote by V/G the affine variety for which $\mathbb{C}[V]^G$ is the coordinate ring. Then the inclusion $\mathbb{C}[V]^G \hookrightarrow \mathbb{C}[V]$ induces a map $\pi: V \longrightarrow V/G$ called the <u>quotient map</u>. The pair $\{\pi, V/G\}$ is called the <u>categorical quotient</u> because it satisfies the following characteristic properties:

(i) the fibers of π a G-invariant;

(ii) if $\pi': V \longrightarrow M$ is a morphism, such that M is in affine variety and the fibers of π' are G-invariant, then there exists a unique map $\psi: V/G \longrightarrow M$ such that $\pi' = \psi \circ \pi$.

Note that V/G is a (weighted) cone (i.e., it has a closed imbedding $V/G \hookrightarrow \mathbb{C}^m$, which is invariant under transformations $(c_1,\ldots,c_m) \longrightarrow (t^{s_1}c_1,\ldots,t^{s_m}c_m)$, $t \in \mathbb{C}$, $s_i > 0$).

Note that $\mathbb{C}[V]^G$ is a polynomial ring $\Leftrightarrow$ the vertex of V/G is a regular point $\Leftrightarrow$ V/G is smooth.

§2.3. Here we prove the following classical fact: V/G <u>parametizes the closed orbits</u>, i.e., for each $x \in V/G$, the fiber $\pi^{-1}(x)$ contains a unique closed orbit.

This follows from the following two facts:

(i) if $M_1, M_2 \subset V$ are two closed disjoint G-invariant subvarieties, then there exists an invariant polynomial P which is identically 0 on M_1 and identically 1 on M_2;

(ii) the map π is surjective.

Then the closed orbit in a fiber is an orbit of minimal dimension in the fiber.

For (i), let P_i's and Q_j's be generators of defining ideals I_1 and I_2 for M_1 and M_2. Then by Hilbert's Nullstelensatz, $\{P_i, Q_j\}_{i,j}$ generate $\mathbb{C}[V]$ is an ideal, hence $\sum_i g_i P_i + \sum_j g_j' Q_j = 1$ for some g_i, g_j'. Denoting the first summand by f_1 and second by f_2, we have:

$$f_1 + f_2 = 1, \quad \text{where} \quad f_1 \in I_1, \quad f_2 \in I_2.$$

Applying $\natural$, we get:

$$f_1^{\natural} + f_2^{\natural} = 1, \quad \text{where} \quad f_i^{\natural} \in I_i.$$

Set $P = f_1^{\natural}$; clearly, $P|_{M_1} = 0$ since $P \in I_1$, so $P|_{M_2} = 1$.

If (ii) fails then there exists a maximal ideal $I \subseteqq \mathbb{C}[V]^G$ which generates $\mathbb{C}[V]$ as an ideal; then we have: $1 = \sum_i f_i P_i$, where $f_i \in \mathbb{C}[V]$, and $P_i \in I$. Applying $\natural$, we get:

$$1 = \sum_i f_i^{\natural} P_i.$$

Hence $I = \mathbb{C}[V]^G$, a contradiction. □

§2.4. Let $p \in V$ be such that the orbit $G(p)$ is closed. We have: $V = T_p \oplus S_p$ (G_p-invariant decomposition). We have by restriction: $\pi\colon S_p \longrightarrow V/G$, and by the universality property, this can be pushed down to the morphism:

$$\pi_p\colon \; S_p/G_p \longrightarrow V/G.$$

It is clear that $\dim S_p/G_p = \dim V/G$. From Luna's slice theorem we deduce the following lemma [5].

Lemma. *The morphism* π_p *induces an isomorphism of completions of local rings of* $\pi(p) \in V/G$ *and the vertex of the cone* S_p/G_p

Note that this lemma says, in particular, that π_p is an analytic isomorphism on some neighbourhood of the vertex of S_p/G_p (in complex topology), i.e., this vertex has the same singularity as $\pi(p) \in V/G$. Moreover, it follows from §2.3 that we get all the singularities of V/G in this way.

§2.5. Here are some precise special cases of our general principle. Let $p \in V$ be such that the orbit $G(p)$ is closed. Set $R = \mathbb{C}[V]^G$ and $R_1 = \mathbb{C}[S_p]^{G_p}$.

Proposition. a) *If* R *is a polynomial ring (resp. complete intersection), then* R_1 *is a polynomial ring (resp. complete intersection) too*.

b) *Let* m *(resp.* m_p*) denote the minimal number of homogeneous generators of* R (*resp*. R_1). *Then* $m \geq m_p$.

c) Set $A = \mathbb{C}[z_1,\dots,z_m]$, $A_1 = \mathbb{C}[z_1,\dots,z_{m_p}]$ and let

$$\dots \longrightarrow A^{r_2} \longrightarrow A^{r_1} \longrightarrow R \longrightarrow 0 \quad \text{and} \quad \dots \longrightarrow A^{r_{2p}} \longrightarrow A^{r_{1p}} \longrightarrow R_1 \longrightarrow 0$$

be the minimal free resolutions for R and R_1. Then

$$r_1 \geq r_{1p},\ r_2 \geq r_{2p}, \dots .$$

Proof. follows from §2.4 and the following remarks. From the point of view of the properties we are interested in

(i) the behaviour of a local ring is the same as the one of its completion;

(ii) the behaviour of the local ring of the vertex of a cone is the same as the one of the coordinate ring of the cone;

(iii) the local ring of the vertex is the "worst" among all the local rings of a cone.

All these statements about local rings are quite simple and can be found e.g. in [14]. □

§2.6. One often can find slice representations for which G_p is a finite group. Then one can apply the Shepard-Todd-Chevalley theorem to check that R is not a polynomial ring. In order to check that R is not a complete intersection one can apply the following result [16].

Proposition. Let G be a finite linear group operating on a vector space V of dimension n. Suppose that $\mathbb{C}[V]^G$ has m generators and that the ideal of relations (i.e., the kernel of the surjection $\mathbb{C}[z_1,\dots,z_m] \longrightarrow \mathbb{C}[V]^G$) is generated by $m - n + s$ elements (note that $m \geq n$ and $s \geq 0$). Then G is generated by those σ such that $\operatorname{rank}(\sigma - I) \leq s + 2$. In particular, if V/G is a complete intersection, then $G = \langle \sigma \in G \mid \operatorname{rank}(\sigma - I) \leq 2 \rangle$.

Proof. Let F_g denote the fixed point set of $g \in G$. Denote by Z the union of all $F_g \subset V$ such that $\operatorname{codim}_V F_g \geq s + 3$. Then G acts on $X := V\setminus Z$ and $X/G = (V/G)\setminus(Z/G)$. Note that V/G is simply connected since, being a cone, it is contractible to the vertex. Furthermore, X/G is simply connected by the following fact, proved by Goresky and Macpherson. Let M be a closed affine subvariety in $\mathbb{C}^m$ of dimension n and suppose that the ideal of M is generated by $m - n + s$ elements. Then if M is simply connected, $M\setminus Y$ is simply connected for any sibvariety $Y \subset M$ of codimension $\geq s + 3$.

So, G acts on a connected variety X such taht X/G is simply connected. But then $G = \langle G_x | x \in X \rangle$. Indeed, let G_1 denote the right-hand side. Then G/G_1 acts on X/G_1 such that $g \neq e$ has no fixed points. Since $X/G = (X/G_1)/(G/G_1)$, we deduce that $G/G_1 = e$. This completes the proof. □

Note that the same proof gives the "only if" part of the Shepard-Todd-Chevalley theorem.

<u>Remark</u>. If $G \subset SL_2(\mathbb{C})$, then $\mathbb{C}^2/G$ is always a complete intersection (F. Klein). But for $G = \{\begin{pmatrix} \varepsilon & 0 \\ 0 & \varepsilon \end{pmatrix}\}$, where ε is a cube root of 1 , $\mathbb{C}^2/G$ is not a complete intersection. Indeed, $u_1 = x^3$, $u_2 = x^2y$, $u_3 = xy^2$, $u_4 = y^3$ is a minimal system of generating invariants, and a minimal system of relations is: $u_1u_4 = u_2u_3$, $u_1u_3 = u_2^2$, $u_2u_4 = u_3^2$. This is the simpliest counter-example to the converse statement of the proposition.

§2.7. In order to apply the slice method one should be able to check that an orbit $G(x)$ is closed. For this one can use the Hilbert-Mumford-Richardson criterion, which I will not discuss here, or the following

<u>Proposition</u> [8]. <u>Let</u> G <u>be a reductive group operating on a vector space</u> V, $p \in V$ <u>and</u> $H \subset G_p$ <u>be a reductive subgroup</u>. <u>The normalizer</u> N <u>of</u> H <u>in</u> G <u>acts on the fixed point set</u> L <u>of</u> H <u>in</u> V. <u>The orbit</u> $G(p)$ <u>in</u> V <u>is closed iff the orbit</u> $N(p)$ <u>in</u> L <u>is closed</u>.

Note also that $G(p)$ is closed iff $G^0(p)$ is closed, where G^0 is the connected component of the unity of G.

§2.8. Now we turn to an example of the action of the group $G = SL_2(\mathbb{C})$ on the space V of binary forms fo degree d by

$$\begin{pmatrix} \alpha & \beta \\ \gamma & \delta \end{pmatrix} P(x,y) = P(\alpha x + \beta y, \gamma x + \delta y).$$

Fix the following basis of V_d:

$$v_0 = x^d, \ v_2 = x^{d-1}y, \dots, \ v_{d-1} = xy^{d-1}, \ v_d = y^d.$$

We consider separately the cases d odd and even.

a) d odd and ≥ 3. Set $p = x^{d-1}y + xy^{d-1}$ and $\varepsilon = \exp \frac{2\pi i}{d-2}$.

Then $G_p = \{A_k = \begin{pmatrix} \varepsilon^k & 0 \\ 0 & \varepsilon^{-k} \end{pmatrix}$, $k - 1, \dots, d-2\}$ is a cyclic group of order $d - 2$. The fixed point set of G_p is $L = \mathbb{C}v_1 + \mathbb{C}v_{d-1}$; the connected component of the unity of the normalizer of G_p is $N^0 = \{\begin{pmatrix} t & 0 \\ 0 & t^{-1} \end{pmatrix}$, $t \in \mathbb{C}^*\}$. The orbit $N^0(p)$ is clearly closed. Hence by §2.7, the orbit

$G(p)$ is closed.

The tangent space T_p to $G.p$ is

$$\left[\mathbb{C}x\frac{\partial}{\partial y}+\mathbb{C}y\frac{\partial}{\partial x}+\mathbb{C}(x\frac{\partial}{\partial x}-y\frac{\partial}{\partial y})\right](p)\ ,$$

hence the eigenvalues of A on T_p are : $\varepsilon^2, \varepsilon^{-2}$ and 1.

On the other hand, we have:

$$A_1(v_j) = \varepsilon^{(d-2j)}v_j.$$

Hence the eigenvalues of A_1 on $S_p = V/T_p$ are:

$$1, \varepsilon, \varepsilon^2, \varepsilon^3, \ldots, \varepsilon^{(d-3)}.$$

So, according to our principle, the representation of a cyclic group $H = \langle A_1\rangle$ of order $d - 2$ on $\mathbb{C}^{d-2}$ acting by $A_1(e_j) = \varepsilon^{-j}e_j$ in some basis $e_1,\ldots,e_{d-2}$, is "better" than the action of $SL_2(\mathbb{C})$ on the space of binary forms of odd degree $d > 1$.

Let $f_1,\ldots,f_{d-2}$ be the basis dual to $e_1,\ldots,e_{d-2}$. Then the monomials $f_1^{k_1}\ldots f_{d-2}^{k_{d-2}}$ such that

$$(*)\quad \sum_j jk_j \equiv 0 \bmod (d-2)$$

form a basis of the space of invariant polynomials for this action of H.

An integral solution of (*) is called positive if all $k_i \geq 0$ and not all of them = 0; a positive solution is called indecomposable if it is not a sum of two positive solutions. It is clear that the minimal number of generating invariant polynomials for the action of H on $\mathbb{C}^{d-2}$ is equal to the number of indecomposable positive solutions of (*). I do not know how to compute this number. However, (as observed by R. Stanley) it is clear that if a solution $(k_1,\ldots,k_{d-2})$ is positive and the left-hand side of (*) is equal to $d - 2$, then this is an indecomposable solution. Also it is clear that $(d-2)(\delta_{1i},\ldots,\delta_{d-2,i})$ is an indecomposable solution provided that i and $d-2$ are relatively prime and $i \neq 1$. This gives us the following estimate: (number of positive indecomposable solutions of (*)) $\geq p(d-2) + \phi(d-2) - 1$, where $p(k)$ is the classical partition function and $\phi(k)$ is the number of $1 \leq j \leq k$ relatively prime to k.

For the discussion of the number of relations we need the following defintiion. Let $R = \mathbb{C}[z_1,\ldots,z_m]/I$ be a finitely generated ring, where m is the minimal number of generators; let $n (\leq m)$ be the dimension of R. We say that R requires at least s extra equations if the minimal system of generators of the ideal I contains at least

$m - n + s$ elements.

It is clear by §2.6 that for the action of the cyclic group H the ring of invariants requires at least $d-5$ extra relations. Again, applying the principle, this gives an estimate for the number of relations between invariants of binary forms.

The obtained results are sumarized in this following

Proposition. Let R be the ring of invariant polynomials for the action of $SL_2(\mathbb{C})$ on the space of binary forms of degree $d > 1$, d odd. Then the minimal number of generators of R is $\geq p(d-2)+\phi(d-2)-1$ and R requires at least $d - 5$ extra relations.

A complete information about degrees of polynomials in a minimal system of homogeneous generators of R (it is easy to see that these are well-defined numbers) and the generating relations are known (for odd d) only for $d \leq 5$. Namely, for $d = 3$, R is generated by a homogeneous polynomial of degree 4; for $d = 5$, R is generated by homogeneous polynomials of degrees 4,8,12 and 18 and there is exactly one generating relation.

Note that for $d = 3$ and 5 our low bounds are exact. However, for $d \geq 7$ the low bounds given by the slice method are far from being exact. For instance taking $p = x^7 + y^7$ for $d = 7$ gives the best low bound, which is 17, for the minimal number of generating invariants; it is known, however, that this number lies between 28 and 33 [15].

b) d even and ≥ 4. We take $p = x^d + y^d$ and let $\varepsilon = \exp \frac{2\pi i}{d}$. Then $G_p = \langle \begin{pmatrix} \varepsilon & 0 \\ 0 & \varepsilon^{-1} \end{pmatrix}, \begin{pmatrix} 0 & 1 \\ -1 & 0 \end{pmatrix} \rangle$. The same argument as in a) shows that the orbit $G(p)$ is closed, and our principle gives similar low bounds. In particular, we get that R requires at least $\frac{3}{4}d - 5$ (resp. $\frac{1}{4}(3s + 2) - 5$) extra relations if $4|d$ (resp. $4|d + 2$).

One has a complete information about R (for even d) only for $d \leq 8$. Namely for $d = 2$, R is generated by one polynomial of degree 2; for $d = 4$, R is freely generated by polynomials for degree 2 and 3; for $d = 6$, R is generated by polynomials of degree 2,4,6, 10 and 15 and there is exactly one generating relation; for $d = 8$ R is generated by polynomials of degree $2,3,4,\ldots,10$ and requires two extra relations.

One can see that for $d = 4$, 6 and 8 our low bounds are exact.

§2.9. It follows from the results of §2.8 that for the action of $SL_2(\mathbb{C})$ on the space V_d of binary forms of degree d, the ring R of invariant polynomials is a complete intersection iff $d \leq 6$ (and is a polynomial ring iff $d \le 4$).

Similarly, one can apply §2.7 to the classification of reductive linear groups for which the ring of invariants is a complete intersection. As an example, let us prove the following

Proposition. For the action of $SL_n(\mathbb{C})$ $(n > 1)$ on the space $S^d(\mathbb{C}^n)$ the ring of invariants R is a complete intersection iff either $d \le 2$, or $n = 2$ and $d \le 6$, or $n = d = 3$, or $n = 4$, $d = 3$. Moreover if R is a complete intersection but is not a polynomial ring, then $(n,d) = (2,5)$ or $(2,6)$ or $(4,3)$ and R is the coordinate ring of a hypersurface.

Proof. The case $(4,3)$ was worked out by Salmon about a hundred years ago [12]. He showed that R is generated by invariants of degree 8, 16,24,32,40 and 100 with one generating relation. It is well known that in the case $(3,3)$, R is a polynomial ring generated by invarinats of degree 4 and 6. The case $d \le 2$ is obvious.

In order to show that in the remaining cases R is not a complete intersection take $p = \sum_{i=1}^{n} z_i^d \in S^d(\mathbb{C}^n)$. Then as in §2.8 we show that the orbit of p is closed. Using §§2.5 and 2.6 we deduce that R is not a complete intersection in all cases in question except $(3,4)$. In the last case one should take $p = z_1^3 z_2 + z_2^3 z_3 + z_3^3 z_1$. □

§2.10. Let $G = SL_n(\mathbb{C})$ and V be the direct sum of $m \ge n$ copies of the natural representation of SL_n on $\mathbb{C}^n$. Let $p = (v_1,\dots,v_m) \in V$, $p \neq 0$. Then the orbit $G(p)$ is closed iff $\operatorname{rank}(v_1 \dots v_m) = n$; in this case $G_p = \{e\}$. So all non-trivial slice representations are nice. However, if $m > n + 1$, the point $\pi(0)$ is (the only) singular point of V/G.

In other words, for these representations the slice method does not simplify the problem. In fact the slice principle works best of all for irreducible representations.

References.

1. Bernstein I.N., Gelfand I.M., Ponomarev V.A., Coxeter functors and Gabriel's theorem, Russian Math. Surverys 28, 17-32(1973).

2. Gabriel P., Unzerdegbare Darstellunger I., Man. Math. 6, 71-103 (1972).

3. Kac V.G., Infinite root systems, representations of graphs and invariant theory, Invent. Math. 56, 57-92(1980).

4. Kac V.G., Infinite root systems, representations of graphs and invariant theory II, Journal of Algebra 77, 141-162 (1982).

5. Kac V.G., Popov V.L., Vinberg E.B., Sur les groupes lineares algebrique dont l'algebre des invariants est libre, C.R. Acad. Sci., Paris 283, 875-878(1976).

6. Kraft H. Parametrisierung der Konjugationklassen in $s\ell_n$, Math. Ann. 234, 209-220(1978).

7. Luna D., Slices étales, Bull. Soc. Math. France, Memoire 33, 81-105(1973).

8. Luna D., Adhérences d'orbites et invariants, Invent. Math. 29, 231-238(1975).

9. Macdonald I.G., Symmetric functions and Hall polynomials, Clarendon press, Oxford, (1979).

10. Nazarova L.A., Representations of quivers of infinite type, Math. U.S.S.R.-Izvestija Ser math. 7, 752-791(1973).

11. Ringel C.M., Representaions of K-species and bimodules, J. of Algebra, 41, 269-302(1976).

12. Salmon G., A treatise of the analytic geometry of three dimensions, v.2, New York, Chelsea, 1958-1965.

13. Serre J.-P., Cohomologie Galoisienne, Lecture Notes in Math. 5 (1965).

14. Serre J.-P., Algèbre Locale, Lecture Notes in Math 11(1975).

15. Sylvester J.J., Tables of the generating functions and ground forms of the binary duodecimic, with some general remarks..., Amer. J. Math. 4, 41-61(1881).

16. Kac V.G., Watanabe K., Finite linear groups whose ring of invariants is a complete intersection, Bull. Amer. Math. Soc. 6 ,221-223(1982).

Victor G. Kac
M.I.T.
Cambridge, MA 02139

INVARIANTS OF Z/pZ IN CHARACTERISTIC p.

Gert Almkvist

1. Representations of $G = Z/pZ$.

Let G denote the cyclic group with p elements ($p > 2$ a prime), written multiplicatively. Let further k be a field of characteristic p. Let V be a finite dimensional vector space over k. A representation of G is a group homomorphism

$$\rho : G \to \mathrm{Aut}V.$$

Then V can be considered as a $k[G]$-module via the action

$$g\cdot x = \rho(g)(x)$$

for $g \in G$ and $x \in V$.

The group ring $k[G] \simeq k[X]/(X^p - 1)$ is artinian and self-injective (i.e. it is injective considered as a module over itself).

A module V is indecomposable if $V = V_1 \oplus V_2 \Rightarrow V_1 = 0$ or $V_2 = 0$.

Then it is not hard to show the following:

1. Every module is a direct sum of indecomposable modules.

2. There exist only finitely many (non-isomorphic) indecomposables $V_1, V_2, \ldots, V_p$ where $\dim_k V_n = n$.
   Observe that $V_1 = k$ with the trivial G-action and $V_p = k[G]$ is the free module of rank 1.

3. $V_n \simeq k[X]/(X-1)^n$ (let the generator of G act as multiplication by X).

Definition 1.1: The representation ring R_G is the free abelian group with basis $V_1, V_2, \ldots, V_p$ and multiplication defined by $V \otimes_k W$, (G acts by $g\cdot(v\otimes w) = (gv)\otimes(gw)$).

Then $V_1 = 1$ is the identity and we have the multiplication table:

$$V_2V_n = V_{n+1} + V_{n-1} \quad \text{if } 1 < n < p$$

$$V_2V_p = 2V_p .$$

Hence R_G is generated by V_2 over Z.

More precisely:

<u>Theorem 1.2</u>: The map $V_2 \to X$ induces a ring isomorphism

$$R_G \simeq Z[X] \big/ (X-2)U_{p-1}(X/2)$$

where $U_{p-1}(\cos\varphi) = \frac{\sin p\varphi}{\sin\varphi}$ is the second Chebyshev polynomial of degree $p-1$.

<u>Definition 1.3</u>: Let V be a module. Denote by $V^G = \{x \in V;\ gx = x \text{ for all } g \in G\}$ the <u>submodule of G-invariant elements</u>.

<u>Proposition 1.4</u>: $V_n^G = V_1$.

---

2. Invariants of $G = Z/pZ$.

<u>Definition 2.1</u>: A homogeneous polynomial $f \in k[x_0, x_1, \ldots, x_n]$ is <u>G-invariant</u> if $f(x_0, x_1+x_0, x_2+2x_1+x_0, \ldots, x_n+\binom{n}{1}x_{n-1}+\binom{n}{2}x_{n-2}+\ldots+x_0) = f(x_0, x_1, \ldots, x_n)$

(i.e. if V has basis $x_0, x_1, \ldots, x_n$ and G has generator σ then σ acts as

$$\sigma = \begin{pmatrix} 1 & 0 & 0 & 0 & \ldots & 0 \\ 1 & 1 & 0 & 0 & \ldots & 0 \\ 1 & 2 & 1 & 0 & \ldots & 0 \\ - & - & - & & & \\ 1 & \binom{n}{1} & \binom{n}{2} & - & - - & 1 \end{pmatrix}$$

If we make the substitution $X \to 1 + X$ it agrees with the action in section 1.

Example 2.2: Let $p = 5$, $n = 3$ and the degree $r = 3$. Then the polynomials

$$x_0^3$$
$$x_0^2x_2 - x_0x_1^2$$
$$x_0^2x_3 + 2x_0x_1x_2 + 2x_1^3$$
$$(2x_1^2-2x_0x_2)x_3 + x_1x_2^2 - x_0^2x_1$$

form a basis for the space of homogeneous G-invariants of degree 3, (observe that the first two polynomials agree with the classical $SL(2,\mathbb{C})$-invariants. This is no accident (see [2])).

Main problem of invariant theory: Find the generators and the relations for the graded ring of G-invariants $k[x_0,x_1,\dots,x_n]^G$. This is very difficult in general. The only non-trivial easy case is the following.

Example 2.3: Let $n = 2$. Then the ring of invariants is the hypersurface

$$k[x_0,x_1,x_2]^G = k[u_0,u_1,u_2,u_3]\Big/\{u_2^2-u_0^pu_3-u_1(u_1^{\frac{p-1}{2}}-u_0^{p-1})^2\}$$

where

$$u_0 = x_0$$
$$u_1 = x_1^2 - x_2x_0$$
$$u_2 = x_1^p - x_0^{p-1}x_1$$
$$u_3 = \prod_{j=0}^{p-1}(x_2+2jx_1+j^2x_0)$$

Example 2.4: Let $p = 5$ and $n = 3$ (as in Example 2.2). Then $R = k[x_0,x_1,x_2,x_3]^G$ has twelve generators and at least 16 relations (see [2] and [8]).

If $R = \bigoplus_{r \geq 0} R_r$ is a graded k-algebra where all $\dim_k R_r < \infty$ let

$$F(R,t) = \sum_{r \geq 0} \dim_k R_r t^r$$

denote the Hilbert series of R.

A more reasonable problem is to find the Hilbert series of the ring of invariants.

Example 2.4 (cont.)

$$F(R,t) = \frac{1 - 2t - 2t^2}{(1-t)^3(1-t^5)} = 1 + t + 2\,t^2 + 4t^3 + \ldots$$

Let V_{n+1} have basis $x_0, x_1, \ldots, x_n$ and let

$$S^r V_{n+1} = \begin{cases} \text{all homogeneous polynomials of} \\ \text{degree } r \text{ in } x_0, x_1, \ldots, x_n \end{cases}$$

be the symmetric product. Then $S^r V_{n+1}$ becomes a G-module via the action

$$g(x_0^{i_0} x_1^{i_n} \ldots x_n^{i_n}) = (gx_0)^{i_0} (gx_1)^{i_1} \ldots (gx_n)^{i_n} .$$

We have

$$k[x_0, x_1, \ldots, x_n]^G = \bigoplus_{r \geq 0} (S^r V_{n+1})^G.$$

If $S^r V_{n+1} = \sum_{j=1}^{p} c_j V_j$ then $\dim_k (S^r V_{n+1}) = \sum_{j=1}^{p} c_j$

because $V_n^G = V_1 = k$.

Theorem 2.5: If n is even then

$$F(k[x_0, \ldots, x_n]^G, t) = \frac{1}{p} \sum_{\gamma \in \mu_p} \prod_{j=0}^{n} \frac{1}{1-\gamma^{n-2j} t}$$

where μ_p = the group of p-th roots of unity.

Remark 2.6: If n is odd there is a more complicated formula (the proof is worse too, see [1] and [3]).

Remark 2.7: The formula in 2.5 looks like Molien's Theorem. Indeed if

$$G' = \left\{ \begin{pmatrix} \gamma^n & & & 0 \\ & \gamma^{n-2} & & \\ & & \ddots & \\ 0 & & & \gamma^{-n} \end{pmatrix} ; \gamma \in \mu_p \right\}$$

acts on $\mathbb{C}[x_0,\dots,x_n]$ then $\mathbb{C}[x_0,\dots,x_n]^{G'}$ has the same Hilbert series. But the rings are far from isomorphic.

Reciprocity Theorem 2.8: If n is even then

$$F(k[x_0,\dots,x_n]^G, 1/t) = (-t)^{n+1} F(k[x_0,\dots,x_n]^G, t)$$

(if n is odd there is no such result).

Amusing(?) Remark 2.9: Put $\gamma = e^{2\pi i\nu/p}$, $\nu = 0,1,2,\dots,p-1$ and let $p \to \infty$ in 2.5,

$$\frac{1}{p} \sum_{\gamma\in\mu_p} \prod_{j=0} \frac{1}{1-\gamma^{n-2j}t} = \sum_{\nu=0}^{p-1} \frac{1}{\prod_{j=0}^{n} (1-te^{2\pi i\nu(n-2j)/p})} \cdot \frac{1}{p} \to$$

$$\to \frac{1}{2\pi} \int_{-\pi}^{\pi} \frac{1 + \cos\varphi}{\prod_{j=0}^{n} (1-te^{i(n-2j)\varphi})} d\varphi = \sum_{r\geq 0} b_r t^r = F(C_n, t)$$

(the numerator $1 + \cos\varphi$ instead of 1 makes the formula valid also for odd n). Here b_r = number of linearly independent (classical $SL(2,\mathbb{C})$-) covariants of a binary form of degree n, with leading term of degree r (see [2] for an explanation).

Theorem 2.10: If n is even then

$$F(k[x_0,\dots,x_n]^G, t) =$$

$$= \sum_{0\leq j<n/2} (-1)^j \phi_{n-2j} \left\{ \frac{1+t^p}{1-t^p} \cdot \frac{t^{j(j+1)}}{(1-t)^2\dots(1-t^{2(n-j)})(1-t^2)\dots(1-t^{2j})} \right\} .$$

Here $\phi_s : \mathbb{Q}(t) \to \mathbb{Q}(t^s)$ is defined by

$$(\phi_s(h)(t^s) = \frac{1}{s} \sum_{\gamma \in \mu_s} h(\gamma t)$$

(for a more complicated formula when n is odd see [4]).

Example 2.11: Let $n = 4$. Then

$$F(k[x_0,x_1,x_2,x_3,x_4]^G,t) = \frac{(1+t^3)(1+t^p) + 2t^{\frac{p+1}{2}}(1+t)^2}{(1-t)(1-t^2)^2(1-t^3)(1-t^p)}$$

(formulas are also computed for $n = 5$ and 6 for certain classes of primes in [4]).

Ellingsrud and Skjelbred [9] have shown that depth $k[x_0,\dots x_n]^G = 3$. Since $\dim k[x_0,\dots x_n]^G = n + 1$ we have that $k[x_0,\dots,x_n]^G$ is never Cohen-Macaulay if $n > 2$ (but it is a UFD; see also [7]).

3. Computations in R_G.

The proofs are done in the representation R_G (or rather in the extension $R_G[\mu]$ where $V_2 = \mu + \mu^{-1}$). Also computations are much simplified if one disregards the free modules (i.e. consider the ring $\tilde{R}_G = R_G/(V_p)$).

Definition 3.1: (a) $\lambda_t(V) = \sum_{i\geq 0} \Lambda^i V\, t^i$ in $R_G[t]$

(b) $\sigma_t(V) = \sum_{i\geq 0} S^i V t^i$ in $RG[[t]]$.

Reciprocity Theorem 3.2:

$$\sigma_{1/t}(V_{n+1}) = \begin{cases} (-t)^{n+1}\sigma_t(V_{n+1}) & \text{if } n \text{ is even} \\ (-t)^{n+1}(V_p - V_{p-1})\sigma_t(V_{n+1}) & \text{if } n \text{ is odd}. \end{cases}$$

This is the main result from which all other reciprocity and symmetry theorems follow. [3]

Theorem 3.3 (Symmetry).

(a) $S^r V_{n+1} \simeq S^n V_{r+1}$ if $r,n < p$ ("Hermite's reciprocity law")

(b) $S^r V_{n+1} \simeq \Lambda^r V_{n+r}$ if $n+r < p$

(c) $S^{p-n-r-1} V_{n+1} - S^r V_{n+1} = \frac{1}{p}\left\{\binom{p-r-1}{n} - \binom{r+n}{n}\right\} V_p$ if n is even

$S^{p-n-r-1} V_{n+1} + S^r V_{n+1} = \frac{1}{p}\left\{\binom{p-r-1}{n} + \binom{r+n}{n}\right\} V_p$ if n is odd.

Theorem 3.4

(a) $\lambda_{-t}(V_{n+1})\sigma_t(V_{n+1}) = 1$ if n is even

(b) $\lambda_{-t}(V_{n+1})\sigma_t(V_{n+1}) = \dfrac{1-(V_p-V_{p-1})t^p}{1 - t^p}$ if n is odd.

4. Applications to Combinatorics.

Let $A(m,n,r)$ = number of partitions of m into at most r parts all of size $\leq n$.

Put $V(m,n,r) = A(m,n,r) - A(m-1,n,r)$

Let $A(m,n,r) = 0$ if either $n < 0$, $m > nr$ or $m \notin Z$.

If $S^r V_{n+1} = \sum_{j=1}^{p} c_j V_j$ where $r,n, < p$

then

$$c_j = \sum_{\nu} V\left(\frac{nr+1-j}{2} + \nu p, n, r\right).$$

Hence all results about $S^r V_{n+1}$ (like symmetries) can be formulated in the language of partitions without any reference to group representations.

Let $\begin{bmatrix} n+r \\ r \end{bmatrix}(t) = \dfrac{(1-t^{n+r}) \ \dots \ (1-t^{n+1})}{(1-t) \ \dots \ (1-t^{r})}$ be the Gaussian polynomial.

Definition 4.1: A polynomial $a_0 + a_1 t + \dots + a_n t^n$ in $Z[t]$ is

(a) symmetric if $a_j = a_{n-j}$

(b) unimodal if it is symmetric and $0 < a_0 \leq a_1 \leq \dots \leq a_{[n/2]}$.

Theorem 4.2: Let

$$\begin{bmatrix} n+r \\ r \end{bmatrix}(t) = q(t)(1+t+t^2+ \dots + t^{p-1}) + t^s f(t)$$

where q and f are symmetric with $f(0) \neq 0$.

Then f(t) is unimodal (see [6]) .

Acknowledgements: My thanks go to R. Fossum who got me interested in invariant theory. Many results are also due to him (see [1]). Furthermore I want to thank L. Avramov and R.P. Stanley for stimulating discussions.

References.

1. G. Almkvist-R. Fossum: Decompositions ... "Sém d'alg., Paul Dubreil 1976-77", Springer Lecture Notes No 641, 1-111.

2. G. Almkvist: Invariants, mostly old ones, Pac. J. Math. 86(1980), 1-13.

3. G. Almkvist: Representations of Z/pZ,..., J.Alg. 68(1981),1-27.

4. G. Almkvist: Some formulas in invariant theory, to appear in J.Alg.

5. G. Almkvist: Rings of invariants, Preprint Lund 1981:2.

6. G. Almkvist: Representations of SL(2,ℂ) and unimodal polynomials, Preprint Lund 1982.

7. M.J. Bertin: Anneaux d'invariants d'anneaux de polynômes en caractéristiques p, CR 264(1967), 653-656.

8. L.E. Dickson: On invariants and the theory of numbers, Dover 1966.

9. G. Ellingsrud-T. Skjelbred: Profondeur d'anneaux d'invariants en caractéristique p, Oslo 1978, No. 1.

University of Lund
Box 725
S-220 07 LUND
Sweden

SYMMETRY AND FLAG MANIFOLDS

by

Alain Lascoux & Marcel-Paul Schützenberger

In spite of its links with the symmetric group, the study of flag varieties has not yet fully used the customary technics (permutoèdre, Ehresman's order, Lehmer's code) of the theory of symmetric functions.

To the so-called Schubert cycles are associated polynomials, which are no other than Schur functions in the case of Grassmann varieties, and which can be studied through the help of symmetrizing operators, acting both on the cohomology ring, and the Grothendieck ring as special cases. Conversely, the study of representations of the symmetric group benefits from the geometrical intuition coming from the action of the symmetric group on the flag variety.

As an example, we indicate how to effectively compute the projective degrees of Schubert cycles. A note submitted to the Académie des Sciences apply these methods to the calculation of the Chern classes of the flag variety - as for its harmonic functions, the theory of which requires some properties of the plactic monoïd, they will be the subject of a separate article.

Half of the authors warmly thanks Mittag Leffler Institute & the University of Stockholm for their hospitality, the C.I.M.E. Foundation for providing the opportunity of displaying the symmetrizing operators, as well as A.Björner, D.Laksov and F.Gherardelli for their interest in this work.

<u>Caution</u> : the operators are placed on the right.

§ 1 Symmetrizing operators.

It is always delicate to distinguish between a permutation and its inverse, or between right and left multiplication for the symmetric group, if one does not take a set of "values" and a set of "places". To avoid misunderstandings, we shall consider permutations of $n+1$ elements as operators on the ring of polynomials $\mathbb{Z}[a,b,\dots]$, $\{a,b,\dots\}$ being a totally ordered alphabet of cardinal $n+1$.

Starting from the special element $a^E = a^n\, b^{n-1}\, c^{n-2} \dots$, one uses the transpositions $\sigma_{ab}, \sigma_{bc} \dots$ of consecutive letters to generate all monomials (written in the lexicographic order) whose multidegrees are a reordering of $\{0,1,\dots,n\}$.

This process gives us a ranked poset, as shown in the following figure, the permutations being considered as paths (directed downwards) in the graph of the poset (the graph is called the "permutoëdre"). The permutoëdre gives us all the reduced decompositions of the elements of the symmetric group, the length of a permutation being the length of any corresponding path; ω denotes the permutation of greatest length.

For example, the symmetric group on three letters gives

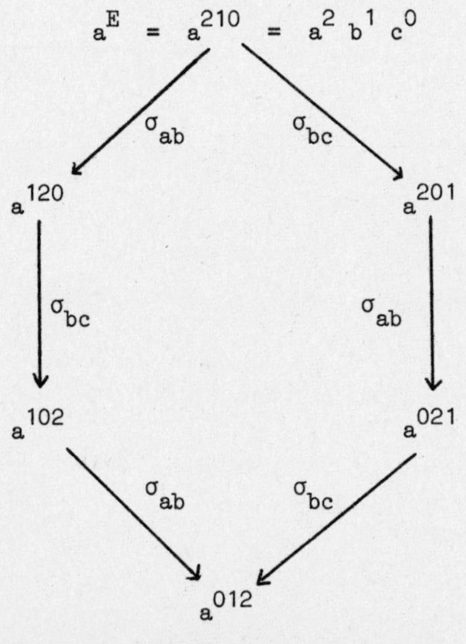

FIGURE 1

Moore's relations say that two paths having the same end points can be obtained by a sequence of elementary transformations

<u>1.1</u> $\sigma_{ab} \cdot \sigma_{bc} \cdot \sigma_{ab} = \sigma_{bc}\, \sigma_{ab}\, \sigma_{bc}$

a,b,c being any triple of consecutive letters

<u>1.2</u> $\sigma_{ab} \cdot \sigma_{de} = \sigma_{de}\, \sigma_{ab}$

if $\{a,b\} \cap \{d,e\} = \emptyset$

The conventions are such that if $\sigma . \sigma' . \sigma'' \ldots$ is a path from a^E to $a^I = a^{i_0} b^{i_1} \ldots$, then with $\omega w = i_0+1,\ i_1+1,\ \ldots,\ i_n+1$, one has $a^E\, \omega w = a^I$, $a^I\, w^{-1} = a^0 b^1 c^2 \ldots$; less trivial operators on $\mathbb{Z}[a,b,\ldots]$ appear when applying Jacobi's procedure to generate symmetric polynomials through alternating polynomials.

Define ∂_{ab} to be the operator

<u>1.3</u> $\partial_{ab} : f(a,b,\ldots) \rightsquigarrow [f(a,b,c\ldots) - f(b,a,c\ldots)]/(b-a)$

and similarly for all pairs of consecutive letters, where f is an arbitrary polynomial (or rational) function.

We can interpret any path of the permutoëdre as a product of operators ∂_{ab}. Checking the relation similar to 1.1 (1.2 is trivial in this case), one gets

<u>1.4</u> <u>Lemma</u>. <u>The product of operators corresponding to a path from</u> ω <u>to</u> w <u>is independant of the choice of the path and depends only upon</u> w . <u>It is denoted</u> $\partial_{\omega w}$.

The operators ∂_w are not always adequate because they systematically decrease the degrees. To preserve the degree, one defines

<u>1.5</u> $\pi_{ab} : f \rightsquigarrow (af)\, \partial_{ab}$

$\pi_{bc} : f \rightsquigarrow (bf)\, \partial_{bc}$

. . . .

and one checks that these new operators still verify relations 1.1 and 1.2, so that a product of operators corresponding to a path depends only upon the end points : $\pi_{\omega w}$ is given by a path from ω to w , π_w by a path from w^{-1} to $123 \ldots$.

Having at hand three operators verifying the same relations 1.1 and 1.2, one cannot resist in putting them in a single family.

Let p,q,r be fixed integers.

Define

1.6 $$D_{ab}(p,q,r) : f \rightsquigarrow (f)(p\partial_{ab} + q\pi_{ab} + r\sigma_{ab})$$

and similarly for all pairs of consecutive letters.

It is a simple, but not totally trivial verification, that conditions 1.1 and 1.2 are still fulfilled, so that one can write $D_w(p,q,r)$.

To accelerate computations one may remark that symmetric functions are scalars with respect to the D_w :

1.7 $$f,g \in \mathbb{Z}[a,b,\ldots] \ , \ f\sigma_{ab} = f \Rightarrow (fg)D_{ab} = gD_{ab}f \ .$$

In fact $D_\omega(p,q,o)$ is a symmetrizer in the whole alphabet, i.e. $\forall f \in \mathbb{Z}[a,b,\ldots] \ , \ \forall w, \ [f\, D_\omega(p,q,o)]w = f\, D_\omega(p,q,o)$.

One can show that, up to a change of variables, the operators $D_w(p,q,o)$ are the most general symmetrization operators verifying certain natural conditions, and thus we cannot find a family with more parameters containing them.

More precisely concerning $D_\omega(p,q,o)$, one has:

1.8 $$f\, D_\omega(p,q,o) = \sum \ (f\ \Delta_{pq})w$$

sum on all permutations, Δ_{pq} being the generalization of Vandermonde's determinant:

$$\Delta_{pq} \ = \ \Pi_{x<y} \ (p + qx)/(x - y) \ .$$

1.9 Remark. $f\pi_\omega = (fa^E)\partial_\omega = \sum (-1)^{\ell(w)}(fa^E)w \,/\, \sum (-1)^{\ell(w)} a^E w$;

$a^I\pi_\omega$ is the classical Schur function of index $i_n, i_{n-1}, \ldots, i_0$ (cf Macdonald).

Thus the operators $D_\omega(1,0,0) = \partial_\omega$ and $D_\omega(0,1,0)$ are essentially the same, and formula 1.8 becomes in this case Jacobi's expression of Schur functions, Weyl's character formula for the linear group and Bott's theorem for the cohomology of line bundles on flag manifolds.

We did not use the square D^2_{ab} of an operator; in fact, one has

1.10 $$D^2_{ab} = q\, D_{ab} + r(q+r)$$

so that one is really working with a representation of the Hecke algebra of the symmetric group.

§ 2 Schubert polynomials.

As the action of ∂_ω, or π_ω as well, transforms a monomial into a Schur function, the operators D_w will give generalizations of Schur functions.

Following Demazure, and independently, Bernstein-Gelfand and Gelfand, we shall define, for every permutation w, the polynomial X_w by

2.1 $$X_w = a^E\, \partial_{\omega w} ,$$

a^E being the monomial $a^n\, b^{n-1}\, c^{n-2} \ldots$. Thus, the X_w are obtained just by pushing down $X_\omega = a^E$ along the edges of the permutoëdre.

Figure 2 gives the result in the case of four letters.

One could generate the X_w through the π_w ; this more complicated process leads to a combinatorial representation of the X_w .

FIGURE 2

POLYNÔMES DE SCHUBERT

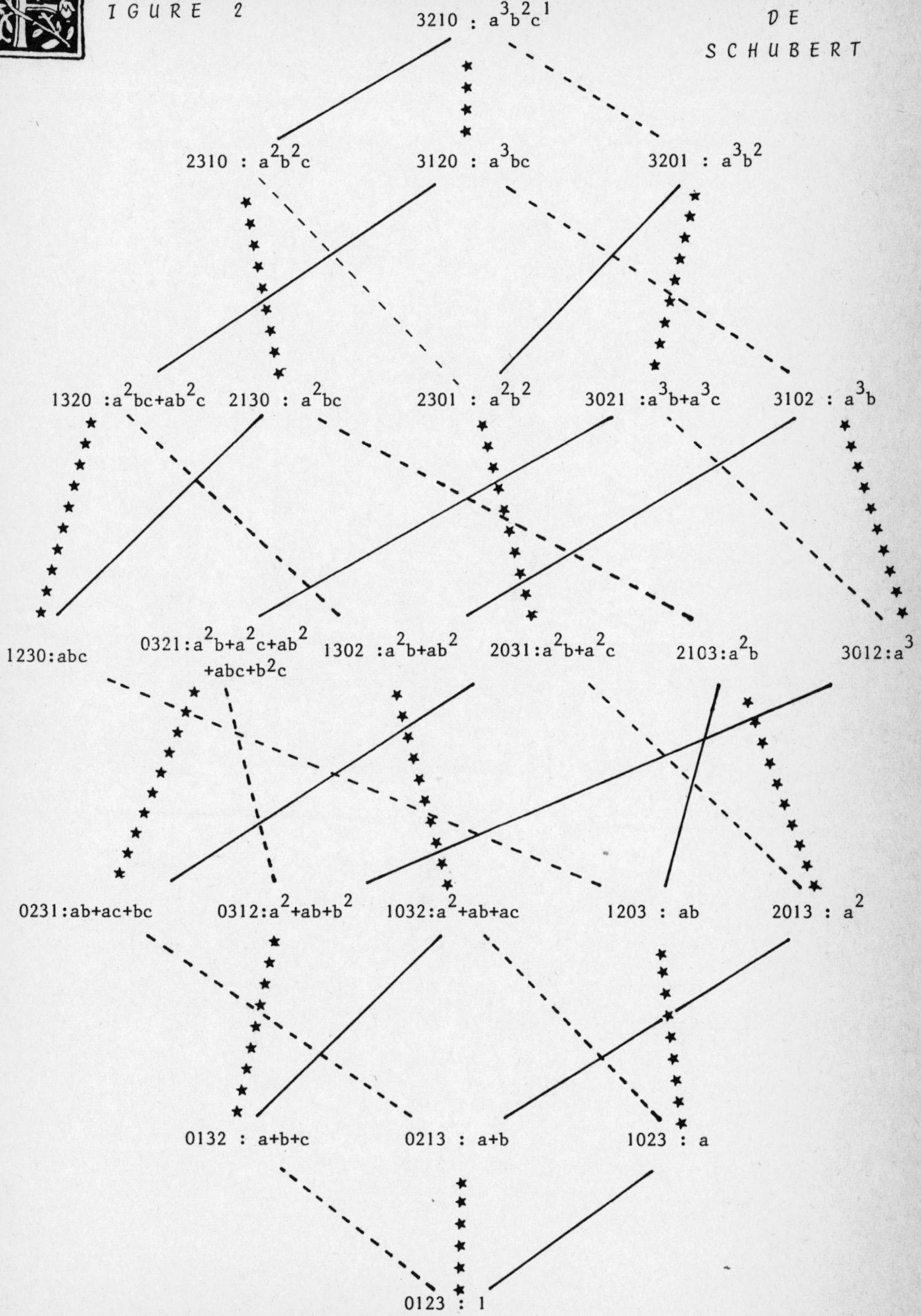

One notices on the example that X_w is a polynomial of degree $\ell(w)$ with positive coefficients, and that X_w is symmetrical in the i-th and i+1-th letters of A if and only if $w_i < w_{i+1}$; in other words, the shape of w (i.e. the sequence of its up's and down's) gives the symmetries of the polynomial X_w . For those permutations which are called grassmannian permutations (i.e. the ones which have only one descent), then X_w is a Schur function in the first letters of A, e.g. X_{2413} , which has to be symmetrical in a,b and also in c,d is indeed the Schur function

$$S_{12}(a+b) = a^2b + ab^2$$

(we identify an alphabet $A = \{a,b,\ldots\}$ and $S_1(A) = a+b+\ldots$) .

As for Schur functions, the first problem will be to multiply two polynomials. The simplest case is due to Monk, but first we need to enlarge the permutoëdre. For each pair of permutations (v,w) , such that v and w differ only by a transposition: $v = \ldots v_i \ldots v_j \ldots \curvearrowright w = \ldots v_j \ldots v_i \ldots$ and that $\ell(w) = \ell(v) + 1$, one draws between v and w j-i edges, of respective "colors" (i+1, i), ..., (j, j-1) (remember that on the permutoëdre, an edge of color (i+1, i) meant the transposition of the letters at place i and i+1).

The graph so obtained, when one forgets about the colours and the multiplicities of the edges, is due to Ehresman, and more generally, for Coxeter groups, to Bruhat. Let us call it the coloured Ehresmanoëdre.

Now, choose one colour (i+1, i) , and consider the monocolour subgraph $\Gamma_{i+1\ i}$ obtained by erasing the edges of colour different from the choosen one.

Then, writing i+1 i for the permutation 1 ... i-1 i+1 i i+2 ... , one has Monk's formula

$$\text{2.2} \qquad X_{i+1\ i} \cdot\cdot X_v = \sum X_w$$

sum on all $w : \ell(w) = \ell(v) + 1$, vw is an edge of $\Gamma_{i+1\ i}$, i.e. there is an edge of colour (i+1 i) between v and w .

It is not too difficult to verify by induction this formula. The remarkable fact is that there is no multiplicity in this multiplication. Pieri's formula asserts that the multiplication of a Schur function by a special Schur function of any degree (i.e. elementary or complete symmetric functions, cf. Macdonald) produces no multiplicities. The same thing happens more generally for Schubert polynomials, i.e. the multiplication of a Schubert polynomial by a special one gives rise to no multiplicities (cf. L & S for the description of the w coming in the summation). Thus Monk's formula is the initial degree one-case of the general Pieri's formula.

As a by-product, one obtains a commutation property which is valid for all finite Coxeter groups.

Let $C_{21}, C_{32} \ldots$ be the matrices of the directed graphs $\Gamma_{21}, \Gamma_{32}, \ldots,$ i.e. we put 1 at the place (v,w) of the matrix $C_{i+1\,i}$ if $\ell(v) < \ell(w)$, and vw is an edge of $\Gamma_{i+1\,i}$, and 0 otherwise. Then one has

2.3 Lemma: The matrices $C_{i+1\,i}$ commute.

Proof.: As the multiplication of Schubert polynomials by $X_{i+1\,i}$ is described by the matrix $C_{i+1\,i}$. The lemma is equivalent to the fact that the product $X_{i+1\,i} \cdot X_{j+1\,j} \cdot X_v$ is equal to $X_{j+1\,j} \cdot X_{i+1\,i} \cdot X_v$ for every v.

This specific property of Bruhat order on Coxeter groups has to be proved in itself without reference to multiplication of Schubert polynomials.

§ 3 Cohomology of the flag manifold.

The reader who wants to use Figure 2 to multiply $a^3\left(=X_{4123}\right)$ by $a\left(=X_{2134}\right)$ finds no edge with colour 21 from 4123 upwards. So he must disagree with Monk's formula as stated above. But if he enlarges the alphabet by just one letter, he certainly obtains that

$$a^3 \cdot a = X_{41235} \cdot X_{21345} = X_{51234} = a^4 .$$

More generally, to use Monk's formula for the symmetric group W_n, one must for safety reasons imbed it into W_{n+1}.

Alternatively one can also notice that

$$a^4 = S_4(a+b+c+d) - (b+c+d)\, S_3(a+b+c+d) + (bc+bd+cd)\, S_2(a+b+c+d) - bcd(a+b+c+d)$$

which has the consequence that a^4 belongs to the ideal generated by the polynomials summetrical in a,b,c,d.

Definition: The cohomology ring of the flag manifold associated to the symmetric group W_{n+1} is

<u>3.1</u> $$H = \mathbb{Z}[a,b,\dots] / I$$

where I is the ideal generated by the symmetric polynomials (with no constant term!) in all the variables (in other words, the ideal generated by the invariants of W_{n+1}).

It is easy to show by induction on n that H has two natural $\mathbb{Z}$-bases:

i) the monomials $a^I = a^{i_1} b^{i_2} \dots$ with $0 \leq i \leq E$

ii) the Schubert polynomials X_w (the class of a Schubert polynomial in H is called a <u>Schubert cycle</u>).

Notice that H is of rank 1 in the maximal degree $\ell(\omega) = n(n+1)/2$ and that $X_\omega = a^E$.

Now, Monk's formula is perfectly valid: passing from an alphabet of $n+1$ letters to n, one annihilates exactly the Schubert polynomials X_w

FIGURE 3

MULTIPLICATION PAR a + b DANS L'ANNEAU DE COHOMOLOGIE

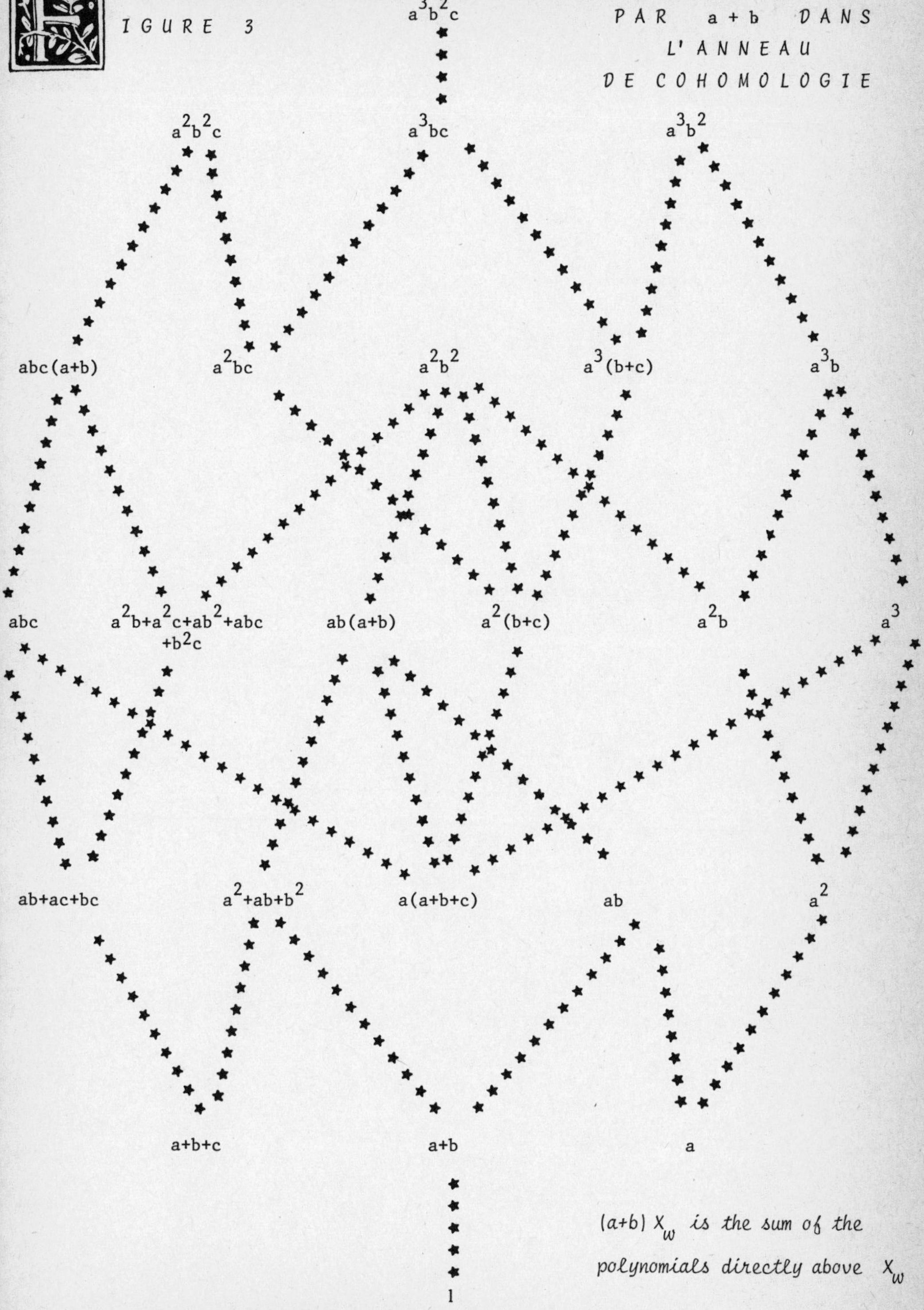

(a+b) X_w is the sum of the polynomials directly above X_w

for those w such that $w_{n+1} \neq n+1$.

But another difficulty comes: how can we see that two polynomials are equivalent modulo the invariants of W_{n+1} ?

Ehresman, generalizing classical results on Grassmann varieties, has shown that the multiplication in H does induce a pairing on the basis of Schubert cycles: for w, w' such that $\ell(w) + \ell(w') = \ell(\omega)$, then

<u>3.2</u> $$X_w \cdot X_{w'} \equiv a^E \text{ or } 0$$

according to $w' = \omega w$ or not.

This result is due to Chevalley for arbitrary Coxeter groups.

Since the operators D_w preserve the ideal I ($(fg)D_{ab} = g\, D_{ab} \cdot f$ if $f\,\sigma_{ab} = f$), they are indeed operators on H .

Using $a^E \partial_\omega = 1$, one gets:

<u>3.3</u> <u>Let</u> P <u>be a homogeneous polynomial of degree</u> $\ell(\omega)$. <u>Then</u>

$$P \equiv (P\, \partial_\omega)\, a^E \mod I .$$

Thus, combining with 3.2, the decomposition of a polynomial in the basis X_w is given by

<u>3.4</u> $$P \equiv \sum (P \cdot X_{\omega w})\partial_\omega\, \varepsilon \cdot X_w$$

sum on all permutations w , the <u>augmentation morphism</u> $\varepsilon : \mathbb{Z}[a,b,\ldots] \to \mathbb{Z}$, $a\varepsilon = b\varepsilon = \ldots = 0$ taking care of the decomposition of P into its homogeneous components.

Now, if one does not have at hand the explicit expression of the Schubert polynomials, one must improve the method to be able to determine when two polynomials are equivalent modulo the ideal I . This will be done in § 6.

§ 4 Projective degree of Schubert cycles.

Consider a graded ring H , call the graduation codimension, and assume that H is of rank 1 in maximal codimension (assumed different from infinity!) : $H^{max} \simeq \mathbb{Z}$.

Let Y in H be an element of codimension 1, and X of codimension d . Then the degree of X relative to Y is the image in $\mathbb{Z}$ of $X \cdot Y^{max-d}$.

When H is the cohomology ring of a projective variety, one chooses an imbedding in a projective space and Y is the class of the intersection with an hyperplane.

In our case, for the natural embedding of the flag variety, which is due to Plücker, Y is equal to the sum of all Schubert polynomials of codimension 1 :

4.1 $$Y = X_{2134\ldots} + X_{1324\ldots} + X_{1243\ldots} + \ldots = na + (n-1)b + (n-2)c + \ldots$$

(To distinguish between the degree of X relative to Y and the degree of X as a polynomial, we call the first projective degree and the second codimension.)

To compute the projective degrees, it is sufficient to know them for the Schubert cycles. In the case of grassmannians (a certain quotient of H) one obtains the degrees of the irreducible representations of the symmetric group, so these projective degrees should be interesting by themselves, regardless of their geometrical interpretations.

We have already done most of the work: as the multiplication by $X_{i+1\,i}$ corresponds to the edges of colour (i+1 i) in the coloured Ehresmanoëdre, one gets:

4.2 Proposition. The projective degree of X_w is the number of paths from w to ω in the coloured Ehresmanoëdre.

FIGURE 4

ORDRE D'EHRESMAN SUR LE GROUPE SYMETRIQUE

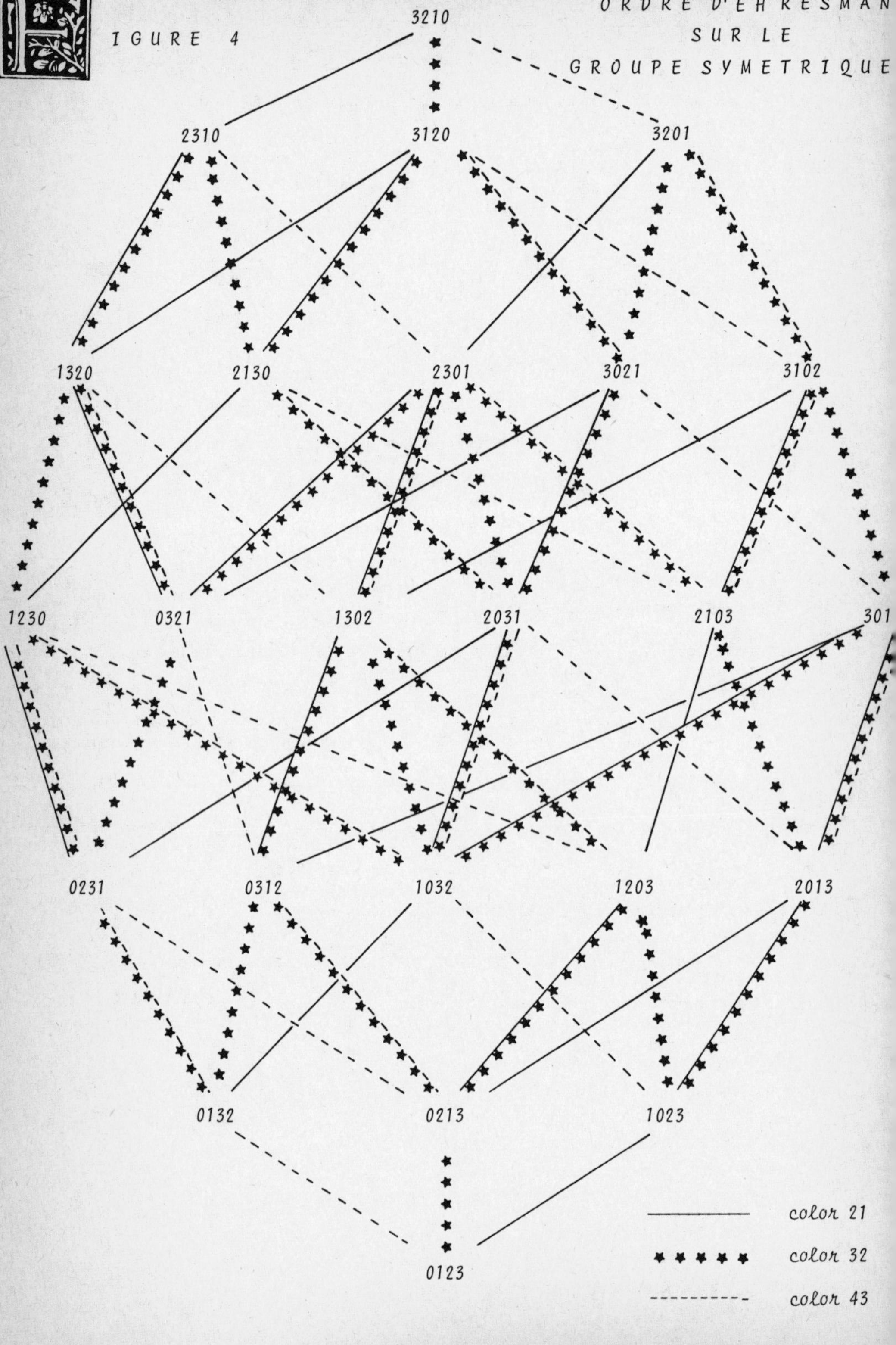

FIGURE 5

DEGRÉ PROJECTIF DES VARIÉTÉS DE SCHUBERT

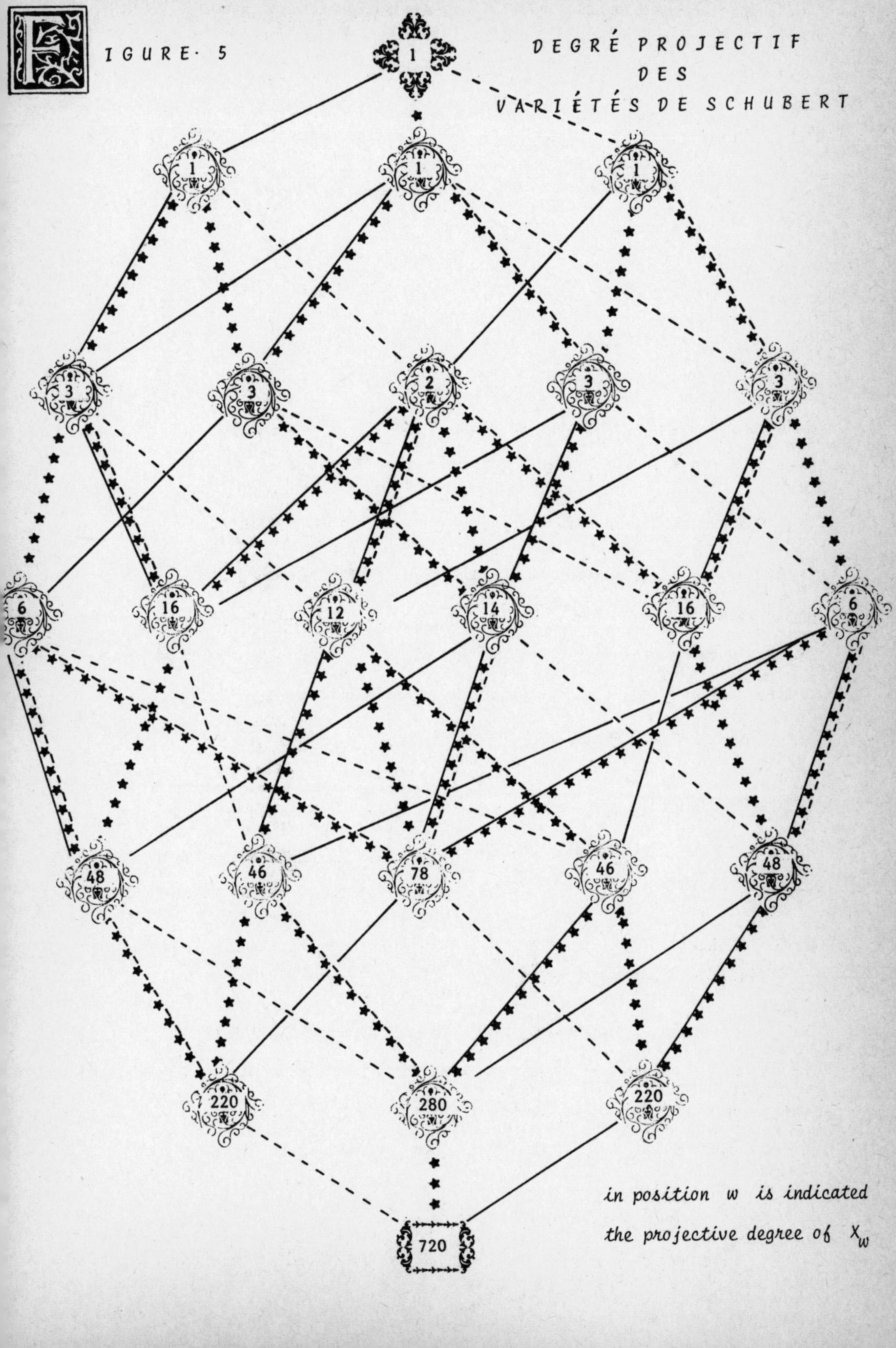

in position w *is indicated the projective degree of* X_w

This proposition is equivalent to the following induction:

<u>4.3</u> $\quad \text{proj.deg}\ X_w = \sum_v m(w,v)\ \text{proj.deg}\ X_v$

sum on all permutations v: $\ell(v) = \ell(w) + 1$, with $m(w,v)$ = number of edges from w to v .

Another formulation is:

<u>4.4</u> $\quad (1-Y)^{-1} = (1 - (na + (n-1)b + \ldots))^{-1} \equiv \sum \text{proj.deg}\ (X_w)\ X_{\omega w}$ in H .

For example, in the case $n = 2$

$(1 - (2a+b)^{-1} \equiv 1 + a + (a+b) + 3a^2 + 3ab + 6a^2b$ (and so the proj.deg are 1,1,1,3,3,6) taking into account that in H , $a^2 + ab + b^2 \equiv 0$, $a^3 \equiv b^3 \equiv a^2b + ab^2 \equiv 0$ modulo the symmetric functions in a,b,c.

One can show this way that proj.deg $X_{123\ldots} = \left(n(n+1)/2\right)!$

Since more information is contained in the Ehresmanoëdre, one can do better than only counting the paths, by reading the paths as words of colours. So denote colours (21), (32), ... by α, β, ..., and read a path as a sequence of colours, i.e. a word in the Greek alphabet.

Define the <u>non-commutative degree</u> of X_w as the sum of the words given by all the paths from w to ω . Then the commutation property 2.3 insures that this non commutative degree is a "Partie reconnaisable" (terminology from the theory of monoids) i.e. is <u>invariant by permutation</u>: whenever you meet the word $\alpha\alpha\beta\gamma$, you have also the word $\gamma\alpha\beta\alpha$ with the same frequency.

<u>4.5</u> <u>Example</u>. For X_{3214} , one gets the noncommutative degree

$(\alpha+\beta+\gamma)(\beta\gamma+\gamma\beta+\gamma\gamma) + (\beta+\gamma)(\alpha\gamma+\gamma\alpha) + \gamma(\alpha\beta+\beta\alpha+\beta\beta)$.

Thus, the non commutative degree is given by restricting to the <u>increasing words</u>; in the above examples, the degree is obtained by permutation of (6) $\alpha\beta\gamma$ + (3) $\alpha\gamma\gamma$ + (3) $\beta\beta\gamma$ + (3) $\beta\gamma\gamma$ + (1) $\gamma\gamma\gamma$

(inside the parenthesis, we have indicated how many words are associated in the non-commutative degree).

In other words, if $\varphi : \mathbb{Z}\left[[\alpha,\beta,\ldots]\right] \to \mathbb{Z}[\alpha,\beta,\ldots]$ is the natural morphism (the <u>evaluation</u>) from the ring of non-commuting variables to the ring of polynomials, then the non-commutative degree is the inverse image of a polynomial $Z_{\omega,w}$ that we call the <u>colour-degree</u> of X_w .

More generally, one defines the polynomials $Z_{v,w}$, when $\ell(v) \geq \ell(w)$, to be the sum of all increasing paths from v to w (this will correspond to the degree of the intersection of two Schubert cycles); put $Z_{v,w} = 0$ if $\ell(v) < \ell(w)$.

If moreover, one defines M_α to be the matrix: the entry (v,w) is α or 0 , according as $\ell(v) = \ell(w) + 1$ and there is an edge of colour α between v and w , or not, and similarly for M_β, M_γ ..., one obtains from 2.3 the commutation of the matrices M_α, M_β, ...

<u>Exercise</u>. Prove that $Z_{n+1\ 12..n,\ 12..n+1} = \sum \alpha^i \beta^j \gamma^k \ldots$, sum on all different monomials of total degree n , with $i \leq n$, $j \leq n-1$, $k \leq n-2$, ...

e.g. $Z_{4123,\ 1234} = \alpha(\alpha\alpha + \alpha\beta + \beta\beta + \alpha\gamma + \beta\gamma)$.

§ 5 The G-polynomials.

Instead of taking the cohomology ring of the flag manifold as we did in § 3, it is more fruitful to take another quotent of the ring of polynomials, which is called the <u>Grothendieck ring of the flag manifold</u>; denote the variables by L_a, L_b, ... to distinguish from the preceeding case, and keep the same notations for the operators ∂_{ab}, π_{ab}, ..., as no ambiguity is to be feared.

Call θ the <u>specialization</u> ring-morphism: $L_a\theta = L_b\theta = \ldots = 1$ and let $\mathcal{J}$ be the ideal generated by the relations

<u>5.1</u> $\forall\ f \in \mathbb{Z}[L_a,L_b,\ldots]$, $f\,\pi_\omega \equiv f\,\pi_\omega\theta$

i.e. the totally symmetric polynomials are equalled to their value for $L_a = L_b = \dots = 1$.

5.2 Definition. The Grothendieck ring of the flag manifold is

$$K = \mathbb{Z}[L_a, L_b, \dots] / J .$$

The properties of this ring are strongly linked with those of symmetric functions, whose theory has been formalized in the theory of λ-rings.

As J is invariant under the action of the D_w , these operators still act on K . Of course, taking relations 5.1 instead of 3.1 do not change the $\mathbb{Z}$-bases of the quotient ring, so that one has

5.2 The set of monomials $L^I = L_a^{i_1} L_b^{i_2} \dots$, for $0 \leq I \leq E$. is a $\mathbb{Z}$-basis of K .

5.3 The Schubert polynomials (in the alphabet $L_a, L_b, \dots$) are a $\mathbb{Z}$-basis of K .

As $L_a L_b L_c \dots \equiv 1$, we see that the ring K contains the inverse of the variables $L_a, L_b, \dots$. The inversion of $L_a, L_b, \dots$ extends to an involution morphism of K which is called duality by reference to vector bundles ($L_a, L_b, \dots$ are the tautological line bundles of the flag manifold).

It is convenient to introduce new variables $x = 1 - L_a^{-1}$, $y = 1 - L_b^{-1}$, $z = 1 - L_c^{-1}$ .... The symmetrizers associated to $x, y, \dots$ are related to those of $L_a, L_b, \dots$; One checks

5.4 $$\forall f \in K , \quad f\, \pi_{ab} = (f - fy)\partial_{xy}$$

and similarly for all pairs of consecutive letters, which, incidentally, shows that a change of variables induces a non trivial transformation of the symmetrizers. As in § 2, we choose a maximal element

$$G_\omega = x^E = x^n y^{n-1} z^{n-2} \dots$$

and we define the G-polynomial indexed by the permutation w by

5.5 $$G_w = G_\omega \pi_{\omega w} .$$

Figure 6 gives the case $n = 3$.

It is clear from lemma 5.4 that the homogeneous part of smallest degree ($= \ell(w)$) of G_w is the Schubert polynomial X_w (in the variables $x,y,\ldots$ instead of $a,b,\ldots$). Thus the Schubert polynomials are nothing but the leading term of the G_w.

Schubert polynomials (in $x,y,\ldots$) being a basis of K, one can express the G_w in term of the X_w, or conversely, the X_w in term of the G_w, the matrix being triangular.

e.g. for $(w) = 4$, one has

$$X_{2413} = G_{2413} + G_{3412} \; ; \qquad X_{2341} = G_{2341} \; ;$$

$$X_{3142} = G_{3142} + G_{3241} \; ; \qquad X_{3214} = G_{3214} \; ;$$

$$X_{1432} = G_{1432} + 2G_{2431} + G_{3421} + G_{3412} \; ; \qquad X_{4123} = G_{4123} .$$

To express the multiplication in the basis of G_w is more complicated than in the basis X_w. We shall give elsewhere the corresponding "Pieri-formula". For example, one reads on figure 5 that

$$G_{1324}\, G_{1324} = (x+y-xy)^2 = G_{2314} + G_{1423} - G_{2413} .$$

(We previously had $X_{1324} \cdot X_{1324} = X_{2314} + X_{1423}$; here we had to substract the supremum of 2314 and 1423 which is 2413 ; bigger intervals are involved in general.)

To understand the link between the two rings H and K, one must recall the existence and properties of Chern classes:

Denote by $1 + H^+$ the multiplicative monoid of polynomials with constant term 1 (and coefficients in Q). There exists a homomorphism:

$$c : K \to 1 + H^+$$

FIGURE 6

ANNEAU DE GROTHENDIECK DES VARIÉTÉS DE DRAPEAUX

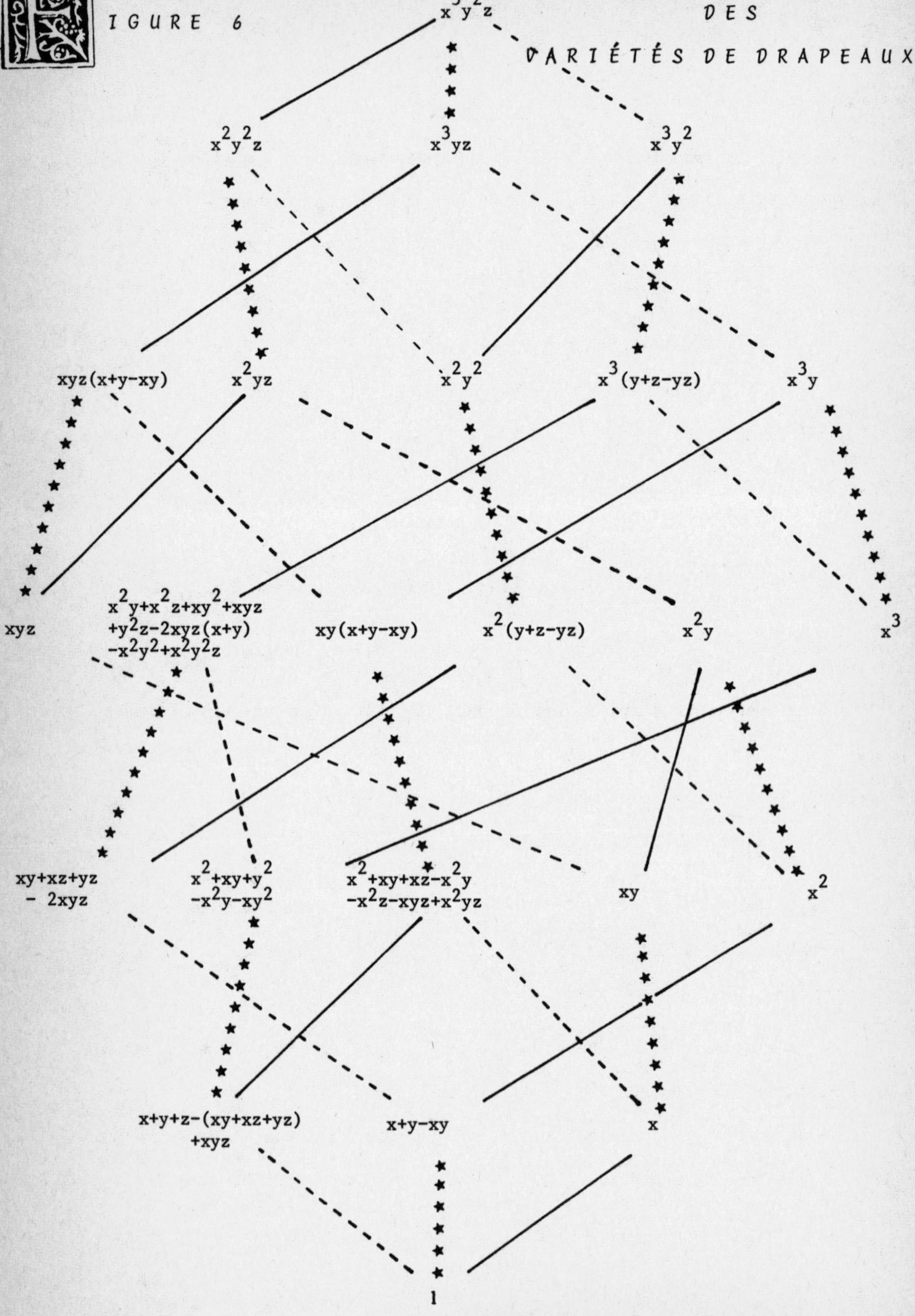

such that $c(1) = 1$, $c(E+F) = c(E)\cdot c(F)$. On the basis L^I , it takes the values

$$c(L_a^{\,i}\, L_b^{\,j}\, L_c^{\,k} \ldots) = 1 + ia + jb + kc + \ldots .$$

(Of course the multiplication in K induces a "product" in $1 + H^+$, for whose explicit description one needs Schur functions - see Macdonald.)

Now one can check

$$c(G_w) = 1 - (-1)^{\ell(w)} \; (\ell(w)-1)! \; X_w + X'$$

where X' is a polynomial of degree $> \ell(w)$. (cf. SGA6, exp. 0 formule 1.18).

If G is a sum $\Sigma\, n_w G_w$ with w of constant length $\ell(w) = d$ (one says $G \in K^d$)

$$c(G) = 1 - (-1)^d \; (d-1)! \; \Sigma\, n_w X_w + X'$$

and

$$c(G\pi_{ab}) = 1 - (-1)^{d-1} \, (d-2)! \; \Sigma\, n_w X_w \partial_{ab} + X' \partial_{ab}$$

so that one sees that:

5.6 Proposition: $-(d-1/^{-1} \partial_{ab}$ is the image by the Chern homomorphism of π_{ab} acting on K^d .

§ 6 Quadratic form on the cohomology ring.

Most of the preceeding description relies heavily upon the natural bases of the cohomology or Grothendieck ring of the flag variety. To be able to compute without restriction in these rings, one must be able to express a general element (in a finite time) in the bases already defined.

The operators corresponding to the permutation of greatest length are the most effective rool for this purpose. It amounts to define on each of the spaces H and K a quadratic form.

We consider sequences as vectors in $\mathbb{Z}^{n+1}$ and thus can write $I \pm J$; through the identification $I \curvearrowright a^I$, the symmetric group acts on sequences: $I \rightsquigarrow Iw$;

recall that E is the sequence $n, n-1, \ldots, 0$.

Now, when $-E \leq I \leq E\omega$, one checks from formula 1.9 that

<u>6.1</u> $$a^I \pi_\omega = a^{I+E} \partial_\omega \begin{cases} = (-1)^{\ell(w)} & \text{if there exists } w \text{ such that } I + E = Ew \\ = 0 & \text{otherwise.} \end{cases}$$

Moreover, $a^I \pi_\omega = (a^{-I}\omega)\pi_\omega$, and $a^I \pi_\omega = 0$ if $i_1 + \ldots + i_{n+1} (= |I|) \neq 0$.

E.g. $I = -3102 \Rightarrow I+E = 0312$ is a permutation of E and so is $-I\omega + E$ $(= -2\ 0\ -1\ 3 \ + \ 3\ 2\ 1\ 0 \ = \ 1\ 2\ 0\ 3)$.

Owing to this symmetry between I and $-I\omega$, one defines a scalar product on H by its values on the basis a^I (for $0 \leq I \leq E$) :

<u>6.2</u> $$(a^I, a^J) = a^{I\omega - J}\, \pi_\omega \,.$$

For example, for four letters and degree 3

2100	1110	2010	1200	3000	0210	I \ J
0	0	0	0	0	−1	2100
0	0	0	0	−1	0	1110
0	0	0	−1	1	1	1200
0	0	−1	0	1	1	2010
0	−1	1	1	−1	0	3000
−1	0	1	1	0	−1	0210

On the example one sees that the quadratic form, for a given weight, is positive or negative definite, and triangular for an appropriate ordering of the monomials.

Instead of describing the ordering, which directly comes from the interpretation of sequences I such that $0 \leq I \leq E$ as coding permutations, one can do better and give the adjoint basis of a^I.

<u>6.3</u> Let $P_I = \Pi_{0 \leq p \leq n} \Lambda_{i_p} (A_{n-p})$.

A_p being the alphabet of the first p letters, and Λ_i the elementary symmetric function of degree i ; then one has

<u>6.4</u> <u>Proposition: The family</u> $(-1)^{|I|}\{P_I\}$ <u>with</u> $0 \leq I \leq E$, <u>is the adjoint basis of</u> $\{a^I\}$.

For example, for $n=3$, $I = 2010$, $P_I = \Lambda_2 (a+b+c) \Lambda_1(a) = a^2b + a^2c + abc$, and one checks from the preceeding table that

$$(a^I, P_I) = -1 \quad , \quad (a^J, P_I) = 0 \quad \text{if} \quad J \neq I .$$

<u>6.5</u> <u>Corollary</u> (<u>Bott-Rota's straightening</u>).

<u>If</u> Y <u>is an homogeneous polynomial of degree</u> d , <u>then in</u> H , $(-1)^d Y \equiv$ $\equiv \Sigma (Y, P_I) a^I \equiv \Sigma (Y, a^I) P_I$ <u>sum on all sequences I such that</u> $0 \leq I \leq E$.

Thanks to 6.1, this straightening is an efficient way of decomposing in H the class of a polynomial. For the decomposition in the basis X_w , we already have 3.4.

We note that $X_w \partial_v = 0$ if $\ell(v) = \ell(w)$ and $v \neq w^{-1}$, because $X_w \partial_v = X_\omega \partial_{\omega w} \partial_v$, so that either $\omega w v = \omega$, or $\ell(\omega w v) < \ell(\omega) \Leftrightarrow \partial_{\omega w} \partial_v$ can be written $\partial_{w'} \partial_u \partial_u \partial_{w''}$ $(= 0)$. Thus we have the other way of decomposing a polynomial Y .

<u>6.6</u> $$Y \equiv \Sigma \, (Y \cdot \partial_{w^{-1}}) \varepsilon \, X_w$$

sum on all permutations w , which $P\varepsilon$ = term of degree o of P , as in 3.4.

§ 7 Quadratic form on the Grothendieck ring.

As Schubert polynomials are still a basis of the Grothendieck ring K , one could still keep the scalar product for which the Schubert polynomials are an orthonormal basis. This would not fit well with the action of the operators on K .

Remembering that π_ω sends K to its subring $\mathbb{Z}$, one can define

7.1 $\quad \forall\ P, Q \in K\ ,\ \langle P, Q \rangle = (PQ)\,\pi_\omega\ ;$

on the basis G_w , the quadratic form takes only the values 0 or 1 .

For example, for S_3 , the multiplication table is

	G_{123}	G_{213}	G_{132}	G_{231}	G_{312}	G_{321}
G_{123}	G_{123}	G_{213}	G_{132}	G_{231}	G_{312}	G_{321}
G_{213}	G_{213}	G_{312}	$G_{312}+G_{231}-G_{321}$	G_{321}	0	0
G_{132}	G_{132}	$G_{312}+G_{231}-G_{321}$	G_{231}	0	G_{321}	0
G_{231}	G_{231}	G_{321}	0	0	0	0
G_{312}	G_{312}	0	G_{321}	0	0	0
G_{321}	G_{321}	0	0	0	0	0

and the quadratic form is the image of this table by π_ω (noting that $\forall w\ ,\ G_w\,\pi_\omega \equiv 1$):

1	1	1	1	1	1
1	1	1	1	0	0
1	1	1	0	1	0
1	1	0	0	0	0
1	0	1	0	0	0
1	0	0	0	0	0 .

We shall not prove here the two following propositions which generalize 3.4 and 6.6.

7.2 Proposition. For any w, let H_w be the sum of G-polynomials $\Sigma_{v \geq w} (-1)^{\ell(v)-\ell(w)} G_v$.

Then $\{H_{\omega w}\}$ is the adjoint basis of $\{G_w\}$ (with respect to $\langle\,,\,\rangle$).

For example,

$$\langle G_{132} - G_{231} - G_{312} + G_{321}\,,\, G_w \rangle = 0 \quad \text{except for } w = 312 = \omega \cdot 132 .$$

7.3 Proposition. Let θ be the specialization morphism $L_a\,\theta = L_b\,\theta = \ldots = 1$. Then $\forall\, P \in K$, $\forall w$, $\langle Y, G_w \rangle = Y\, \pi_{w^{-1}}\, \theta$.

As for every w, $\langle G_w, 1 \rangle = (G_w \cdot 1)\, \pi_\omega = 1$ the fact that $\langle H_w, 1 \rangle = 0$ generalizes the property of the Moebius function (for the Bruhat order) to be ± 1.

§ 8 Applications.

We have mainly described the tools to study the cohomology or Grothendieck ring of the flag manifold. Many questions arising from the theory of groups or algebraic geometry can be then easily studied.

8.1 The representation of the symmetric group W_{n+1} on H or K. One must note that as $\mathbb{Z}$-modules, H and K are isomorphic to the regular representation of W_{n+1} but that the degree gives us an extra information; in fact, the multiplicity of an irreducible representation can be considered as a polynomial (which happens to be a Kostka-Foulkes polynomial coming in the theory of representation of the finite linear groups). More generally, De Concini et Procesi have studied the quotients of H associated to the variety of flags fixed by a given unipotent matrix.

8.2 Enumerative geometry on the flag manifold.

We have only given the projective degree of a Schubert cycle in § 4.

One needs also the postulation of the cycle X_w with respect to a line bundle L^I : by definition, it is $\Sigma (-1)^i \dim H^i (\mathcal{O}_w, L^I)$; once given the rules of the translation, it simply becomes $(L^I G_w) \pi_\omega$, which is also equal to $L^I \pi_{w-1} \theta$, as asserted in 7.3.

The Chern classes of a variety are the first invariants of it that one tries to get. In the case of the flag manifold, the tangent bundle T has class

$$L_a \cdot L_b^{-1} + L_a \cdot L_c^{-1} + L_b L_c^{-1} + \ldots \quad \text{in } K$$

so that its Chern class is

$$c(T) = (1+a-b)\,(1+a-c)(1+b-c) \ldots$$

and it remains to compute $c(T)$ in the basis of Schubert cycles. This will be done elsewhere.

8.3 Representations of the linear group $G\ell(\mathbb{C}^{n+1})$.

One can consider the ring of invariants of W_{n+1}: $\mathbb{Z}[a,1/a,b,1/b,\ldots]^W$ to be the ring of formal sums of representations of $G\ell(\mathbb{C}^{n+1})$.

Bott's theorem evaluates in this ring, for any line bundle L^I, $I \in \mathbb{Z}^{n+1}$, and any i , the representation $H^i(X, L^I)$, X being the flag manifold.

We have obtained here a little less:

$$\Sigma (-1)^i H^i(X, L^I) = (L^I)\, \pi_\omega ,$$

(in fact, all the $H^i(X, L^I)$ are $\{0\}$ except at most one, so that the two computations are not very different).

One can also look for syzygies of the Schubert variety corresponding to w , i.e. try to get a complex of locally free bundles which "solves" the ring of the Schubert variety. The class of the complex in $\mathbb{Z}[a,1/a,b,1/b,..]$ is given by

$$[\,(1-L_a^{-1})^n\,(1-L_b)^{n-1} \ldots]\; \pi_{\omega w}$$

but, of course there remains to describe the morphisms inside the complex. This, we shall not do.

8.4 Root systems and Coxeter groups.

Most of the properties of the operators D_w can be extended to other finite Coxeter groups, as shown first by Demazure and independently Bernstein, Gelfand-Gelfand.

In the case of the symmetric group, if α, β, ..., are the simple roots, and ρ half the sum of positive roots, then e^α, e^β, ... are respectively $L_a L_b^{-1}$, $L_b L_c^{-1}$, ..., and e^ρ is equal to L^E up to a power of $L_a L_b L_c \ldots$. If I is weakly decreasing (L^I is dominant), then Weyl's character formula for the corresponding irreducible representation E_I is

$$ch(E_I) = \Sigma(-1)^w (L^I L^E)w \;/\; \Sigma(-1)^w L^E w$$

and as we have remarqued in 1.9, it can be written

$$ch(E_I) = (L^I L^E)\,\partial_\omega \;;$$

an equivalent result, using instead the operator π_ω , is

$$ch(E_I) = L^I \pi_\omega$$

which is in fact Bott's formula.

8.5 Determinants.

For $m : 0 \leq m \leq n+1$, one can consider The Grassmann variety of subvector spaces of dim $m+1$ of $\mathbb{C}^{n+1}$. The associated cohomology ring is the subring $H^{W' \times W''}$ of H invariant under the product of symmetric groups $W' \times W''$ (W' being the group on the first $m+1$ letters, W'', on the remaining letters).

A $\mathbb{Z}$-basis of $H^{W' \times W''}$ is the set of Schubert polynomials X_w for the w of minimum length in their class modulo $W' \times W''$. In this case X_w is a Schur function on the alphabet of the first $m+1$ letters, and all the properties of the cohomology ring of the Grassmann variety (or of the Grothendieck ring) can be translated in term of Schur functions. In particular, the determinantal expression of Schur function gives rise to a determinantal expression for X_w (due to Giambelli), for G_w, for the postulation (due to Hodge), etc...

Unfortunately, not all permutations in general give determinants. We have given in [L & S] several characterizations of those permutations for which X_w, G_w, ... are determinants (permutations vexillaires); for them, the computations are very similar to the ones in the more special case of Grassmannvariety.

8.6 Combinatorics.

A combinatorial and powerful description of Schur functions is given by Ferrers diagrams and Young tableaux. One can similarly associate to any permutation a diagram (due to Riguet), and fill it according to rules deduced from Pieri-Monk's formula 2.2. This seems to be an interesting generalization of Young tableaux and the plactic monoid.

BERNSTEIN I.N., GELFAND I.M. and GELFAND S.I., *Russian Math.Surv.28, 1973, p.1-26.*

DEMAZURE M., *Inv. Math. 21, 1973, p. 287-301.*

EHRESMAN C., *Annals of Maths. 35, 1934, p. 396-443.*

HILLER H., *Geometry of Coxeter groups (Pitman, 1982).*

LASCOUX A. and SCHÜTZENBERGER M.P., C.R. Acad. Sc. *Paris 294, 1982, p. 447.*

MACDONALD I.G., *Symmetric and Hall polynomials (Oxford Math. Mono., 1979).*

ON SOME RESTRICTION THEOREMS FOR SEMISTABLE BUNDLES

Vikram B. Mehta

Department of Mathematics
University of Bombay
Bombay 400 098

Introduction

In this survey article, we shall mainly be concerned with the following problem: given a semistable vector bundle on a smooth projective variety X, describe its restriction to a general hyperplane section X. In characteristic zero, one can prove that the instability degree of the restriction is bounded by the degree of X (Thm. 2.1). The characteristic zero assumption is used to prove that a certain morphism is constant if its differential is zero. This is precisely what fails in char p, where we have to factor the given morphism by its Frobenius transform [2].

The use of the "standard construction" was used by Van de Ven in studying uniform bundles, and then by Grauert-Mullich for rank 2 bundles on $\mathbb{P}^2$ and then by Forster--Hirschowitz-Schneider and Maruyama for bundles of arbitrary rank on any variety.

Another method which can be used is to restrict the bundle to hypersurfaces of high degree [11]. This enables one to prove that the restriction to a hypersurface of sufficiently high degree is semistable. We shall also sketch a proof of a theorem of Maruyama [9] where he proves that if $\mathrm{rk}V < \dim X$, then the restriction of V to any hyperplane section of X is again semistable. This enables us among other things, to prove immediately that the set of semistable bundles on X of rank 2 and fixed Chern Classes is bounded, for arbitrary X.

We shall not touch upon the topics of uniform bundles on $\mathbb{P}^n$, for which we refer the reader to [4] for the classification in char. 0 and to [2,6] for characteristic $p > 0$. Also we shall not mention the restriction theorems proved for special bundles on $\mathbb{P}^3$, such as the results of Barth [1] on null-correlation bundles in Char. 0 and those of Ein [2] in Char. $p > 0$. The reader is also urged to consult [3] which appeared too late for inclusion in this survey.

Throughout this paper, we use the notions of semistability and stability in the sense of Mumford-Takemoto [9, 11]. For the sake of convenience we recall these notions as well as those relating to the Harder-Narasimhan forunstable filtration of a vector bundle on a variety [7, 9].

Definition 0.1: Let X be smooth projective of dim n and $\theta_X(1)$ be very ample on X. For a coherent torsion-free sheaf V on X, define $\deg V = c_1(V)\cdot\theta_X(1)^{n-1}$ and $\mu(V) = \deg V/\mathrm{rk}V$. Define V stable (resp. semistable) if for all $W \subset V$, we have $\mu(w) < \mu(v)$ (resp. $\le \mu(V)$).

Suppose V is not semistable. Then there exists a subsheaf V_1 of V with $\mu(V_1) = \sup_{W\subset V} \{\mu(w)\} = \mu_o$ say, and V_1 having maximal rank among subsheaves W of V with $\mu(W) = \mu_o$.
Further V_1 is unique and infinitesimally unique, i.e.
$\mathrm{Hom}(V_1, V/V_1) = 0$ [7]. Now look at V/V_1. Then there exists a subsheaf V_2 of V such that $V_1 \subset V_2 \subset V$ and V_2/V_1 is the maximal subsheaf, in the above sense, of V/V_1. Continue this process to get a flag

$$0 = V_o \subset V_1 \subset \ldots \subset V_r = V$$

with the properties:

1) Each V_i/V_{i-1} is semistable, $1 \le i \le r$.
2) $\mu(V_i/V_{i-1}) > \mu(V_{i+1}/V_i)$, $1 \le i \le r-1$.
3) The above flag is unique.

We put $\mu_i(V) = \mu(V_i/V_{i-1})$ and we call V_1 the β-subsheaf of V. We also $\mu_{max}(V) = \mu_1$ and $\mu_{min}(V) = \mu_r$. The above filtration (1) is called the Harder--Narasimhan filtration of V. By the uniquenes of the flag, it is defined over any field of definition of V.

Let $f : X \to S$ be a smooth projective family of varieties and V a vector bundle on X such that $V_K|X_K$ is unstable, where K is the quotient field of S. Then there exists a nonempty open subset U of S and a filtration

$$0 = W_o \subset W_1 \subset \ldots \subset W_r = V/f^{-1}(U)$$

such that for every point $x \in U$, $0 = W_{o,x} \subset W_{1,x} \subset W_{r,x}$ is the unstable filtration of $V/f^{-1}(x)$.

Section I

First we introduce some notation. Let X be a smooth projective variety of dimension n. Let $(\alpha_1,\ldots,\alpha_{n-1})$ be a sequence of integers with each $\alpha_i \geq 2$. For any $m > 0$, put $\underline{m} = (\alpha_1^m, \ldots, \alpha_{n-1}^m)$. Put $S_{\underline{m}}$ = the multi-projective space of homogeneous polynomials of multi-degree $\alpha_1^m, \ldots, \alpha_{n-1}^m$, let $Z_{\underline{m}}$ be the correspondence variety:

$$\begin{array}{ccc} Z_{\underline{m}} & \xrightarrow{q_{\underline{m}}} & S_{\underline{m}} \\ {\scriptstyle p_{\underline{m}}}\downarrow & & \\ X & & \end{array}$$

For a closed point $x \in S_{\underline{m}}$, we call $q_{\underline{m}}^{-1}(x)$ a curve of type $\underline{m}$.
Let $K_{\underline{m}}$ be the quotient field of $S_{\underline{m}}$ and we define $Y_{\underline{m}}$, the <u>generic complete intersection</u> curve of type $\underline{m}$, by the fibre product:

$$\begin{array}{ccc} Y_{\underline{m}} & \longrightarrow & K_{\underline{m}} \\ \downarrow & & \downarrow \\ Z_{\underline{m}} & \longrightarrow & S_{\underline{m}} \end{array}$$

Now we can state

Theorem 1.1 [11]: Let V be a semistable vector bundle on X. Then $\exists$ m_o s.t. $V/Y_{\underline{m}_o}$ is semistable.

We sketch a proof. First assume that $\mathrm{rk}V = 2$, $\dim X = 2$ and $\mathrm{Pic}\, X = \mathbb{Z}$. Suppose $V/Y_{\underline{m}}$ is not semistable $\forall$ $m > 0$. Let $L_{\underline{m}} \in \mathrm{Pic}(Y_{\underline{m}})$ contradict the semi-stability of $V/Y_{\underline{m}}$. Then $M_{\underline{m}} \in \mathrm{Pic}(X)$ such that $M_{\underline{m}}/Y_{\underline{m}} = L_{\underline{m}}$.

This is shown by taking a Lefschetz pencil of hyperplane sections of X for $m >> 0$ and then deducing the existence of an imbedded curve in X of type $\underline{m}$ which is integral and singular. Or, one can appeal to a theorem of Weil [13].

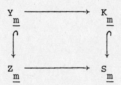

Next, we take a ruve $C_{\underline{m+1}}$ of type $\underline{m+1}$ and degenerate it, over a discrete valuation ring A, to α_1 curves $C_{\underline{m}}^{(i)}$ of type $\underline{m}$, $1 \leq i \leq \alpha_1$. Let $D \to A$ be the family and extend $L_{\underline{m+1}}$, which is defined on the general fibre of D, to a line bundle L_A on D and which is contained in V/D. Restrict L_A to the components of the special fibre. By comparing degrees, we get the important inequality:

$$\deg L_{\underline{m+1}} \le \alpha_1 \deg L_{\underline{m}} \quad . \tag{1}$$

This shows that $\{\deg M_{\underline{m}}\}$ is bounded above. Since $\mathrm{Pic}\,X = \mathbb{Z}$, there exists a unique $M \in \mathrm{Pic}\,X$ such that $M/Y_{\underline{m}}$ is the β-subbundle of $V/Y_{\underline{m}}$ for all $m >> 0$. By Enriques-Severi, the inclusion $M/Y_{\underline{m}} \subset V/Y_{\underline{m}}$ lifts to an inclusion $M \subset V$ on X, contradicting the semistability of V on X.

Now assume $\mathrm{rk}V$, $\dim X$ are arbitrary. Let $W_{\underline{m}}$ be the β-subbundle of $V/Y_{\underline{m}}$, which may be assumed to be of constant rank. Again by degenerating $C_{\underline{m+1}}$ to $\alpha = \prod_{i=1}^{n-1} \alpha_i$ curves of type $\underline{m}$, we get an analogous inequality:

$$\deg W_{\underline{m+1}} \le \alpha \deg W_{\underline{m}} \tag{2}$$

Again this shows the existence of a unique $M \in \mathrm{Pic}(X)$ such that $M/Y_{\underline{m}} = \det W_{\underline{m}}$ for all $m >> 0$. Now one can construct, for each x in $S_{\underline{m}}$ for $m >> 0$, a subsheaf W of V on X such that $W/q_{\underline{m}}^{-1}(x) = W_{\underline{m}}/q_{\underline{m}}^{-1}(x)$. Hence $\exists$ m_o such that $V/Y_{\underline{m}_o}$ is semistable.

By using inequality (2) one can prove that $V/Y_{\underline{m}}$ is semistable $\forall$ $m \ge m_o$. So if V_1 and V_2 are two semistable bundles on X, then both V_1 and V_2 are semistable on $Y_{\underline{m}}$ for some m. In particular, if we are in characteristic zero, then $V_1 \otimes V_2$ is also semistable on $Y_{\underline{m}}$, and hence also on X. This has also been proved by Maruyama and Ramanathan-Ramanan ([8], [12]).

Section II

In this Section we consider the following situation: Let $X \subset \mathbb{P}^n(\mathbb{C})$ be a smooth projective variety, $\dim X = r$. Let V be a semistable vector bundle on X. Put $t = n-r+1$ and let $G = G(t,n)$ be the Grassmann variety of t-planes in $\mathbb{P}^n$. Consider the following diagram

$$\begin{array}{ccc} Z & \xrightarrow{\ g\ } & G \\ {\scriptstyle f}\downarrow & & \\ X & & \end{array}$$

where Z is the correspondence variety, $Z = \{(x,W) \in X \times G / x \in W\}$. We consider $f^*(V)$ as a family of vector bundles on the fibres of g. We may assume $\exists$ an open set U of G and a filtration

$$0 = W_o \subset W_1 \subset \dots \subset W_s = f(V)/g^{-1}(U)$$

such that for all $x \in U$,

$$0 = W_{o,x} \subset W_{1,x} \subset \dots \subset W_{s,x} = V/g^{-1}(x)$$

is the Harder-Narasimhan filtration for $V/g^{-1}(x)$.

Now we have the following, which is due independently to Forster-Hirschowitz--Schneider and Maruyama [5,8]:

Theorem 2.1: We have $\mu(W_i) - \mu(W_{i+1}) \leq d =$ degree X, $1 \leq i \leq s-1$.

Proof: Suppose i such that $\mu(W_i) - \mu(W_{i+1}) > d$. If rank $W_i = p$, W_i gives us a map $\sigma : g^{-1}(U) \longrightarrow G_p[f^*(V)|g^{-1}(U)] = H$ say. Let $h : H \to X$. Denote by T_Z the tangent bundle of Z and by T_f and T_h the relative tangent bundles of f and h respectively. The map σ defines $d\sigma : T_z \longrightarrow T_h$. We restrict $d\sigma$ to $g^{-1}(x)$, $x \in U$. Now the restriction of $T_f/g^{-1}(x)$ can be identified with $\Omega^1(1)^{\oplus n-t}/g^{-1}(x)$, where $\Omega^1(1)$ is the bundle of twisted 1-forms on the t-space corresponding to x. Denote $T_f/g^{-1}(x)$ by B. Now $\mu_{min}(B) \geq -d$ and $\mu_{max}(B) \leq 0$. And $T_h/g^{-1}(x) = (W_i \otimes V/W_i)$, which, by assumption on the $\mu_i'^s$ is seen to be filtered by semistable bundles whose μ's are $< -d$. (Here, we use the fact that tensor products of semistable bundles over $\mathbb{C}$ is semistable). Hence $d\sigma : T_f \longrightarrow T_h$ is zero for every $x \in U$. Thus σ provides us with a map $\tilde{\sigma}: X \to H$ and hence W_i descends to a subsheaf of V on X, contradicting the semistability of V.

Remark 2.2: [8] Assume in the above theorem $X = \mathbb{P}^n$ and we restrict a semistable V on X to a general $\mathbb{P}^r \subset \mathbb{P}^n$. Then if $0 = V_o \subset V_1 \subset \dots \subset V_s = V/\mathbb{P}^r$ is the Harder-Narasimhan filtration for $V/\mathbb{P}^r$, we have $\mu_i(V) - \mu_{i+1}(V) \leq \frac{1}{r}$, $1 \leq i \leq s$. For the proof, we note that if $G = G(r,n)$, and $x \in G(r,n)$ then $T_f/G^{-1}(x)$ is semi-stable and $\mu[T_f/g^{-1}(x)] = \frac{-1}{r}$. Hence if $\mu_i(V) - \mu_{i+1}(V) > \frac{1}{r}$ for some i, V_i descends to a subsheaf of V, contradicting the semistability of V.

Remark 2.3: Assume that X is an r-dimensional complete intersection in $\mathbb{P}^n$, of multi-degree $(d_1,\ldots,d_{n-r})$. Let $d = \deg X = \Pi\, d_i$. Then for any semistable V on X and for any general $\mathbb{P}^t \subset \mathbb{P}^n$ with $t+r \geq n+1$, if $0 = V_o \subset V_1 \ldots \subset V_s = V/\mathbb{P}^t \cdot X$ is the Harder-Narasimhan filtration of $V/\mathbb{P}^t \cdot X$, we have $\mu_i(V) - \mu_{i+1}(V) \leq \frac{d}{t}$. The proof follows from Remark 2.1 and (1) of Section I. One has to show that $\mu_{min}[T_f/\mathbb{P}^t \cdot X] \geq d\, \mu_{min}[T_f/\mathbb{P}^t \cdot \mathbb{P}^r]$.

Remark 2.4: In [15] Schneider has proved that if V is a rank 3 semistable bundle on $\mathbb{P}^3(\mathbb{C})$ with $\deg V = 0$ then V restricted to a general $\mathbb{P}^2$ is again semistable. Making use of Maruyama's theorem [9, thm. 3.1] and 2.2 above, we can generalize this as follow:

Let V be rank n semistable on $\mathbb{P}^n(\mathbb{C})$ with $\deg V \neq -1$ or $1-n$. Then V restricted to a general $\mathbb{P}^{n-1}$ is again semistable. For, if the restriction is not semistable, then by [10], the Harder-Narasimhan filtration of $V/\mathbb{P}^{n-1}$ is

$$0 \to W_i \subset V/\mathbb{P}^{n-1} \to W_2 \to 0$$

with rank $W_1 = 1$ or $n-1$. Further, we must have $\mu_1(V) - \mu_2(V) \leq \frac{1}{n-1}$, which is impossible.

Remark 2.5: In nonzero characteristic, the above does not hold as Ein [2] has constructed bundles of rank 2 on $\mathbb{P}^2$ of degree zero whose restrictions to a general $\mathbb{P}^1$ is not semistable. However, since the moduli space of semistable sheaves of rank 2 and degree 0 on $\mathbb{P}^2$ is irreducible [10], one can easily deduce that there exists a nonempty open subset of the moduli space such that the corresponding bundles have semistable reductions.

Section III

Now we sketch a proof of the following, due to Maruyama [9]:

Theorem 3.1: Let V be a semistable vector bundle on a smooth projective variety X. If $\mathrm{rk}E < \dim X$, then for a general $Y \in |\theta_X(1)|$, E/Y is also semistable.

First we need a construction: Let $X \subset \mathbb{P}^n$ and let G_1 and G_2 be the Grassmannians of codimension 1 and codimension 2 linear spaces in $\mathbb{P}^n$, respectively.

Define $\tilde{X}$ by:

$$\tilde{X} = \{(x,V,W) \in X \times G_1 \times G_2 \mid x \in V \supset W\}$$

Define $\tilde{Y}$ by: $\tilde{Y} = \{(x,V,W) \in \tilde{X} / x \in W\}$.

Define a correspondence variety T between G_1 and G_2 by $T = \{(V,W) \in G_1 \times G_2 / V \supset W\}$.

Then we have:

$$\begin{array}{ccccc} \tilde{Y} & \subset & \tilde{X} & \xrightarrow{q} & T \\ & & {\scriptstyle P}\downarrow & & \\ & & X & & \end{array}$$

Suppose E is a vector bundle on X and suppose $E|X_K$ and $E|Y_K$ are both unsemistable, where K is the quotient field of T.

Let $0 = E_o \subset E_1 \subset \ldots \subset E_m = P^*(V)/\tilde{X}$ and

$0 = F_o \subset F_1 \subset \ldots \subset F_n = P^*(V)/\tilde{Y}$ be the Harder-Narasimhan filtrations of $P(V)/\tilde{X}$ and $P(V)/\tilde{Y}$, respectively. Under the above conditions, we have

Lemma 3.2: If for some i, $1 \le i \le m-1$, $E_i/Y \simeq F_j$, for some j, $1 \le j \le n-1$, then there exists a subsheaf M of E such that $M/\tilde{X} \simeq E_i$. In particular, E is not semistable.

Proof: Let $x \in G_2$ and consider $p_2^{-1}(x)$, where p_1 and p_2 are the two projections: $T \to G_1$ and $T \to G_2$. Then $p_2^{-1}(x)$ is a $\mathbb{P}^1$, imbedded via p_1 in G_1. Put $Z = q^{-1}(\mathbb{P}^1)$. Then Z is the blow-up of X along Y, where Y is the Scheme-theoretic intersection of X with the Codim 2 linear space in $\mathbb{P}^n$ corresponding to x. We have the diagram:

$$\begin{array}{ccccccc} p^{-1}(Y) & = & Y \times \mathbb{P}^1 & \subset & Z & \xrightarrow{q} & \mathbb{P}^1 \\ & & \downarrow & & \downarrow {\scriptstyle p} & & \\ & & Y & \subset & X & & \end{array}$$

Consider E_i/Z, $E_i/Y \times \mathbb{P}^1$ and $F_j/Y \times \mathbb{P}^1$. By assumption, the last two bundles on $Y \times (t)$ are isomorphic for every $t \in \mathbb{P}^1$. If rank $E_i = s$, then E_i/Z gives a map $\sigma : Z \to G_s(E)$, the Grassmannian of rank s subbundles of E. By restriction

σ to $Y \times \mathbb{P}^1$, one sees that is independent of the $\mathbb{P}^1$ factor, hence σ descends to a map $\tilde{\sigma} : X \to G_s(E)$, which defines M. Hence E is not semistable.

Now the proof of Theorem 3.1 follows by induction: Let Y be a hyperplane section of X. If E/Y is not semistable and if $0 = E_o \subset \ldots \subset E_t = E/Y$ is the Harder-Narasimhan filtration then $rkE_i < rkE$, $1 \le i \le t-1$. So far a hyperplane section Z of Y, E_i/Z is semistable $\forall$ i. In particular $0 = E_o/Z \subset \ldots \subset E/Z$ is the Harder-Narasimhan filtration of E/Z. Then by Lemma 3.2 there is a subsheaf $\bar{E}$ of E with $\bar{E}/Y = E_i$. Clearly $\bar{E}$ contradicts the semistability of E.

In [2] Ein proves the following:

Let E be rank 2 stable on $\mathbb{P}^2$ in char $p > 0$. Let $\deg E = -1$ and assume that for a general line $L \subset \mathbb{P}^2$, $E/L = \theta_L(a) + \theta_L(-a-1)$, with $a \ge 0$. Then $a \le \frac{1}{2}[(\frac{4\,c_2(E) - 1}{3})^{1/2} - 1]$. And if $\deg E = 0$ then $a \le (c_2(E)/3)^{1/2}$. He proves this by factoring the map from the correspondence variety to $\mathbb{P}(E)$ by its Frobenius transform. See also [10, Lemma 7.3] and [14, Thm. 8.2] where similar results are proved using the Riemann-Roch Theorem.

REFERENCES

[1] Barth, W.: "Some properties of stable rank-2 vector bundles on $\mathbb{P}^n$", Math. Ann. 226, 125-150 (1977).

[2] Ein, L.: "Stable Vector Bundles on projective spaces in Char. $p > 0$", Math. Ann. 254, 53-72 (1980).

[3] Ein, L., Hartshorne, R., Vogelaar, H.: "Restriction Theorems for Stable Rank 3 Vector Bundles on $\mathbb{P}^n$", Math. Ann. 259, 541-569 (1982).

[4] Elencwajg, Hirschowitz, Schneider: "Les Fibrés Uniforms de Rang au Plus n sur $\mathbb{P}^n(\mathbb{C})$ Sont Ceux Qu'on Croit", (Preprint).

[5] Forster, Hirschowitz, Schneider: "Type De Scindage Généralisé Pour Les Fibrés Stables", (Preprint).

[6] Lang, H.: "On Stable and Uniform Rank - 2 Vector Bundles on $\mathbb{P}^2$ in characteristic P", Manuscripta Math. 29, 11-28 (1979).

[7] Langton, S.: "Valuative criteria for families of vector bundles on algebraic varieties", Ann. Math. 101, 88-110 (1975).

[8] Maruyama, M.: "The Theorem of Grauert-Mullich-Spindler", Math. Ann. 255, 317-333 (1981).

[9] Maruyama, M.: "Boundedness of Semistable Sheaves of Small Ranks", Nagoya Math. J. 78, 65-94 (1980).

[10] Maruyama, M.: "Moduli of Stable Sheaves II", J. Math. Kyoto Univ., 18 (1978).

[11] Mehta, V.B., Ramanathan, A.: "Semistable Sheaves on Projective Varieties and their Restriction to Curves", Math. Ann. 258, 213-224 (1982).

[12] Ramanan, S., Ramanathan, A.: "Some remarks on the Unstability flag", (Preprint).

[13] Weil, A.: "Sur les Critère d'équivalence en géometrie algébrique", Math. Ann. 128, 95-127.

[14] Hartshorne, R.: "Stable Vector Bundles of rank 2 on $\mathbb{P}^3$", Math. Ann. 238, 229-280 (1978).

[15] Schneider, M.: Chernklassen semi-stabilen Vektorraum bündel Vom Rang 3 auf Hyperebenen des Projectiven Raumes.Grelle J. 323, 177-192 (1981).

Acknowledgements: During the preparation of this paper the author was a Visiting Professor at the University of Naples, supported by the Consiglio Nazionale delle Ricerche, Italy. He is grateful to the University of Naples and the C.N.R. for their hospitality.

FONDAZIONE C.I.M.E.
CENTRO INTERNAZIONALE MATEMATICO ESTIVO
INTERNATIONAL MATHEMATICAL SUMMER CENTER

"Complete Intersections"

in the subject of the First 1983 C.I.M.E. Session.

The Session, sponsored by the Consiglio Nazionale delle Ricerche and the Ministero della Pubblica Istruzione, will take place under the scientific direction of Prof. SILVIO GRECO (Politecnico di Torino, Italy) at the Azienda Regionale delle Terme, Acireale (Catania), Italy, ***from June 13 to June 21, 1983.***

Courses

a) ***Complete intersections in affine-algebraic spaces and Stein spaces.*** (8 lectures in English).
Prof. Otto FORSTER (Ludwig-Maximilians-Universität, München, BRD).

1. Estimate of the number of generators of ideals in non-local rings. Proof of the Forster-Eisenbud-Evans conjecture.
2. Estimate of the number of equations necessary to describe algebraic (analytic) sets. Proof of the theorem of Storch-Eisenbud-Evans.
3. The role of the normal bundle.
4. Topological conditions for ideal-theoretical complete intersections in Stein spaces.
5. The Ferrand construction. Set theoretical complete intersections.

References

1. BANICA-FORSTER, Complete intersection in Stein manifolds. Manuscr. Math. 37 (1982), 343-356.
2. EISENBUD-EVANS, Every algebraic set in n-space is the intersection of n hypersurfaces. Inv. Math. 19 (1973), 107-112.
3. FERRAND, Courbes gauches et fibrés de rang 2. CR Acad. Sci. Paris 281 (1975), 345-347.
4. FORSTER, Uber die Anzahl der Erzeugenden eines Ideals in einem Noetherschen Ring. Math. Z. 84 (1964), 80-87.
5. FORSTER-RAMSPOTT, Analytische Modulgarben und Endromisbündel. Inv. Math. 2 (1966), 145-170.
6. KUNZ, Einführung in die kommutative Algebra und algebraische Geometrie, Kap. V., Vieweg 1980.
7. MOHAN KUMAR, On two conjectures about polynomial rings. Inv. Math. 46 (1978), 225-236.
8. SATHAYE, On the Forster-Eisenbud-Evans conjecture. Inv. Math. 46 (1978), 211-224.
9. SCHNEIDER, Vollständige, fast-vollständige und mengentheoretischvollständige Durchschnitte in Steinschen Mannigfaltigkeiten. Math. Ann. 260 (1982), 151-174.
10. STORCH, Bemerkung zu einem Satz von M. Kneser. Arch. Math. 23 (1972), 403-404.
11. SWAN, The number of generators of a module. Math. Z. 102 (1967). 318-322.
12. SZPIRO, Equations defining space curves. Tata Institute Bombay, Springer 1979.

b) ***Work of Zak and others on the geometry of projective space.*** (8 lectures in English).
Prof. Robert LAZARSFELD (Harvard University, USA).

A conjecture of Hartshorne, to the effect that any smooth subvariety of sufficienty small codimension in projective space must be a complete intersection, has sparked a considerable body of work over the past decade. We will survey some of these results, focusing on Zak's recent solution of a related problem of Hartshorne's on linear normality. Specifically, the course will be organized as follows:

1. Historical Introduction; theorems of Barth, Fulton-Hansen, et. al.
2. Work of Zak.
3. Further results; open problems.

References:

1. R. HARTSHORNE, Varieties of small codimension in projective space, Bull. A.M.S. 80 (1974), 1017-1032.
2. W. FULTON and R. LAZARSFELD, Connectivity and its applications in algebraic geometry, in Libgober and Wagreich (eds), Algebraic geometry, Proceedings, Chicago Circle (1980), Lecture notes in math. no. 862, Springer Verlag.

c) ***Complete intersections in weighted projective spaces.*** (4 lectures in English).
Prof. Lorenzo ROBBIANO (Università di Genova, Italy).

The purpose of this course is to give a brief account ρf some results relating classical theorems on complete intersections in projective spaces to new results in weighted projective spaces.

The first part will treat some basic facts on weighted projective spaces, while the second one will be concerned with more specialized facts, such as Lefschetz-type theorems. In particular, problems of factoriality and semifactoriality will be studied.

Basic references

1. C. DELORME, Espaces projectifs anisotropes, Bull. Soc. Math. France 103 (1975).
2. M. DEMAZURE, Anneaux gradués normaux, in Séminaire Demazure-Giraud-Teissier, Singularités des surfaces, Ecole Polytechnique 1979.
3. I. DOLGACHEV, Weighted projective varieties. Mimeographed notes. Moscow State University 1975/76.
4. R.M. FOSSUM, The divisor class group of a Krull domain, Ergeb. Math. Grenz. Bd. 74, Springer Berlin 1973.
5. S. MORI, On a generalization of complete intersections, J. Math. Kyoto Univ. 15 (1975).

d) ***On set-theoretic complete intersections.*** (4 lectures in English).
Prof. Giuseppe VALLA (Università di Genova, Italy).

The aim of this course is to give a comprehensive approach to some of the research frontiers in the topic of algebraic varieties which are set-theoretic complete intersections.

Focusing on the special case of affine or projective algebraic curves over a field of characteristic zero, the course will develop to include the most important and recent results on this subject, such as the theorems, given by D. Ferrand and M. Kumar, on affine curves which are locally complete intersections.

The final part of the course will be devoted to make some hints at the case of projective space curves.

Reference

1. J.P. SERRE, Sur le modules projectifs, Sem. Dubreil-Pisot 14 (1960/61).
2. P. MURTHY, Complete intersections, Conference on Commutative Algebra 1975, Queen's Unviersity, 196-211.
3. M. KUMAR, On two conjectures about polynomial rings, Inv. Math. 46 (1978), 225-236.

Seminars

A number of seminars and special lectures will be offered during the Session.

FONDAZIONE C.I.M.E.
CENTRO INTERNAZIONALE MATEMATICO ESTIVO
INTERNATIONAL MATHEMATICAL SUMMER CENTER

"Bifurcation Theory and Applications"

in the subject of the Second 1983 C.I.M.E. Session.

The Session, sponsored by the Consiglio Nazionale delle Ricerche and the Ministero della Pubblica Istruzione, will take place under the scientific direction of Prof. LUIGI SALVADORI (Università di Trento, Italy) at Villa «La Querceta», Montecatini Terme (Pistoia), Italy, ***from June 24 to July 2, 1983.***

Courses

a) ***Bifurcation Phenomena in Biomathematics***. (6 lectures in English).
Prof. Stavros BUSENBERG (Harvey Mudd College, USA).

Lecture 1: Origins of bifurcation problems in biomathematics.
Nonlinear interactions in population dynamics, nerve pulse propagation, cell growth and morphogenesis.

Lecture 2: Bifurcation and stability in models with monotone properties.
Global bifurcation and stability of constant, periodic and almost periodic solutions. Applications to epidemic and other population models.

Lecture 3: Hopf type bifurcation.
Models in population dynamics and metabolic control with Hopf bifurcations. Periodic, quasiperiodic and chaotic behavior.

Lecture 4: Linear and nonlinear diffusion.
Spatial diffusion and pattern formation. Chemotaxis, strain guided diffusion and morphogenesis.

Lecture 5: Separable age-dependent processes.
A method for decomposing the equations of age-structured processes. Bifurcation phenomena in age-dependent population and cell growth models.

Lecture 6: Diffusion in age-dependent processes.
Spatial diffusion in age-dependent population and cell growth models. Bifurcation of spatially heterogeneous solutions and the development of spatial structure.

References

General background: Mathematics of Biology, M. Iannelli editor, CIME ciclo 1979, Liguori, Napoli (1981). Hoppensteadt, F., Mathematical Theory of Population, Demographics, Genetics and Epidemics, SIAM, Philadelphia (1975).

Population and epidemic models: Busenberg, S. and Cooke, K., «The effect of integral conditions in certain equations modelling epidemics and population growth», J. Math. Biol. 10 (1980), 13-22. Busenberg, S. and Cooke, K., «Models of vertically trasmitted diseases with sequential continuous dynamics», in Nonlinear Phenomena in Mathematical Science, V. Lakshmikantham, editor, Academic Press, New York (1982). Lajmanovich, A. and Yorke, J., «A deterministic model for gonorrhea in a nonhomogeneous population», Math, Biosc. 28 (1976), 221-236.

Diffusion an age-dependence: Busenberg, S. and Travis, C., «Epidemic models with spatial spread due to population migration», J. Math. Biol. (1983) (in press). Busenberg S. and Iannelli, M., «A class of nonlinear diffusion problems in age-dependent population dynamics», Nonlinear Analysis MTA, (1983) (in press). Okubo, A., Diffusion and Ecological Problems: Mathematical Models, Springer Verlag, New York (1980).

b) ***Bifurcation of periodic solutions near equilibria of Hamiltonian systems***. (6 lectures in English).
Prof. I.J. DUISTERMAAT (State University of Utrecht, NL).

A short outline of the content

The problem of finding periodic solutions is formulated as an equation in the loop space. Using Lyapunov-Schmidt reduction this is equivalent to finding zeros of a vectorfield in a finite dimensional space, with a built-in circle invariance. Variants of this procedure work for discrete dynamical systems and for finding homoclinic orbits. If the original system is Hamiltonian then the reduced problem amounts to finding critical points of a circle-invariant function. Near equilibrium points the problem is solved approximately up to any order using Birkhoff normal forms. If only two degrees of freedom are in resonance then, in the generic case, one can bring the equations into an exact normal form, using Wasserman's group invariant version of Mather's theory. If the system itself has two degrees of freedom then this also gives useful information about the other solutions near the equilibrium points. The course will be concluded with a discussion of the situation when more than two degrees of freedom are in resonance.

References

1. D.S. SCHMIDT, Periodic solutions near a reasonant equilibrium of a Hamiltonian system, Celestial Mechanics 9 (1974), 81-103.
2. J. MOSER, Periodic orbits near an equilibrium and a theorem by Alan Weinstein, Comm. Pure Appl. Math. 29 (1976), 727-747.
3. G. WASSERMAN, Classification of singularities with compact abelian symmetry, Regensburger Math. Schriften 1, 1977.

c) ***Topics in Bifurcation Theory***. (6 lectures in English).
Prof. Jack K. HALE (Brown University, USA).

The topics include:

Bifurcation from an equilibrium point with one zero or two pure imaginary roots, relations between the bifurcation function and the center manifold, and the extent to which the theory is valid in infinite dimensions. Nonlocal results and some codimension two bifurcations in R^2 and the role of symmetry. Generic theory, dynamic behavior and stable equilibria in a parabolic equation. Nonlinear oscillations and chaotic behavior in functional differential equations.

The basic references are:

1. S.N. CHOW and J.K. HALE, Methods of Bifurcation Theory, Grundlehren der Math. Wiss. 251, Springer-Verlag, 1982.
2. J.K. HALE, Topics in Dynamic Bifurcation Theory, NSF-CBMS Lectures 27, Am. Math. Soc., Providence, R.I. 1981.

d) ***Bifurcation and transition to turbulence in hydrodynamics***. (6 lectures in English).
Prof. Gérard IOOSS (Université de Nice, F).

Outline of the contents:

Physical motivation - Experimental results.
Navier-Stokes equations as a dynamical system, regularity properties of the solution. Poincaré map.
Specific examples: Taylor problem, plane Bénard problem.
Bifurcations which break symmetries, rotating waves, quasiperiodic solutions, frequency lockings.
Routes for transition to turbulence. Conjectures, open problems.

Basic literature references:

1. V.I. ARNOLD, Chapitres supplémentaires de la théorie des équations différentielles ordinaires, ed. MIR, Moscou 1980.
2. S.N. CHOW, J.K. HALE, Methods of bifurcation theory, Springer Verlag, 1982.
3. G. IOOSS, Arch. Rat. Mech. Anal., 64, 4 (1977), 339-369.
4. G. IOOSS, Bifurcation of maps and applications, North Holland Math. Stud. 36, 1979.
5. D.D. JOSEPH, Stability of fluid motions, vol. I and II, Springer Tracts in Phil., vol. 27, 28, 1976.
6. T. KATO, Perturbation theory for linear operators, Springer Verlag, 1966.
7. O.A. LADYZENSKAYA, The mathematical theory of viscous incompressible flow, Gordon and Breach, 1969.
8. J. MARSDEN, M. Mc CRACKEN, The Hopf bifurcation and its applications, Math. Applied Sciences, 1, Springer Verlag 1976.

Seminars

A number of seminars and special lectures will be offered during the Session.

FONDAZIONE C.I.M.E.
CENTRO INTERNAZIONALE MATEMATICO ESTIVO
INTERNATIONAL MATHEMATICAL SUMMER CENTER

"Numerical Methods in Fluid Dynamics"

in the subject of the Third 1983 C.I.M.E. Session.

The Session, sponsored by the Consiglio Nazionale delle Ricerche and the Ministero della Pubblica Istruzione, will take place under the scientific direction of Prof. FRANCO BREZZI (Università di Pavia, Italy) at «Villa Olmo», Como, Italy, ***from July 4 to July 12, 1983.***

Courses

a) ***Finite Elements Method for Compressible and Incompressible Fluids.*** (6 lectures in English).
Prof. Roland GLOWINSKI (INRIA, France).

1. Finite Elements Method for the Stokes problem.
2. Nonlinear least-square and applications to fluid flow problems.
3. Alternating directions methods for Navier-Stokes equations.
4. Upwinding methods for transonic flows.

References

1. GIRAULT, RAVIART, Finite Element Approximation of Navier-Stokes equations. Lecture Notes in Math. n. 749 (1979), Springer.
2. GALLAGHER, NORRIE, ODEN, ZIENKIEWICZ (Eds.), Finite Elements in Fluids. Vol. IV. J. Wiley, 1982.
3. GLOWINSKI, Numerical Methods in Nonlinear Variational Problems. Cap. VII, Springer, 1983.

b) ***Spectral methods for partial differentiation equations of fluid dynamics.*** (6 lectures in English).
Prof. David GOTTLIEB (NASA, USA).

Lecture 1 : Presentation of spectral methods - Fourier Chebyshev and others, survey of approximation results.
Lecture 2+3: Stability and convergence of spectral methods for parabolic and hyperbolic P.D.E.'s.
Lecture 4 : Time marching and iterative techniques.
Lecture 5 : Application - incompressible flows.
Lecture 6 : Application - compressible flows.

Literature

1. D. GOTTLIEB & S.A. ORSZAG, Numerical Analysis of Spectral Methods, Theory and Applications C.B.M.S.-S.I.A.M. No. 26, 1977.
2. B. MERCIER, Analyse numérique des Méthodes Spectrales. Note CEA-N-2278 Commissariat à l'Energie Atomique, Centre d'études de Limiel.

c) ***Transonic flow calculations for aircrafts.*** (6 lectures in English).
Prof. Antony JAMESON (Princeton University, USA)

— Review of mathematical models
— Potential flow methods
— Multigrid acceleration
— Solution of the Euler equations in 2 and 3 dimensions.

References

1. K.W. MORTON, R.P. RICHTMYER, Difference methods for initial value problems, New York, 1967.
2. A. BRANDT, Math. Comp. 31 (1977).

3. A. JAMESON, Comm. Pure & Appl. Math., 27 (1974).
4. A. JAMESON, «Steady state solution of Euler equation for Transonic Flow» in «Transonic, shock, and multidimensional flows: Advances in scientific computing», R.E. Meyer ed., Academic Press 1982.

d) ***An Analysis of Particle Methods***. (6 lectures in English).
Prof. P.A. RAVIART (Université P. et M. Curie, Paris).

By particle methods, one usually means numerical methods where some dependent variables of the problem are approximated by a sum of delta functions. This course intends to provide the mathematical basis of these methods which play an increasing role in Fluid Mechanics and in Physics. The following topics will be discussed:

1. Particle approximation of linear hyperbolic equations
2. Numerical approximation of Euler equations in two and three dimensions by vortex and vortex in cell methods.
3. Particle approximation of Vlasov-Poisson equations in plasma physics.

References

1. J.T. BEALE & A. MAJDA, Vortex methods I: Convergence in three dimensions, Math. Comp. 32 (1982), 1-27.
2. J.T. BEALE & A. MAJDA, Vortex methods II: Higher order accuracy in two and three dimensions, Math. Comp. 32 (1982), 29-52.
3. G.H. COTTET, Méthodes particulaires pour l'équation d'Euler dans le plan, Thèse de 3ème cycle, Université Pierre & Marie Curie, Paris 1982.
4. G.H. COTTET & P.A. RAVIART, Particle methods for the one-dimensional Vlasov-Poisson equations, rapport interne 82027, Laboratoire d'Analyse Numérique, Université Pierre & Marie Curie, Paris (to appear in SIAM J. Num. Anal.).
5. R.W. HOCKNEY & J.W. EASTWOOD, Computer simulation using particles, McGraw-Hill, New York, 1981.
6. A. LEONARD, Vortex methods for flow simulation, J. Comp. Physics, 37 (1980), 289-335.

Seminars

A number of seminars and special lectures will be offered during the Session.

Vol. 845: A. Tannenbaum, Invariance and System Theory: Algebraic and Geometric Aspects. X, 161 pages. 1981.

Vol. 846: Ordinary and Partial Differential Equations, Proceedings. Edited by W. N. Everitt and B. D. Sleeman. XIV, 384 pages. 1981.

Vol. 847: U. Koschorke, Vector Fields and Other Vector Bundle Morphisms – A Singularity Approach. IV, 304 pages. 1981.

Vol. 848: Algebra, Carbondale 1980. Proceedings. Ed. by R. K. Amayo. VI, 298 pages. 1981.

Vol. 849: P. Major, Multiple Wiener-Itô Integrals. VII, 127 pages. 1981.

Vol. 850: Séminaire de Probabilités XV. 1979/80. Avec table générale des exposés de 1966/67 à 1978/79. Edited by J. Azéma and M. Yor. IV, 704 pages. 1981.

Vol. 851: Stochastic Integrals. Proceedings, 1980. Edited by D. Williams. IX, 540 pages. 1981.

Vol. 852: L. Schwartz, Geometry and Probability in Banach Spaces. X, 101 pages. 1981.

Vol. 853: N. Boboc, G. Bucur, A. Cornea, Order and Convexity in Potential Theory: H-Cones. IV, 286 pages. 1981.

Vol. 854: Algebraic K-Theory. Evanston 1980. Proceedings. Edited by E. M. Friedlander and M. R. Stein. V, 517 pages. 1981.

Vol. 855: Semigroups. Proceedings 1978. Edited by H. Jürgensen, M. Petrich and H. J. Weinert. V, 221 pages. 1981.

Vol. 856: R. Lascar, Propagation des Singularités des Solutions d'Equations Pseudo-Différentielles à Caractéristiques de Multiplicités Variables. VIII, 237 pages. 1981.

Vol. 857: M. Miyanishi. Non-complete Algebraic Surfaces. XVIII, 244 pages. 1981.

Vol. 858: E. A. Coddington, H. S. V. de Snoo: Regular Boundary Value Problems Associated with Pairs of Ordinary Differential Expressions. V, 225 pages. 1981.

Vol. 859: Logic Year 1979–80. Proceedings. Edited by M. Lerman, J. Schmerl and R. Soare. VIII, 326 pages. 1981.

Vol. 860: Probability in Banach Spaces III. Proceedings, 1980. Edited by A. Beck. VI, 329 pages. 1981.

Vol. 861: Analytical Methods in Probability Theory. Proceedings 1980. Edited by D. Dugué, E. Lukacs, V. K. Rohatgi. X, 183 pages. 1981.

Vol. 862: Algebraic Geometry. Proceedings 1980. Edited by A. Libgober and P. Wagreich. V, 281 pages. 1981.

Vol. 863: Processus Aléatoires à Deux Indices. Proceedings, 1980. Edited by H. Korezlioglu, G. Mazziotto and J. Szpirglas. V, 274 pages. 1981.

Vol. 864: Complex Analysis and Spectral Theory. Proceedings, 1979/80. Edited by V. P. Havin and N. K. Nikol'skii, VI, 480 pages. 1981.

Vol. 865: R. W. Bruggeman, Fourier Coefficients of Automorphic Forms. III, 201 pages. 1981.

Vol. 866: J.-M. Bismut, Mécanique Aléatoire. XVI, 563 pages. 1981.

Vol. 867: Séminaire d'Algèbre Paul Dubreil et Marie-Paule Malliavin. Proceedings, 1980. Edited by M.-P. Malliavin. V, 476 pages. 1981.

Vol. 868: Surfaces Algébriques. Proceedings 1976–78. Edited by J. Giraud, L. Illusie et M. Raynaud. V, 314 pages. 1981.

Vol. 869: A. V. Zelevinsky, Representations of Finite Classical Groups. IV, 184 pages. 1981.

Vol. 870: Shape Theory and Geometric Topology. Proceedings, 1981. Edited by S. Mardešić and J. Segal. V, 265 pages. 1981.

Vol. 871: Continuous Lattices. Proceedings, 1979. Edited by B. Banaschewski and R.-E. Hoffmann. X, 413 pages. 1981.

Vol. 872: Set Theory and Model Theory. Proceedings, 1979. Edited by R. B. Jensen and A. Prestel. V, 174 pages. 1981.

Vol. 873: Constructive Mathematics, Proceedings, 1980. Edited by F. Richman. VII, 347 pages. 1981.

Vol. 874: Abelian Group Theory. Proceedings, 1981. Edited by R. Göbel and E. Walker. XXI, 447 pages. 1981.

Vol. 875: H. Zieschang, Finite Groups of Mapping Classes of Surfaces. VIII, 340 pages. 1981.

Vol. 876: J. P. Bickel, N. El Karoui and M. Yor. Ecole d'Eté de Probabilités de Saint-Flour IX – 1979. Edited by P. L. Hennequin. XI, 280 pages. 1981.

Vol. 877: J. Erven, B.-J. Falkowski, Low Order Cohomology and Applications. VI, 126 pages. 1981.

Vol. 878: Numerical Solution of Nonlinear Equations. Proceedings, 1980. Edited by E. L. Allgower, K. Glashoff, and H.-O. Peitgen. XIV, 440 pages. 1981.

Vol. 879: V. V. Sazonov, Normal Approximation – Some Recent Advances. VII, 105 pages. 1981.

Vol. 880: Non Commutative Harmonic Analysis and Lie Groups. Proceedings, 1980. Edited by J. Carmona and M. Vergne. IV, 553 pages. 1981.

Vol. 881: R. Lutz, M. Goze, Nonstandard Analysis. XIV, 261 pages. 1981.

Vol. 882: Integral Representations and Applications. Proceedings, 1980. Edited by K. Roggenkamp. XII, 479 pages. 1981.

Vol. 883: Cylindric Set Algebras. By L. Henkin, J. D. Monk, A. Tarski, H. Andréka, and I. Németi. VII, 323 pages. 1981.

Vol. 884: Combinatorial Mathematics VIII. Proceedings, 1980. Edited by K. L. McAvaney. XIII, 359 pages. 1981.

Vol. 885: Combinatorics and Graph Theory. Edited by S. B. Rao. Proceedings, 1980. VII, 500 pages. 1981.

Vol. 886: Fixed Point Theory. Proceedings, 1980. Edited by E. Fadell and G. Fournier. XII, 511 pages. 1981.

Vol. 887: F. van Oystaeyen, A. Verschoren, Non-commutative Algebraic Geometry, VI, 404 pages. 1981.

Vol. 888: Padé Approximation and its Applications. Proceedings, 1980. Edited by M. G. de Bruin and H. van Rossum. VI, 383 pages. 1981.

Vol. 889: J. Bourgain, New Classes of $\mathcal{L}^p$-Spaces. V, 143 pages. 1981.

Vol. 890: Model Theory and Arithmetic. Proceedings, 1979/80. Edited by C. Berline, K. McAloon, and J.-P. Ressayre. VI, 306 pages. 1981.

Vol. 891: Logic Symposia, Hakone, 1979, 1980. Proceedings, 1979, 1980. Edited by G. H. Müller, G. Takeuti, and T. Tugué. XI, 394 pages. 1981.

Vol. 892: H. Cajar, Billingsley Dimension in Probability Spaces. III, 106 pages. 1981.

Vol. 893: Geometries and Groups. Proceedings. Edited by M. Aigner and D. Jungnickel. X, 250 pages. 1981.

Vol. 894: Geometry Symposium. Utrecht 1980, Proceedings. Edited by E. Looijenga, D. Siersma, and F. Takens. V, 153 pages. 1981.

Vol. 895: J.A. Hillman, Alexander Ideals of Links. V, 178 pages. 1981.

Vol. 896: B. Angéniol, Familles de Cycles Algébriques – Schéma de Chow. VI, 140 pages. 1981.

Vol. 897: W. Buchholz, S. Feferman, W. Pohlers, W. Sieg, Iterated Inductive Definitions and Subsystems of Analysis: Recent Proof-Theoretical Studies. V, 383 pages. 1981.

Vol. 898: Dynamical Systems and Turbulence, Warwick, 1980. Proceedings. Edited by D. Rand and L.-S. Young. VI, 390 pages. 1981.

Vol. 899: Analytic Number Theory. Proceedings, 1980. Edited by M.I. Knopp. X, 478 pages. 1981.

Vol. 900: P. Deligne, J. S. Milne, A. Ogus, and K.-Y. Shih, Hodge Cycles, Motives, and Shimura Varieties. V, 414 pages. 1982.

Vol. 901: Séminaire Bourbaki vol. 1980/81 Exposés 561–578. III, 299 pages. 1981.

Vol. 902: F. Dumortier, P.R. Rodrigues, and R. Roussarie, Germs of Diffeomorphisms in the Plane. IV, 197 pages. 1981.

Vol. 903: Representations of Algebras. Proceedings, 1980. Edited by M. Auslander and E. Lluis. XV, 371 pages. 1981.

Vol. 904: K. Donner, Extension of Positive Operators and Korovkin Theorems. XII, 182 pages. 1982.

Vol. 905: Differential Geometric Methods in Mathematical Physics. Proceedings, 1980. Edited by H.-D. Doebner, S.J. Andersson, and H.R. Petry. VI, 309 pages. 1982.

Vol. 906: Séminaire de Théorie du Potentiel, Paris, No. 6. Proceedings. Edité par F. Hirsch et G. Mokobodzki. IV, 328 pages. 1982.

Vol. 907: P. Schenzel, Dualisierende Komplexe in der lokalen Algebra und Buchsbaum-Ringe. VII, 161 Seiten. 1982.

Vol. 908: Harmonic Analysis. Proceedings, 1981. Edited by F. Ricci and G. Weiss. V, 325 pages. 1982.

Vol. 909: Numerical Analysis. Proceedings, 1981. Edited by J.P. Hennart. VII, 247 pages. 1982.

Vol. 910: S.S. Abhyankar, Weighted Expansions for Canonical Desingularization. VII, 236 pages. 1982.

Vol. 911: O.G. Jørsboe, L. Mejlbro, The Carleson-Hunt Theorem on Fourier Series. IV, 123 pages. 1982.

Vol. 912: Numerical Analysis. Proceedings, 1981. Edited by G. A. Watson. XIII, 245 pages. 1982.

Vol. 913: O. Tammi, Extremum Problems for Bounded Univalent Functions II. VI, 168 pages. 1982.

Vol. 914: M. L. Warshauer, The Witt Group of Degree k Maps and Asymmetric Inner Product Spaces. IV, 269 pages. 1982.

Vol. 915: Categorical Aspects of Topology and Analysis. Proceedings, 1981. Edited by B. Banaschewski. XI, 385 pages. 1982.

Vol. 916: K.-U. Grusa, Zweidimensionale, interpolierende Lg-Splines und ihre Anwendungen. VIII, 238 Seiten. 1982.

Vol. 917: Brauer Groups in Ring Theory and Algebraic Geometry. Proceedings, 1981. Edited by F. van Oystaeyen and A. Verschoren. VIII, 300 pages. 1982.

Vol. 918: Z. Semadeni, Schauder Bases in Banach Spaces of Continuous Functions. V, 136 pages. 1982.

Vol. 919: Séminaire Pierre Lelong – Henri Skoda (Analyse) Années 1980/81 et Colloque de Wimereux, Mai 1981. Proceedings. Edité par P. Lelong et H. Skoda. VII, 383 pages. 1982.

Vol. 920: Séminaire de Probabilités XVI, 1980/81. Proceedings. Edité par J. Azéma et M. Yor. V, 622 pages. 1982.

Vol. 921: Séminaire de Probabilités XVI, 1980/81. Supplément: Géométrie Différentielle Stochastique. Proceedings. Edité par J. Azéma et M. Yor. III, 285 pages. 1982.

Vol. 922: B. Dacorogna, Weak Continuity and Weak Lower Semicontinuity of Non-Linear Functionals. V, 120 pages. 1982.

Vol. 923: Functional Analysis in Markov Processes. Proceedings, 1981. Edited by M. Fukushima. V, 307 pages. 1982.

Vol. 924: Séminaire d'Algèbre Paul Dubreil et Marie-Paule Malliavin. Proceedings, 1981. Edité par M.-P. Malliavin. V, 461 pages. 1982.

Vol. 925: The Riemann Problem, Complete Integrability and Arithmetic Applications. Proceedings, 1979-1980. Edited by D. Chudnovsky and G. Chudnovsky. VI, 373 pages. 1982.

Vol. 926: Geometric Techniques in Gauge Theories. Proceedings, 1981. Edited by R. Martini and E.M.de Jager. IX, 219 pages. 1982.

Vol. 927: Y. Z. Flicker, The Trace Formula and Base Change for GL (3). XII, 204 pages. 1982.

Vol. 928: Probability Measures on Groups. Proceedings 1981. Edited by H. Heyer. X, 477 pages. 1982.

Vol. 929: Ecole d'Eté de Probabilités de Saint-Flour X – 1980. Proceedings, 1980. Edited by P.L. Hennequin. X, 313 pages. 1982.

Vol. 930: P. Berthelot, L. Breen, et W. Messing, Théorie de Dieudonné Cristalline II. XI, 261 pages. 1982.

Vol. 931: D.M. Arnold, Finite Rank Torsion Free Abelian Groups and Rings. VII, 191 pages. 1982.

Vol. 932: Analytic Theory of Continued Fractions. Proceedings, 1981. Edited by W.B. Jones, W.J. Thron, and H. Waadeland. VI, 240 pages. 1982.

Vol. 933: Lie Algebras and Related Topics. Proceedings, 1981. Edited by D. Winter. VI, 236 pages. 1982.

Vol. 934: M. Sakai, Quadrature Domains. IV, 133 pages. 1982.

Vol. 935: R. Sot, Simple Morphisms in Algebraic Geometry. IV, 146 pages. 1982.

Vol. 936: S.M. Khaleelulla, Counterexamples in Topological Vector Spaces. XXI, 179 pages. 1982.

Vol. 937: E. Combet, Intégrales Exponentielles. VIII, 114 pages. 1982.

Vol. 938: Number Theory. Proceedings, 1981. Edited by K. Alladi. IX, 177 pages. 1982.

Vol. 939: Martingale Theory in Harmonic Analysis and Banach Spaces. Proceedings, 1981. Edited by J.-A. Chao and W.A. Woyczyński. VIII, 225 pages. 1982.

Vol. 940: S. Shelah, Proper Forcing. XXIX, 496 pages. 1982.

Vol. 941: A. Legrand, Homotopie des Espaces de Sections. VII, 132 pages. 1982.

Vol. 942: Theory and Applications of Singular Perturbations. Proceedings, 1981. Edited by W. Eckhaus and E.M. de Jager. V, 363 pages. 1982.

Vol. 943: V. Ancona, G. Tomassini, Modifications Analytiques. IV, 120 pages. 1982.

Vol. 944: Representations of Algebras. Workshop Proceedings, 1980. Edited by M. Auslander and E. Lluis. V, 258 pages. 1982.

Vol. 945: Measure Theory. Oberwolfach 1981, Proceedings. Edited by D. Kölzow and D. Maharam-Stone. XV, 431 pages. 1982.

Vol. 946: N. Spaltenstein, Classes Unipotentes et Sous-groupes de Borel. IX, 259 pages. 1982.

Vol. 947: Algebraic Threefolds. Proceedings, 1981. Edited by A. Conte. VII, 315 pages. 1982.

Vol. 948: Functional Analysis. Proceedings, 1981. Edited by D. Butković, H. Kraljević, and S. Kurepa. X, 239 pages. 1982.

Vol. 949: Harmonic Maps. Proceedings, 1980. Edited by R.J. Knill, M. Kalka and H.C.J. Sealey. V, 158 pages. 1982.

Vol. 950: Complex Analysis. Proceedings, 1980. Edited by J. Eells. IV, 428 pages. 1982.

Vol. 951: Advances in Non-Commutative Ring Theory. Proceedings, 1981. Edited by P.J. Fleury. V, 142 pages. 1982.

Vol. 952: Combinatorial Mathematics IX. Proceedings, 1981. Edited by E. Billington, S. Oates-Williams, and A.P. Street. XI, 443 pages. 1982.

Vol. 953: Iterative Solution of Nonlinear Systems of Equations. Proceedings, 1982. Edited by R. Ansorge, Th. Meis, and W. Törnig. VII, 202 pages. 1982.

"Fabulous! This beautifully written book brings a little-known part of American history to life with characters so real they leap off the pages into readers' hearts and linger there long after the last page is turned. *Along a Storied Trail* is a story to savor, to ponder, and to read again and again."

Amanda Cabot, bestselling author of *Dreams Rekindled*

"Ann Gabhart has woven a tale of a feisty, endearing, and thoroughly memorable character, Tansy Calhoun, as she settles into her route as a WPA packhorse librarian in Depression-era Appalachia. Adventure, a romantic triangle (or two!), plus an unexpected natural disaster roil up, spilling into a dramatic, heart-pounding conclusion. *Along a Storied Trail* might be Gabhart's best book yet."

Suzanne Woods Fisher, author of *The Moonlight School*

"From the very first sentence of *Along a Storied Trail,* Ann H. Gabhart has hand-delivered a tale that will make readers feel right at home. With a voice that is every bit as distinct and special as Gabhart herself, the reader feels as though she's riding along with her new friend Tansy, seeing the beauty of Appalachia even amidst the hard times of the Great Depression, through loss, and adversity. This is a story of resilience that is not only representative of the 1930s, but a story of resilience that we so deeply need in our times."

Susie Finkbeiner, author of *The Nature of Small Birds* and *Stories That Bind Us*

along a storied trail

Books by Ann H. Gabhart

Along a Storied Trail
An Appalachian Summer
River to Redemption
These Healing Hills
Words Spoken True
The Outsider
The Believer
The Seeker
The Blessed
The Gifted
Christmas at Harmony Hill
The Innocent
The Refuge

Heart of Hollyhill

Scent of Lilacs
Orchard of Hope
Summer of Joy

Rosey Corner

Angel Sister
Small Town Girl
Love Comes Home

Hidden Springs Mystery as A. H. Gabhart

Murder at the Courthouse
Murder Comes by Mail
Murder Is No Accident

along a storied trail

ANN H. GABHART

Revell
a division of Baker Publishing Group
Grand Rapids, Michigan

Published by Revell
a division of Baker Publishing Group
PO Box 6287, Grand Rapids, MI 49516-6287
www.revellbooks.com

Printed in the United States of America

Library of Congress Cataloging-in-Publication Data
Names: Gabhart, Ann H., 1947– author.
Title: Along a storied trail / Ann H. Gabhart.
Description: Grand Rapids, Michigan : Revell, a division of Baker Publishing Group, [2021]
Identifiers: LCCN 2020047617 | ISBN 9780800737214 (paperback) | ISBN 9780800739997 (casebound) | ISBN 9781493430390 (ebook)
Subjects: GSAFD: Love stories.
Classification: LCC PS3607.A23 A79 2021 | DDC 813/.6—dc23
LC record available at https://lccn.loc.gov/2020047617

Scripture quotations, whether quoted or paraphrased, are from the King James Version of the Bible.

This book is a work of fiction. Where real people, events, establishments, organizations, or locales appear, they are used fictitiously. All other elements of the novel are drawn from the author's imagination.

Published in association with Books & Such Literary Management, www.booksandsuch.com.

21 22 23 24 25 26 27 7 6 5 4 3 2 1

In memory of my grandparents,

Rose and Herbert Hawkins,

who loved books,

and my aunt

Lorin Bond Houchin,

who kept a list of every book she read.

They passed down a love of reading and books to me.

one

EVERYBODY THOUGHT Tansy Calhoun was heartbroken after Jeremy Simpson threw her over for Jolene Hoskins. Or that she should be. She had to admit her pride was bruised, but the whole thing was simply a shin bump in life. In fact, after pondering it some, she decided she'd gotten the best of things.

Of course, that had been some while back. Three years, when she counted it up. Jolene had a baby now, with another on the way. That could have been Tansy if she'd gone down the path folks thought she was ready to take. Married at seventeen. A mother at eighteen. Worn out by thirty, with more children than she had chairs around her kitchen table. A poor man's riches, some said. That was about the only riches a man was apt to see here in the mountains in 1937.

Family did matter. Tansy wasn't against marrying and having her fair share of children, but she was glad enough to put it off a while. Longer than most around here thought sensible. Marrying young and for forever was the way of life in these Eastern Kentucky hills. A person should marry with the intention of staying married forever, but that could still be a long time even if you waited a while for forever to start.

Her own mother had been eighteen when Tansy was born and

she wasn't the firstborn. Her sister Hilda was nigh on two when Tansy came along. Hilda married young too, but that wasn't because she was following in Ma's footsteps. She had in mind to escape mountain life by marrying the schoolteacher and happily going off with him to live up in Ohio somewhere. Hilda sent a letter now and again and sometimes a book for Tansy, but no word of any babies on the way. Ma worried some about that.

She worried about Tansy even more. Ma was glad enough to still have her home to help with things. She needed the help. Giving birth to Livvy, the least one who was only four, had stolen her health. Did something to her hip so that moving was ever painful. She said Livvy might need to be her last and sounded sad about that, although Tansy thought five a fine number. Well, six. She couldn't forget her little brother Robbie, who got the fever and died when he was seven. He did count. Dear little Robbie. The sweetest of the bunch. Way sweeter than Tansy. But Livvy was turned sweet too.

Took that after their ma. Not their pa. Might be a blessing Pa took off on them last September. Him being gone would be a sure way to save her mother from the ordeal of carrying another baby.

Looking for work, Pa told Ma. No work was to be had here in the mountains. Coal mines had mostly shut down, with nobody having money to buy anything since what they called Black Tuesday happened off in New York City. Tansy still hadn't quite gotten her mind around how rich people losing their money through something they called a ticker tape had leaked down to close mines and bring such hard times to Robins Ridge. Didn't people still need coal to keep the fires in their grates burning? At least those who didn't have trees to cut for firewood.

But Pa never liked the mines anyway. Said working down under the ground stole his breath. Sometimes at night he did seem to do more coughing than breathing.

Ma hadn't wanted him to go. Tansy heard them talking the night before he left. Her bed in the loft was right over theirs, and she often covered her ears with her pillow to keep from hearing more than she should. But this time she'd done the opposite and leaned off the side of the bed to catch every word when she heard the pleading in her mother's voice.

"Joshua, looks to me it would be best you stayin' here. Little Josh can help you do the farming. We can get by."

Pa's voice rumbled in return. "Unless another dry spell comes along. Corn don't grow without rain, Eugenia. You saw that this year, what with the beans drying up in your sass patch and some of the corn ears no more than nubbins."

"We had enough sass to get by." Ma didn't like admitting her vegetable garden didn't supply food enough for their table. "I made pickles. We got some of the early planting for shucky beans. They're hung up all over the attic."

Her father didn't say anything for so long that Tansy about decided he'd let Ma have the last word and gone to sleep. But then he said, "It ain't your fault, Eugenia. Ain't mine either. The Lord just didn't send us no rain."

"The Lord supplies our needs."

Tansy heard the absolute certainty in Ma's voice. She refused to hear complaints against the Lord. Even when Robbie died. She had sat by Robbie's bed night and day, praying fervently for him to get well. But once the boy's breaths stopped, she folded his little hands together, kissed his forehead, and accepted it as the Lord's will. She didn't war against that, like Tansy wanted to. Or like Pa did. He'd gone off and not come back until after the neighbors dug a grave and helped lay Robbie to rest. Ma said that was the only way Pa could take losing a son.

"At times better than other times." Pa sounded worn out. "Lately he must not be paying much mind to what folks is needin' down here."

That went too far for Ma. "I'm thinking you best offer up

a prayer for the Lord to open your eyes to the blessings he sends down to us."

"Maybe so." Pa's voice gentled. "Be that as it may, I see how you never put much on your plate at suppertime. If I head out to find work somewheres, that'll be one less mouth at the table. Tansy and Josh are old enough to take care of things around here."

"Tansy might find a feller and get married."

Tansy couldn't decide if Ma sounded worried or hopeful about that. Maybe resigned to their fates, whatever they were.

Pa made a sound of disgust. "She had a good enough feller and let him get away. That's been over two years back and I haven't seen no boys making paths to our door. The girl turned twenty in July. She ain't liable to find a suitor less'n a widower comes along. Maybe not then if she don't get her head out of them books. Hilda hasn't done us no favors sending storybooks that's got Tansy thinking above herself."

Thinking above herself. Those words had made her want to get up and climb down the steps to tell her father a few things. Like she had a right to think. To read. She didn't slack off helping Ma, but in stolen moments, books took her beyond the mountains. Let her fly like an eagle to take in the view of other places and ways.

That didn't mean she didn't want to roost right where she was. She loved the mountains. Time and again she did let a little daydream tickle her mind. That maybe a prince on a white stallion might ride into her life the way it happened sometimes in fairy tales. Not really a prince, but a man with aplomb. She'd discovered that word a while back and studied out what it might mean. A man who was handsome. Self-confident. Capable. Like the men in the stories she read.

She had lain stiff in her bed while the words she wanted to throw at her father boiled in her head. She never got the chance to say any of them. The next morning he was gone.

Books had always been a bone of contention between them. Once he'd even jerked a book from her to sling into the fire. She had a scar on her hand from yanking the book out of the flames. The page edges were charred, but the words were saved. After that, Tansy hid out to read. In the hayloft. In a tree during the summer. In the loft, crouched beside her bed near the half window as soon as light dawned.

Her pa might change his mind about books if he could see her now. Her love of reading had opened up a way for her to be a packhorse librarian, since it turned out she wasn't alone in thinking books mattered. The government did too and had established a work program to get books to people in these Eastern Kentucky mountains where roads a truck could travel were few and far between. Women like Tansy on mules and horses took books to the people.

Sometimes Tansy thought every prayer she'd ever thought to pray and some she hadn't thought up had been answered the day she got one of the book routes here in Owsley County. Then on top of the pure pleasure of working with books, to get paid for it. Let people feel sorry for her because they thought she was on an old maid's path. She wasn't ready for a rocking chair on a relative's front porch yet. She had time to find a man to love.

But now, she had a horse, access to more books than she had time to read, and money to keep food on her mother's table. She'd felt practically rich the first time she collected her monthly wages. Twenty-eight dollars.

Not that she didn't earn every penny. Her route was over some rough ground, up and down steep hills, through creeks and woods. And she couldn't let the weather stop her. Rain, sleet, or snow, she had promised Madeline Weston to faithfully ride her routes.

Mrs. Weston, the head librarian, had found a room in the back of an old store building in Booneville for their central location. She talked some local men into building shelves along the walls and then sent out a plea for book donations. Tansy had carried the few she had down to the library even before she was hired. Without books, a library was nothing more than a quiet room. Others had brought in books and magazines too, and the headquarters of the women's work programs in London, Kentucky, sent them a few boxes of books. Little by little, the place was beginning to look like a real library.

"We have to make this work," Mrs. Weston told Tansy when she traced out her route. Three other women had routes in other parts of the county. "The people who live in your area will depend on you making your rounds on a regular schedule. Winter coming on doesn't make things easy, but when was anything ever easy for us up here in the hills?"

November had gifted them with some fine fall weather. Riding her route had been a pleasure. Then December was nothing to complain about. But January came in cold and mean.

Rain had poured down all night, and now ice pellets stung Tansy's face as she rode away from the barn to start her route. She was relieved to see snowflakes mixing with the rain, even if that did mean the temperature had dropped. Snow made for better footing for her horse than ice.

Easy or not, Tansy was glad to head down the hill with her saddlebags loaded with books on this first Monday in the new year. She reined in her horse at the edge of Mad Dog Creek. Well named, she thought every time she crossed it. The creek could be a sweet, flowing stream, or it could turn evil when rains brought the tides. Flatlanders thought it odd to call floods *tides*, but that might be because they'd never seen tides of creek water sweeping down the mountains after a downpour.

Tansy stroked her horse's neck. She leased Shadrach, a Morgan gelding, from Preacher Rowlett, who lived at the bottom of Robins Ridge. Their mule was so old he'd give out if she tried to ride him on the miles of her rounds each week. Preacher Rowlett promised Shadrach would be sure-footed on the rough trails, and a person could depend on what Preacher said. Not that he was really a preacher. At least not one with a church pulpit and all, but he did know his Bible and could be counted on to say words at a funeral whenever no bona fide preachers were around.

He was right about Shadrach. The white stockings on his front legs were the only thing fancy about the horse, but Tansy didn't need fancy. She needed willing. Shadrach was that. If she pointed him up the side of a hill, he did his best to climb it, although sometimes Tansy got off and walked to make it easier for him.

"What do you think, boy?" she asked now as the water churned past them. She weighed her options. Heading down the creek for what might be a better crossing place would put her off schedule. Crossing here would be cold and wet.

When she flicked the reins, she wasn't sure if Shadrach was game to give the creek a try or simply resigned. He shook his head slightly and stepped into the water. Tansy gasped as the water sloshed over her boot tops and soaked the wool pants she'd borrowed from her brother. She twisted around to grab her saddlebags of books and lift them above the water.

She let the reins hang loose to give Shadrach his head. The gelding would get through the creek. All she had to do was not fall off. And keep those books dry. Books were hard to come by, and she wasn't about to ruin her allotment.

Her feet were soaked by the time they climbed out of the creek, but the books were dry. After she settled the saddlebags back down on Shadrach's flanks, she stopped the horse to dump water out of her boots. Then nothing for it but to stick her wet feet back in those boots and head up the hill. Her pant legs stiffened and froze in the frigid wind.

She could warm up at her next stop, Jenny Sue Barton's

house. Poor woman had lost her husband a few months back when a tree fell on him. A dark cloud had settled down over her house after that, what with her son, Reuben Jr., falling sick with an ailment nobody had exactly figured out. They'd fetched in a doctor from the next county who didn't have any more answers than the granny healer, Geraldine Abrams, who'd concocted a tonic for the boy. The doctor said to feed him eggs to build up his strength and time would heal him. A boy of six should shake off whatever was ailing him.

Reuben Jr.'s grandmother claimed that was the first town doctor with sense. While he hadn't come right out and said it, he'd tiptoed all around how losing his pa had bruised the boy's heart. Until that healed some, they simply needed to be patient and do what they could to give him something to think about besides missing his pa.

Tansy was glad to help with that. She always stayed long enough at their house to read some from *The Life and Adventures of Robinson Crusoe.* A while reading aloud to bring cheer to the Bartons' worried cabin seemed little enough to do, and as a book woman, she was supposed to find ways to awaken people to the joy of books.

Today, she could dry out her boots while she read more of Crusoe's adventures and taught young Reuben some new words. Learning to read did seem to lift some of the sorrow off the boy. *Snow* might be a good word for him to learn this day.

two

THE WEATHER WAS MEAN. Worse than mean. Wicked. Rain, sleet, snow. All at once. Caleb Barton shivered. He tried to fasten his mind on how hot he'd been last summer down in Tennessee working with the Civilian Conservation Corps to make trails up into the Smoky Mountains. He'd been glad enough to do the work and thankful he'd been sent down south to the mountains instead of stuck in a city hauling rocks for bridge abutments, digging ditches, or whatever. He was happiest with his feet on a mountain.

After being in the Smokies, the Kentucky mountains here in Owsley County didn't seem near as big as they had when he was a boy. The Appalachian Mountains were old, worn down from centuries of standing in this place. He straightened up from the tree he was working up for firewood and peered across the hill. Couldn't see much with the snow clouds hanging low. But he didn't need to see the hills to know what they looked like. Worn down or not, they were home, and he was glad to be standing here on one of them even with the cold seeping through him.

He looked back at the tree he'd felled. He needed to trim off the branches and then saw a piece of the trunk into a size his mule could pull back to the house. Ebenezer, hunched up over to the side, looked longingly in the direction of his barn. The animal was more than ready to be out of the weather, but a work mule knew to wait.

"All right, Ebenezer. I'll quit woolgathering and get this done." Caleb spoke aloud as though the mule would know what he was saying. Ebenezer. He couldn't imagine what had possessed his brother to name his mule Ebenezer. A mule needed a name you could yell. Like Joe or Bud.

But Reuben always did have a fanciful streak. Caleb smiled, thinking about the brother only a couple of years older than him. At the same time, his heart clenched with grief. He did miss Reuben. It didn't seem right him being taken while Caleb was away working with the Corps. Made it not seem real, like Reuben might just come out of the trees and ask Caleb why he didn't bring the two-man saw.

But in November, Reuben had cut his last tree. Nobody for sure knew what happened. Whether Reuben didn't notch the tree right or maybe the tree had a weak side he didn't notice, but something had gone wrong. Tragically wrong.

Caleb's ma said Jenny Sue didn't get worried at first when Reuben didn't come in for supper. But then Ebenezer had come back to the barn. By himself. With night falling. Ma said Jenny Sue was beside herself when she came across the field to Ma's house.

Ma shook her head as she told Caleb about it when he came home after his enlistment time with the Civilian Conservation Corps was up in December.

"That Jenny Sue is a sweet child and pretty as a spring flower," she said, "but fortitude ain't never been her strong suit. Jest as well I was the one to go huntin' for Reuben 'stead of her." Ma swallowed hard but didn't let any tears fall. "'Twas a wonder I come up on him at all in the dark. Figured the Lord must have led me to him. Poor boy. That tree had him pinned tight to the ground. The day had turned cold the way it can once we edge toward winter. Gone to raining some, the kind of rain that ain't exactly sleet but don't feel far from it. So he was chilled to the bone. Didn't hardly know where he was at when I first knelt down beside him. He shook that off and come back to his senses when I put my hand on his cheek. It was cold as death."

Caleb had reached over and taken her hand that felt cold to his touch the way Reuben's cheek must have felt to her that night. "You don't have to tell me everything about it if it's too hard."

"No." She shook her head. "A man should know about his brother dying. The two of you was always nigh on close as twins."

She was right. In fact, he'd felt a strange uneasiness about home even before the letter reached him down there in the Smokies.

"All right then," he said. She needed to say the words. To put him down there with her on the ground by his brother.

They'd been in front of the fireplace, her in the grapevine rocker Reuben made her some years before. He'd always had a way about him, ready to try anything. The rockers had been the hard part and he'd finally given over that job to Caleb. He knew Caleb would figure it out and make it work. Caleb had watched his mother rocking back and forth and felt a flare of anger at Reuben for whatever he'd done to get himself killed.

Then sorrow drowned that out as he waited for her to say more. She stared at the flames leaping up around the wood he'd just laid on the fire. The cabin he and Reuben built for her was small. Just two rooms, but the fireplace was big. Spread halfway across the back wall. It had taken them days to gather enough rocks for the chimney. When he made it home last week, Caleb had been happy to see smoke drifting up out of their chimney, still standing strong on the side of the cabin.

For a bit, Ma had rocked back and forth as if letting the words build in her. "He'd been trapped there for a good spell, but he could still talk a little. Weren't easy for him and I had to lean down close to hear." She paused again. "I can still feel his breath on my ear."

Caleb didn't say anything, but he wanted to be anywhere but sitting in that straight-back chair with the fire heating up his face while he heard how his brother died. Not that shutting out the words would make them any less true. Reuben would still be gone.

Her voice was soft when she started talking again. "I didn't tell Jenny Sue ever'thing he said. Just about how much he cared for her and the children. Having to leave his young'uns was a pain worse than any he was feeling from the tree on him. I hurt for him and tried to get him to hush. Told him we'd pray and the good Lord might lift the tree and he'd find out he weren't hurt so bad after all. I was already prayin' underneath my words, but there is times prayers don't move mountains. Or trees."

"Did he say what happened?"

"I never asked him that. Didn't seem to matter none then with whatever it was havin' done happened."

"I guess not." She was right. Knowing what happened wouldn't bring his brother back, but for some reason it seemed something he needed to know.

"Don't let it trouble you, son. I know how you like to put reasons behind things. That's where you and Reuben were always different. You reasoned it out and he . . ." She paused as a little smile slipped across her face. "And he, he just ran after whatever he saw out ahead."

Caleb didn't say anything to that. He knew Reuben was their mother's favorite. That never bothered Caleb all that much. Reuben was Caleb's favorite too. Everybody's favorite. He said he hadn't ever seen the need to make an enemy. And when he had now and again got on somebody's wrong side through some unintended, often careless cause, he let Caleb take care of settling things down.

Caleb had thought it would be good for him to be gone for a while. That would make Reuben step up and figure ways to take care of his own messes, but Caleb hadn't thought he'd die from one of those times. Caleb was glad when his mother started talking again so he could push away the feeling that if he'd been there, he could have kept his brother from being taken down by a tree. Caleb liked trees. Felt a kinship with the woods on their place, and now one of them had stolen his brother from him.

"Like I said, I didn't tell Jenny Sue Reuben's every last word. At the time I was hopin' with ever'thing in me that there'd be plenty more words for him to share with her. But I reckon you need to hear them."

"Not if they were for Jenny Sue."

"I didn't say they were for her. They was about her. And you." His mother looked straight at Caleb then. "I ain't sure you'll be wantin' to hear them, but I can't not pass them along."

Caleb wanted to shut his ears, but that wouldn't stop his mother's words.

"He said he always thought you were some sweet on Jenny Sue and that he'd take it as a favor if you'd see to her and his young'uns. That maybe you could take his place. Be a daddy to them." She rocked back and forth twice before she went on. "A husband to her."

Caleb held his breath a few seconds and then blew it out slowly. "I'll do what I can for his family, Ma. You know I will. See they have what they need." He looked straight at his mother. "But I'm not marrying my brother's widow. Reuben was deluded thinking I was ever sweet on Jenny Sue. Or that she'd ever be sweet on me."

"That's how Bible people used to do things," his mother

said. "When a brother died, another brother took the man's widow as wife."

"But they weren't worried about how many wives they might have and it was just a way of giving them a roof. Besides, this isn't Bible times. It's the 1930s. Things aren't the same as when Jesus walked on the earth."

"No, I reckon not." His ma looked over at Caleb. "I ain't saying it's something you oughta do, but that don't keep you from needin' to know your brother's last words to you."

"Reuben didn't always think before he spoke."

"True enough, but he had time to think things out that day." She stared down at her lap and folded her apron in pleats. "I told him not to talk nonsense. That he'd be all right as soon as I got some men to carry him back to the house. That me and Jenny Sue would nurse him back to health. Then I had to leave him there in the dark. By hisself. I did leave the lantern. 'Twas the least I could do, but by the time I got back with Frank Sims and his boy from down the way, my Reuben was past nursing. Preacher Rowlett said he'd been healed. Just not the healing we was wanting."

Caleb shook his head now to get the memory of his mother's words out of his thoughts. He needed to get finished with his own tree cutting and not think about his brother pinned to the ground, offering his wife to him. That wasn't something he wanted to even consider.

Not that he had any other woman in mind to marry. He was in love once. If he was honest, he had to admit he still carried feelings for her, but the girl he fell for had fallen for someone else. That's why he'd gone off the mountain to find work. Best thing for all of them. He didn't have to see the girl he loved married to another man, and once he went to work for the Corps, he had money to send home.

He bent back to the task at hand to get firewood for his mother and his brother's wife and children. He hadn't picked one of the big trees to cut. Those needed two men working a crosscut saw. He'd chosen a smaller one. Second growth. Most of the trees were second growth now. Settlers early on had cut the trees for cabins and to clear ground for corn, even if the hills were more suited for woods than crops. Then loggers had come in to cut the big trees and float them down the river to a sawmill. They left the hillsides scarred, but they brought jobs to the area. Back then before the market collapse, America had been building everywhere. Lumber was at a premium. Especially the American chestnuts growing across these hills and all through the eastern states.

Then a fungus rode in on the wind to take down the mighty trees. He'd heard it started in New York. That seemed too far away to be a worry, but the fungus had blown south to find more trees to kill.

Caleb wanted to straighten up and look around again to let his eyes rest on one of the rare chestnuts still standing. You couldn't tell the trees were dying in the winter. But they were. Some dead ones were gray trunks that served as signposts these days. But most everywhere, folks had come into the woods and taken down the chestnut giants. In the Smokies too. No reason to leave them standing to fall down. Might as well get the good out of the trees. But that left a mighty hole in the forests. There. And here.

Caleb sighed and kept working his axe. He might be foolish to mourn a tree. But it wasn't just one of them. It was thousands. He prayed the Lord would leave a few standing. That man would too. The mountains needed the chestnuts, but needs weren't always answered. Prayers either.

Or Reuben would be the one out here cutting wood for his family.

three

TANSY WAS GLAD TO SEE the Barton cabin through the trees as Shadrach plodded across the ridge. She should have gotten off and walked alongside the horse instead of sitting in the saddle while the icy cold crept up her legs. Not a good day to get water down her boots.

Jenny Sue would have a good fire going to keep Reuben Jr. and her little girl, Cindy Sue, warm. The neighbors must be keeping her in firewood. Or Tansy wouldn't doubt that Vesta Barton, Jenny Sue's mother-in-law, had been wielding an axe to fill the woodbox. Ma Vesta, as everybody called her, might be on up in years, but she refused to give in to old age. Folks around were half surprised she hadn't found a way to keep her son, Reuben, alive after that tree fell on him. But some things couldn't be changed just for the wanting of it no matter how strong-willed a person might be. Or how many prayers were sent heavenward.

Tansy glanced up at the sky with an apology for that thought. Preacher Rowlett would tell her all prayers were answered, but some not the way a person hoped. For sure, Ma Vesta's prayers that day weren't answered how she wanted. A person shouldn't demand anything from God, but they could ask. So Tansy sent up a hopeful prayer that her toes wouldn't freeze off. A person needed her toes. All of them.

"Tansy Calhoun, the cold must be freezing your brain," she muttered to herself.

Here she was praying for her toes while thinking about poor Ma Vesta's prayers not saving her son. Time to beg forgiveness. Then again, her mother had taught her the Lord was ready to walk alongside a person through thick and thin. Tansy was more than ready to invite him along with her on her book route.

Her book route. Freezing or not, she still felt blessed to be riding Shadrach with a load of books to take out to the people she'd see today. She thought about the books and magazines in her saddlebags to see if she had one Jenny Sue Barton might like enough to lift her a bit away from her sorrow. The poor woman seemed to be floating in a sea of grief with nothing to catch hold of to steady herself since Reuben got killed.

Jenny Sue and Tansy's sister, Hilda, had been about the same age. Hilda used to complain she couldn't even get Jenny Sue to swing on a grapevine or strip down to her skimpies to go swimming. Hilda was always ready to try anything. Tansy supposed she was the same, or else she wouldn't be riding a horse loaded down with books through a snowstorm.

Still, everybody thought Jenny Sue won the prize when Reuben Barton started courting her. Everybody liked Reuben. He was always smiling, with never a bad word to say about anybody. Hilda might have stayed in the mountains if Reuben had looked at her instead of Jenny Sue. She'd probably deny that now, but Tansy remembered how Hilda had acted when Reuben came around.

Ma told Hilda she ought to warm up to Reuben's younger brother, Caleb, but Hilda laughed at that and paid little attention to Caleb, who seemed to find reason to come by their house even more often than Reuben.

He wasn't like Reuben. Not as ready to smile. Appeared to always be thinking something out. Too serious minded, according to Hilda. She would shake her head at how he even worried about trees when anybody could see they had an overabundance of them here in the mountains.

But Tansy had liked it when he talked about trees. They worried together about the chestnuts dying. She hadn't wanted to believe it when folks in the know about such things claimed they might all die. The American chestnuts were as much part of the mountains as the rocks and cliffs, but they did keep dying. Every time she passed one of the trees with the leaves turning brown, she grieved a little and kept hoping the people were wrong.

Caleb had been off in Tennessee planting trees in those big Smoky Mountains when Reuben died. He hadn't made it home for the funeral. Probably didn't get the news in time and maybe couldn't just up and leave his job with the Civilian Conservation Corps, though some thought he should have.

In these times, a job wasn't something easy to throw over, and it could be he was every bit as excited about planting trees as she was about taking books around to readers. It wasn't like he could bring Reuben back. Ma Vesta said he'd come home when he was needed.

Now Ma Vesta stepped out on the porch when Tansy rode up to Jenny Sue's house. Tansy wasn't surprised to see her there, even if her own cabin was down the hill a ways. She'd aim to be a help to Jenny Sue and the children.

"Put your horse in the barn whilst you're here, Tansy," she said. "Poor critter has icicles hanging off his tail. He can take a bit of rest while you warm up inside. The children are pining to hear more of that story you've been reading to them. Might cheer up Jenny Sue too. She can't shake off the black wearies today."

"Thank you, Ma Vesta. Shadrach will be happy to have some

drying-out time. We got more than a little wet down yonder at the creek." Tansy tried to pull her feet free to dismount, but her boots were frozen to the stirrups.

"Appears you're froze up." Ma Vesta didn't hide her smile. "Want me to fetch the broom to give your foot a whack?"

"Don't go to no bother. I can bang it on the post here." Tansy moved Shadrach closer to the edge of the porch and knocked the stirrup against the wooden post. Once that boot broke free, she flipped her leg over to kick her other foot loose.

"I've heard tell of folks froze in the saddle, but I was thinkin' they meant 'cause they was feared of a snake or a bear." Ma Vesta shook her head.

Tansy had to laugh as she slid off Shadrach. "I'll have a story to tell."

"That you will. Take your horse on out to the barn and I'll warm up some ginger tea. Might keep you from catching your death of pneumonia."

When Tansy got back to the cabin, she was glad to find a place by the fire to pull off her boots and socks and try to rub some warmth back into her toes.

Cindy Sue came running to see what book Tansy had brought her.

"Think you'd like a book with pictures of bunny rabbits?"

The little girl's face lit up. A pretty child, she took after her mother with her blonde hair and light blue eyes.

Cindy Sue clutched the book tight to her chest and waited for what Tansy had for her brother.

"You and Reuben Jr. can look at the bunny pictures together. But I have one for him too." Tansy glanced over at the little boy in the bed across the room. "It's got pictures of some of the words you've been learning, Reuben."

Tansy had made the book the last time she was at the library room in Booneville. She'd felt the Lord was blessing both her and Reuben Jr. each time she came across a picture that matched a word he was learning.

Reuben Jr. didn't have a ready smile. So the sweet smile that slipped across his face when Cindy Sue ran across the room to hand him the book was more than enough reward for the time Tansy spent pasting pictures on heavy paper and then lacing the pages together with string.

Maybe she could make another book with different words tomorrow when she was at the library. A parcel might have come in from the folks in London, Kentucky, who rounded up books and magazines for the mountain libraries. Some of those old magazines were falling apart and couldn't be circulated among readers, but Tansy and the other women found ways to use the pictures and stories to make extra books for their readers.

"Junie, say thankee," Ma Vesta told him.

Reuben Jr. obediently thanked Tansy before he began flipping through the book, whispering the words for the pictures.

His grandmother smiled at him. "He and Cindy Sue about wore out the last books you left for them, but I've got them here for you to take on to the next house." She pointed to two books on the table.

The Barton cabin was bigger than most cabins Tansy visited. Bigger than her own cabin. Ma Vesta had given over the house that had three rooms and a loft to Reuben when he married, with the expectation he and Jenny Sue would need plenty of room for the children they'd have. And now Reuben would never have more than two, although Jenny Sue would no doubt marry again once her grieving time was through.

"I thought I heard you, Tansy," Jenny Sue said as she came in from the bedroom. She was a pretty woman, even with the sorrow pulling down her face. She had a quilt draped around her shoulders, and her blonde hair gave evidence of time spent abed.

"How are you doing, Jenny Sue?" Tansy said.

"No way to be doing good right now." She sat down in the rocker by the fire and pulled the quilt closer. "Cold as a wedge in that back room." She glanced down at Tansy's bare feet. "Ain't you got no socks?"

"They got dunked in the creek. Ma Vesta said I could dry them out by the fire while I was reading to Cindy Sue and Reuben."

"We're calling him Junie now. Seems better than having Reuben's name bouncing around the room." A tear slid out the corner of Jenny Sue's eye. She didn't bother wiping it away, as if it was a common thing. "You like that better, don't you, Junie?"

Jenny Sue glanced over at the boy, who looked up from his book to nod solemnly.

She looked back at Tansy. "It ain't like we're robbing him of his pa's name. He'll ever keep that."

"Junie is a fine name for a junior," Tansy said.

"It was Ma Vesta's idea. Most everything is." Jenny Sue pulled the quilt closer around her shoulders and stared at the fire.

Tansy wasn't sure if Jenny Sue meant the words to sound complaining, but Ma Vesta must have heard them that way. Her shoulders stiffened as she added wood to the fire, but she didn't say anything.

Rumor had it she had never been overly happy with Reuben's pick, but Tansy doubted anybody had ever heard Ma Vesta speak that. Sometimes folks decided things for a person when they didn't have the least bit of evidence to prove it so. Just like Tansy being heartbroken over Jeremy marrying Jolene. Never was any truth to that rumor along the mountain gossip grapevine.

Tansy broke the uncomfortable silence in the cabin by pulling the book about Robinson Crusoe out of her bag. "I can read some if you have the time to listen."

Ma Vesta's face eased back into a smile. "We'd be a right

smart disappointed if you didn't. Wouldn't we, children?" She sat down and reached into her basket of mending. "But first, your poor toes look froze still. Put these on to warm you up." She handed Tansy a pair of socks.

Jenny Sue leaned up in her chair to watch Tansy pull on the first one. "Those were Reuben's. I knitted them for him."

Tansy jerked off the sock and sat up. "Well, my feet aren't that cold if you druther I not wear them." She held out the socks toward Jenny Sue.

"No, no." Jenny Sue waved them away. "My Reuben ain't got no use for them no more. But he did say I did a fine job on them, even if they did squeeze his toes a little." She almost smiled. "Weren't he one to make a person smile?"

"He was that," Tansy agreed.

"Go ahead and put the socks on," Ma Vesta said. "You can bring them back next time you come this way."

"Instead of us book borrowing, you can be sock borrowing." Jenny Sue laughed, but it wasn't a good sound.

Tansy pretended not to hear the hint of hysteria in the woman's laugh as she pulled on the socks. She was grateful for the warmth. "My toes thank you." She smiled at Ma Vesta and then Jenny Sue. "You knitting any now, Jenny Sue? I think the library might have some magazines with knitting patterns I could bring next time."

"I've laid off knitting. Can't keep count of the stitches no more. Last time I tried, it ended up in an awful snarl. Ma Vesta picked it out and rolled it back up in a ball."

"It's still there waiting for you to make something." Ma Vesta pulled a shirt out of her basket and began sewing up a ripped seam without looking up. "Cindy Sue could use a sweater."

"Best you knit it for her then," Jenny Sue said. "I'd be apt to leave off a sleeve."

The two children stopped looking at their books and watched their mother and grandmother with round eyes.

Tansy took things in hand. She didn't have long before she needed to move on to her next stop. She scooted her chair closer to Reuben Jr.'s bed, opened her book to the place she had marked, and began reading. A story always made her feel better. She hoped reading about Robinson Crusoe finding a way to survive after being shipwrecked would help this family find a way to carry on after the shipwreck of losing their father, husband, and son.

While there were some things a book couldn't heal, a story could give you some minutes to escape from what was to what a person could imagine.

four

CALEB WAS ONLY A LITTLE SURPRISED to see a strange horse in the barn when he put Ebenezer in his stall. Ma had told him Reuben Jr. was looking forward to the book woman coming by today to read him a story. He shook his head and changed the name to Junie in his mind. If Ma and Jenny Sue wanted the boy called Junie, the least he could do was remember that.

Might be best for the boy too, who seemed to be closing in on himself and not caring about anything much at all. Except he did seem to like the books the book woman brought him.

The boy had always been a somber child. Ma said some like Caleb himself was at that age. She didn't say out loud that Junie being more like his father would be better, but she didn't have to for both of them to think it so.

"The good Lord makes us the way we are for a reason," his mother used to tell Caleb when he worried about not being like Reuben.

Caleb stroked the horse's nose. He recognized the gelding. One of Preacher Rowlett's Morgan horses. A steady ride on the mountain trails, so a good choice for the book woman.

His mother hadn't said her name. Could be she was somebody brought in and not a mountain girl he might know. Could be his ma just hadn't thought to tell him. Since he'd come home last week, their talk had circled around Reuben.

His mother had written him about the packhorse libraries and how women were riding up into the hills to bring books to folks. Several counties around had the new libraries.

President Roosevelt had come up with some good ways to keep people working. They needed that help here in the mountains where even in the best of times it was a struggle to feed a family. Especially with the dry weather and the chestnuts mostly gone. Back when those trees made up a good portion of the forests, a man could turn his stock loose on the hills and let them fatten up on chestnuts in the fall. Not only that, people gathered up wagonloads of the chestnuts to sell.

Caleb was wasting time grieving over the chestnuts. Some things a man couldn't change. He had plenty to grieve without thinking on trees. He'd lost his brother. He couldn't change that either. What happened happened. But he could think on what he could do to help those left behind. Like making sure his sister-in-law and his mother had wood to keep their fires burning and food enough on their tables.

That somehow circled his thoughts back to those missing chestnuts supplying food for abundant wildlife that could have been meat on a mountaineer's table.

Forget the trees, he told himself firmly as he gave Ebenezer some corn and then hand-fed a nubbin to the Morgan horse. He looked at the log he'd dragged into the open center of the barn. Sawing it into chunks for the fireplace could wait until he carried in some of the wood he'd split the day before. Jenny Sue's woodbox probably needed filling.

Besides, he hadn't been in to see Junie today. The boy was a puzzle he hadn't figured out yet, but he intended to in time. He had the feeling that if he could get the boy to talking more, his mysterious illness might disappear like fog when the sun came up. Maybe not all at once but a gradual clearing. A gradual healing.

Seeing this book woman before she was off to the next house would be good too. Something to read would help pass the long evenings by the fire. His mother had never been a big talker and neither was Caleb. An hour might slip by without the two of them saying a word. While she was busy with her sewing or knitting, he had to be busy with his thoughts. Sometimes it was less troubling to read a book of somebody else's thoughts.

Plus, a book to read to Junie and Cindy Sue would give him a way to spend time with the children without him and Jenny Sue having to fill the air with talk. Jenny Sue might find making conversation with him awkward, or more likely he would find it awkward. Especially since he kept remembering what his mother told him about Reuben thinking Caleb should take his place as Jenny Sue's husband.

He couldn't take Reuben's place and had no intention of trying, no matter how often Ma said he needed to think of what was best for Reuben's children.

What would have been best was Reuben not letting a tree fall on him. Caleb felt guilty thinking that, but a person had no way of hiding his thoughts from himself. However, he could push them back into a dark corner of his mind and cover them up with shame. Reuben didn't want the tree to fall on him. He would want to be the one carrying wood into his wife's woodbox and raising his children.

Caleb dropped the load of wood into the box on the small back porch, then picked up a chunk to carry inside. In this kind of weather, a fire always needed feeding. That was why it was best to get the wood piled up before winter set in. No doubt that was what Reuben was doing when things went wrong.

The book woman was reading aloud when Caleb pushed through the back door. He stopped in his tracks. That couldn't be who it sounded like. He held his breath and listened. The woman must be over by Reuben Jr.'s bed. He couldn't see her through the opening between the sitting room and the kitchen, but he knew that voice.

Tansy Calhoun. It had to be. Her voice had lived inside his head for years. She was the reason he'd left the mountains looking for work. He hadn't wanted to see her married to Jeremy Simpson. And now he was about to come face-to-face with her right in Jenny Sue's front room.

Why in the world hadn't his mother told him the book woman was Tansy Calhoun? Or Tansy Calhoun Simpson now. While Ma always thought it was Tansy's sister, Hilda, he headed over to Robins Ridge to see, she still could have let him know the book woman was somebody he knew. A neighbor girl. A girl who had stolen his heart when she was a kid.

He shut his eyes and remembered the first time he'd noticed she was growing up. Really noticed her. She couldn't have been more than fourteen, with her hair in pigtails. He hadn't been but seventeen himself. Not that sometimes people in the mountains didn't get married at young ages like that. Reuben had been only eighteen and Jenny Sue sixteen when they married. But Caleb thought it better to wait, and then he'd waited too long. Tansy had started keeping company with another. She was still happy to see him when he went by the Calhoun place, but the same as his mother, she thought he was sweet on Hilda.

He should have told her straight out that he wanted to court her. Maybe that would have made a difference. Then again, maybe she'd never had a hankering for him the way he did for her. So he had dillydallied. Too long. Reuben always told him he overthought things. That sometimes it was better to take a flying leap into whatever a man wanted to do and see what might happen next. But half the time what happened next when Reuben did just that was a mess Caleb had to clean up.

He hadn't wanted to make a mess of how he felt about Tansy Calhoun. That was why he left home. The Civilian Conservation Corps turned out to be perfect for him. Let him work out in the open, planting trees and getting to see places he might never have had the opportunity to see. He didn't mind the work, and while it didn't keep thoughts of Tansy from floating into his mind, it did give him a way to work out his sorrow of losing her.

Now it might be better to just set down the chunk of wood and head back out to the barn. Sawing on that log he'd brought in might calm him down. Help him see things in a clearer light.

He hesitated too long. His mother called out, "Come on in here, Caleb, and say hello to Tansy. You remember Tansy Calhoun, don't you? Hilda Calhoun's little sister. Did I tell you she was the book woman?"

"No, Ma. No, you didn't," he managed to say. His voice sounded a little rough, but he had been out in the cold. Reason enough for the flush climbing into his cheeks too.

If his mother noted it, she'd think it was because she'd said something about Hilda. She was sure Hilda getting married was the reason Caleb had headed out of the mountains to find work. He'd told her often enough that Hilda didn't care two figs for him. That she had her eye on Reuben, but his mother didn't choose to believe that. Not with Reuben already married and a daddy. Ma had a way of coming to conclusions that she didn't lightly give up, even if you could prove her wrong. And right now, he was glad she had the wrong sister for him to be heart sick over. Some losses a person didn't care to share.

But in spite of how the thought was grabbing the air out

of his lungs, there had never been any doubt he would step into the next room and rest his eyes on Tansy. Three years had passed since he'd had that pleasure. His mother's words ran back through his head. She'd said Tansy Calhoun. Not Tansy Simpson. But then she'd said Hilda Calhoun too, and he'd been there when Hilda married Bill Raymond some years ago. Sometimes people simply remembered you by the name you were born with. Especially if that person grew up one ridge over.

Tansy had stopped reading. They were all waiting for him. Pulling in a deep breath, he ignored how his heart was thumping as though he'd just split a cord of wood and stepped through the doorway.

He let his gaze slide past his mother and Jenny Sue in front of the fireplace to Tansy over by Junie's bed. A book was open in her lap, and Junie was frowning, not happy to have his story interrupted. Caleb couldn't worry about him right then. His eyes were full of Tansy.

She was every bit as pretty as the last time he'd seen her. A little less of the girl in her face and more of the woman she'd become. She wore a wool hat pulled down over her hair, but a few tendrils of honey-brown hair curled out around the hat's edges. She was in her sock feet. The socks appeared to be dry, while her wool britches looked as wet as his felt.

He wasn't close enough to see if her eyes were more blue or green. Their color changed with what she was wearing or sometimes thinking. But they looked wide and eager, exactly how he remembered.

A smile lit up her face. "Caleb, Ma Vesta didn't tell me you were home."

She looked really pleased to see him, but not in the way he was pleased to see her. A friend. He guessed it might be good if he could think of her in the same way. No more than a neighbor and friend.

"It's good to be home." He tried to tone down his smile, but it was such a pleasure to see her. Even if it was too late to change how things were. "And look at you. One of the pack-horse librarians. You always did like books."

"And you did too, didn't you?"

"He did," Ma spoke up. "Was always after me to let him go borrow one from Preacher Rowlett. That man must have a dozen books."

"And I read them all a dozen times," Caleb said.

"So did I." Tansy laughed. "But now we've got different books to carry out to folks. More books than I can read the one time."

"Might have been better if you'd read the Good Book a dozen times." Ma shook out the shirt she was mending and smoothed it down in her lap before she folded it up.

Tansy's smile didn't waver. "You sound like my mother. She was always telling me that too, but I didn't shirk on reading the Scriptures. I read them too. If you want them, I have some Bible lessons in my pack there." She pointed at a sack of books by the fireplace. "They're some of our popular loaners."

"You left one of them last time," Jenny Sue said. She was wrapped in a quilt in the chair by the fireplace. "Dry as sand-paper. Even nigh on put me to sleep, and that ain't an easy task these days what with how things are. I druther read something with a little more interest to it than somebody preaching in writing."

"Now, Jenny Sue, don't be talking against Tansy's books," Ma said quietly.

"No worries, Ma Vesta. Some people would rather read the Bible and figure out things for themselves," Tansy said. "I've got a magazine you might like, Jenny Sue. And Caleb, maybe you'd like to pick out a book."

"I would. Thank you," Caleb said.

When Tansy closed the book she'd been reading, Junie and

Cindy Sue looked sad. "Sorry, kids. I've got to move along to the next house, but we'll read more next time. Until then you can practice on your words, Junie."

"Me too," Cindy Sue said.

Tansy laughed. "You too. You'll both be reading to me before long."

Caleb wanted to gather in the sound of her laughter to store in his heart. This was crazy. He needed to get himself under control before he did something foolish, like reach to touch her hand as she moved past him to pull a magazine out of her bag. She held it out toward Jenny Sue, who waved her away.

"Leave whatever you want. I'll read it if I feel like it." Jenny Sue pulled the quilt closer around her, as though trying to hide in it.

Ma took the magazine. "If you have one, I wouldn't mind having something with quilt patterns in it."

"I don't have anything like that with me, but I'll see if I can find one to bring next time." Tansy held a book out toward Caleb. "Here's something you might like. About the Revolutionary War. I'm not wrong remembering you liked history, am I?"

Caleb took the book without even looking at the title. If she said he'd like it, he'd like it. "So when do you come back around?"

"In two weeks. I have eight different routes to get to all the people in this area. Of course, I don't always get invited in to read some. Tuesdays I'm at the library in Booneville."

"I didn't know they had a library there," Caleb said.

"We didn't until a few months ago. You get to town, you should go see it. Not a lot of books. We carry most of them out with us on our book routes, but Madeline Weston will be there with some books on the shelves. We're hoping to get more soon."

"Sounds good." Caleb looked down at the book. *The War of 1776.* He ran his hand over the cover, but he wasn't thinking about the book. "Where are you and Jeremy living?"

Her hand froze on her book bag. His mother and Jenny Sue seemed to be holding their breath too, as if he'd said something he shouldn't have.

"I guess I never told you," Ma started.

Tansy came back to life, smiling again as she packed up her books. "It's all right, Ma Vesta." Her smile didn't waver as she stood up and looked straight at him. "I guess when you left I was still seeing Jeremy, but he ended up marrying Jolene Hoskins. They've got a little one now with another on the way, and me, I'm still at the home place. Ma needs my help since Pa took off to find work."

He tried not to smile as if he'd just gotten a load of Christmas candy. The news that she was still living at home could be sad to her, but it made his heart sing. At least the part of Jeremy being out of the picture. And with her pa gone, he'd have a good excuse to go see if they needed help, the same as he was helping Jenny Sue.

Tansy pulled on her boots. "I thank you, Ma Vesta, for the time by the fire and the ginger tea and for letting Shadrach warm in your barn. We've got more stops to make before we head home but maybe no more overfilled creeks." She packed up her books. "I'll bring the socks back when I come this way in a couple of weeks."

Caleb reached for the saddlebags. "Let me carry that for you."

She didn't surrender them. "No need you going back out in the weather. Looks like you could use some time by the fire yourself." She waved at the children. "We'll read more about old Robinson next time, but I left a Doctor Dolittle book for you here along with those others. Maybe your uncle will read it to you." She grinned at Caleb.

"Sounds like fun," he said. "But let me go get your horse. I can sit by the fire later."

He was out the door before she could protest. He hardly felt the snow hitting his face as he headed to the barn. Tansy Calhoun wasn't married. He wanted to jump up in the air and click his heels together at that good news.

five

TANSY FOLLOWED CALEB to the barn. Mrs. Weston told them to be independent and not expect help from those on their routes. But Caleb was through the door before she could get out another no.

He'd always been someone ready to help, to fix things. She was glad to see him here. Sometimes when folks left the mountains, they never seemed to find the way back. Like Hilda, who hadn't come home to visit but once since she got married three years ago. Tansy had thought maybe Caleb would be the same way. That he'd find some flatlander to marry since Hilda had spurned his advances and would end up who knows where. Tansy smiled, thinking about how when he came to see Hilda, she would push him off on her.

That was fine with Tansy. She loved talking to Caleb. They'd spent a lot of hours mourning the chestnut trees and talking about how they hoped and even prayed the trees would quit dying. Other folks might laugh at somebody praying for trees, but it was natural as the sun coming up in the morning for the two of them.

Caleb had changed some in the years he'd been gone. His

face was slimmer and his shoulders broader, but he still had those smiling eyes the color of a summer sky. That was something she'd really liked about him. How when he smiled it seemed to come from deep inside him. His curly sandy hair looked the same, as it sprang out in unruly sprigs. The only times she'd seen it flat on his head were when he tamped it down with pomade. Even then, a few curls would spring free and lap down over his forehead. He was always pushing it back off his face, trying to keep it in control the way he liked to keep his life.

She wondered if he still liked things neat. From the way he rushed out of the house toward the barn with no hat, maybe not. He didn't look a bit concerned about the snow collecting in his unruly curls.

Her own hair had some of those same wayward tendencies, especially after she pulled a knitted hat down over it. That was why she wore plaits and left her hat on when she stopped at houses. She should have stuffed an extra hat in her saddlebags so she could yank off this wet one and pull on a dry one. But she'd been wet before and survived.

The weather didn't seem to bother Caleb at all as he strode across the ground between the house and the barn. He looked strong, capable. A man now, instead of the teenager she remembered.

He'd seen parts of the world she'd only read about in books. Ma Vesta said he'd been working in mountains different from theirs, where mist rose up out of the tree-covered slopes like smoke. Fog could blanket these mountains here sometimes, and storm clouds could sit low on them the way they were now, but that wasn't the same as mountain peaks poking up into the sky to pierce clouds.

She wouldn't mind hearing about the places he'd seen, but she didn't have time for talking today. She'd stayed too long in front of Jenny Sue's fire drying out her feet and reading to

Cindy Sue and Reuben Jr. Or Junie, as they were calling him now. He was such a sweet kid, but so sad that even when he did smile the sorrowing leaked out around it.

Maybe that Doctor Dolittle book she'd left would give him something new to think about. Make him want to try some adventures himself. Not like Doctor Dolittle adventures. Nobody could really talk to animals, but the stories were fun. Just the thing for a boy in need of a giggle or two.

Caleb waited at the barn door for her. "Sorry. I didn't mean to run off ahead of you like that. I forget how fast I can cover ground sometimes."

"No worries. I need to be on the move anyway." Tansy stepped past him to get Shadrach.

He followed her into the barn. "It won't be long before the edge of night, with the way the clouds are hanging low."

She had to smile at his words. He must be thinking she was still that kid he'd known from before he left the mountains. "I can find my way home in the dark."

He looked a little abashed. "I guess you can. But a snowstorm in the dark might make it hard."

"True enough, but Shadrach would know the way if I didn't." It was sweet, him being concerned. She didn't know the last time somebody had worried about her. Everybody thought she was tough. And she was. She had to be. No storybook princess life for her.

"I guess that's so," he said. "That's one of Preacher Rowlett's Morgans, isn't it?"

"It is. I'm leasing Shadrach from him for my book routes."

"Do you like taking books around?"

"You know I do. I get up happy every day."

"Every day?"

"Well, most every day. Hard to be happy every morning. Especially in weather like this." Snow had collected in his hair and she had the sudden urge to brush it away. She would have, back when they used to talk about trees and books, but he was a man now. Not that they couldn't still be friends.

"I guess being a book woman is a perfect job for you. Sort of like planting trees with the Corps was a good job for me."

"It's nice you got to come home. You being here might help Reuben Jr. start feeling better. He really misses his pa." When a shadow crossed Caleb's face, Tansy put her hand on his arm. "I'm sorry. I didn't mean to bring up sad thoughts for you."

"No changing what's happened. I'm glad to be here to help my family. But could be you'll need some help over your way too with your pa gone."

"We're fine. Josh is big enough to shoulder a man's load now."

"Little Josh? That pesky kid I'm remembering?" Caleb's smile was back.

"He's eighteen. Plenty old enough to pull his weight. Some years have spun by since you were around. Things are different."

"They are that."

She turned to fasten her saddlebags back on Shadrach. When the horse blew his breath out in a huff, she smiled. "I think he's ready to move."

"Looks like it." Caleb stood back when Tansy took Shadrach's reins. "How many more stops do you have today?"

"Not many, but there's some miles between them. First, Manders School. They aren't doing lessons right now. The teacher quit and they haven't found a replacement. But sometimes one of the Harrison boys waits for me there. Then I go to Aunt Perdie's. You remember her?"

He laughed. "Course I do. If a person ever once met Perdie Sweet, then that person won't likely forget it. She's kin to you, isn't she?"

"Pa's cousin. Once removed." Tansy led Shadrach out into

the covered center of the barn. The wind had picked up and was blowing the snow sideways. "She'll let me know about it if I'm late." She looked over at Caleb. "I'll see you next time I come this way."

"In a couple of weeks, right?"

"Right. Plenty of time for you to get that book read." She put her foot in the stirrup to lift up into the saddle. With a wave, she headed out into the snow. When she finally got home, she would need a lot of sitting by the fire to warm up.

A couple of miles over the hill and then down into Manders Hollow, she stopped at the schoolhouse. Price Harrison looked half frozen when he came out to meet her.

"Sorry I'm late." She slipped off Shadrach, grabbed her saddlebags, and followed Price back inside. It wasn't much warmer in there, but at least they were out of the wind.

"I weren't sure if you were late or I was early. It's hard to know the time with the clouds hiding the sun." Price was twelve going on thirty. As the oldest of six kids, he had to grow up fast when his father got hurt in the mines a couple of years ago.

"Family doing all right?" Tansy asked as she pulled out a book for him to carry home to his little brothers and sister. She spread out a few others to let him pick out his own.

"'Bout the same. Pa's getting cabin fever, but Ma's afeared to let him outside when the weather's bad. He don't know how bad off he is and wanders some. Ma makes Ella follow after him, but that jest has me troubled for the both of them. Ella ain't but seven. She's liable to get lost in the woods her own self." He picked up a book and then a magazine. "Pa likes looking at these *National Geographic* magazines. Says he can't hardly believe there's really places like they show in the pictures."

"It can make a person wonder," Tansy said. "You walk down here?"

He nodded. "Old Joe come up a little lame last week. It ain't but a couple of miles down here as the crow flies. I brung a poke to put the books in."

Tansy loaded up her books after he made his pick. "You be careful going back up the hill then, and I'll see you in a couple of weeks."

"I'll be here unless they find a new teacher to start school up again. If that happens, I reckon I'll have to let Freddy get the books 'stead of me. He's the next one down from me, but I'll be sorry to not be the one to fetch the books home. It's a rare treat." He ran his hand over the books as he carefully slipped them into his sack.

"Won't you be going to school too?"

"Naw. I done learned most ever'thing I need. How to do my sums and read. I've been helping Ella some with her reading till she's nigh on as good as me. She reads to Pa some. He never had the chance to get any schooling. Had to start working early on."

"Like you," Tansy said.

"Some like that, I guess. But I ain't never had to go down in the mines." The boy shivered and not from the cold. "Could be if the mines start back up I might have to, but right now we're getting by. With our pigs and chickens." He held up the bag with a grin. "And our books."

"You keep on coming down even if school starts up. You can listen to me read some of the stories then. We got in some books about a doctor talking to animals."

"That sounds crazy, although Aunt Perdie over the way claims she can understand squirrel chatter. Last time I seen her, I told her if she could make them understand her, she'd best tell them to watch out for Pa. He might have forgot some things, but he's still a pretty straight shot." Price laughed as they stepped out of the school building. He hoisted the sack up over his shoulder and took off in a trot up the trail to his house.

Tansy watched him a minute before heading on over the next hill to Aunt Perdie's. Seemed everybody was talking about Aunt Perdie today. Truth was, plenty of Aunt Perdie stories floated around the mountains. She hadn't ever married. Said nobody ever came calling to ask her, but if they had, she might just have said no. Leastways once she got past normal marrying age. By then she couldn't see the purpose in sharing a cabin that wasn't all that big with some man who'd want her to do things his way. Not when she was happy doing things her way.

Would that be how Tansy ended up? An old lady in a lonesome cabin after spurning whatever marriage prospects might have come knocking on her door? Youngsters then might be sharing Aunt Tansy stories. Could be they already were talking about that crazy book woman, even if they were glad to see her coming. The thought made her laugh out loud.

At Aunt Perdie's, Tansy tied Shadrach to a handy tree, then climbed up the steps to bang on the cabin door. Aunt Perdie didn't have a porch, just a little slab of roof over the top step.

She was ready to knock again when the door creaked open. "Well, if it ain't the book woman. I'd done about give you out." The woman had a quilt draped around her shoulders. She stretched up to her full five-foot height to stare at Tansy. "You bring me a book?"

"That's why I'm here." Tansy smiled at the woman.

She didn't worry about Aunt Perdie not smiling back. It was anybody's guess what mood the old lady might be in. Sometimes calm as a pond at dawn. Other times ready to frown at everything. Not often smiling. She'd asked her mother how old Aunt Perdie was, but Ma just shook her head and said as far as she knew, Aunt Perdie never told anybody her age.

When Aunt Perdie didn't move back from the door, Tansy asked, "Are you gonna let me in?"

"I reckon as how I have to." Aunt Perdie opened the door wider. "You ain't likely to go away less'n I do."

The same as not knowing her age, nobody knew how Aunt Perdie got by. She did some quilting if folks needed it, but most mountain women did their own stitching. Now and again she whittled out figures of things nobody could name but that Mr. Beatty at the General Store in Booneville sold to tourists passing through now and again. Mostly her neighbor from down the way, Hanley Scroggins, saw to it that Aunt Perdie had firewood and some food in her cupboard.

Tansy's pa had made the trek over to Aunt Perdie's now and again to do what he could for her. But as far as Tansy knew, nobody else claimed kin with her, as though they feared ending up responsible for her.

With Pa gone, Tansy supposed it was up to her to make sure somebody saw to the old woman's needs. She looked around the dark cabin. The windows were covered over with wood planks that kept the wind out. But the place was nearly as cold as the schoolhouse had been. A little flame struggled in the fireplace to keep from going out.

"You shouldn't let your fire die down like that in this weather, Aunt Perdie."

"It's burning some. Figured I'd best save a little wood back and not use it all at once." She pulled the quilt closer around her. "Need me to fetch you a quilt? You ain't looking so warm yourself."

"I'm nigh on frozen, but no need getting me a quilt. I can't stay long," Tansy said. "Are you running low on wood?"

"I didn't get none brung in afore this weather came up, and I guess poor Hanley must have come down sick. He was coughing some last time he was up this way. I give him some of the tonic I made up last fall." She sighed. "I'm hopin' it didn't kill him dead. I drunk some of it and I'm still upright."

A black cat jumped down off the bed in the corner and came to wrap her body around Tansy's legs. "Looks like Prissy's feeling friendly." Tansy reached down to stroke the cat.

"Best watch out. She sometimes pulls that trick when she's wanting to slash out at you. She can be a right smart ornery, but she's good at keeping a body's toes warm if'n she'll curl up on them."

Tansy pulled her hand away. Aunt Perdie was right about the cat. She could go from purring to swiping with her claws in an instant. Something like her owner. She looked in the box where Aunt Perdie kept her wood next to the kitchen door. Two little sticks were all that was there.

She picked them up and handed them to Aunt Perdie. "Here, put this on your fire and I'll go dig some wood out of the snow for you."

"Well, if you have to be warm." Aunt Perdie shook her head, but she took the wood over to lay it on the fire. "What you find is liable to be too wet to burn."

"More reason to bring it in so it can dry out a little."

Tansy went outside before Aunt Perdie could voice any more complaints. The woodpile beside the back door looked mighty low, but Tansy dug some chunks out of it and carried them in to the woodbox. It took several trips to fill it up. By then her fingers felt like they might break if she bent them. She pulled off her gloves and held her hands out toward the fire that was flickering a little more heat out into the room. Aunt Perdie sat in the rocking chair next to the fireplace.

"Did you pick out a book?" Tansy asked.

"Didn't see much need in it, but since you come all this way, I figured I might as well make your trip worth it." She held the book in her hand. "Picked this one because it has big print. My eyes ain't as good as they used to be."

"You can read, can't you, Aunt Perdie?"

"Course I can. You don't think I'm ignorant, do you?" Aunt Perdie didn't wait for Tansy to answer. "My pa could read. He was right prideful about that and so he made sure us young'uns could do the same."

"Did you have many brothers and sisters?"

"I was in the middle. There was seven of us. Some passed on young. Others took off somewhere and we never hear'd from them agin." She rocked up and back. "I was all that was left to take care of Ma once the Lord called Pa home. But then I had to bury her too. It ain't all that good burying your folks." She looked up at Tansy. "You give up one of your brothers, didn't you?"

"Robbie. A couple of years back." Tears still poked at her eyes when she thought about it.

Aunt Perdie nodded. "Hard when they's so young. Your pa, he told me about the boy going on. He weren't easy with it, but I reckon it's hard to be easy with something like that."

Tansy turned away from talking about Pa. "I'll note you borrowed that one then." She looked at the table with an empty plate on it as she pulled on her gloves. "You got anything to eat for supper?"

"Don't you be fretting over me, Tansy girl. I've been doing for myself and gettin' along fine for long years afore you were even born. I reckon I can make it a few more days on my own." She waved Tansy away. "Now you git on out of here toward home. Night will be comin' on early."

"The snow will keep it light."

"Not so much less'n the clouds give way to the moon."

Tansy considered giving the old lady a hug, but she figured she'd be some like her cat and more apt to scratch at her than to appreciate the hug. "I'll come by in a few days and see how things are going with you."

"Don't trouble yourself. I'll be fine." She rocked back and forth a couple of times. "Good of you to come bringing them books."

"It's my job."

"Yep. And now you know where the door is at."

That was Aunt Perdie. Nothing to do but smile and head outside.

As she rode away from Aunt Perdie's cabin, she thought about the books she'd given out that day and the ones she'd gathered back in. She hoped she'd have a few minutes to sit by the fire and read a chapter or two out of one of them before time for bed.

six

PERDITA SWEET PEEKED through a crack in the planks over her window to watch Tansy Calhoun ride off. The saddlebags of books bounced on her poor horse's back. Rather, Hiram Rowlett's horse. Foolish of him to let Tansy wear out the animal, packing books around to folks who ought to be working instead of reading.

Then again, Hiram Rowlett had always been too ready to lend anybody something. He'd even once upon a time lent her a listening ear now and again. She hadn't seen him in who knew how long. Would hurt her head trying to count up the months. More probably years.

At one time she and Hiram Rowlett had been right good friends. Nothing more than that, although her mother had had hopes about their pairing. Ma tried. She was always coming up with some need to bring Hiram to the house, but he was a few years younger than Perdita. Ma had never been too good at figures, and by then, she couldn't keep things in her mind. They just flittered away like a bird feather in the wind. So could be, she forgot that Hiram was younger and all. He wasn't about to consider marrying an old woman like Perdita when he was looking for a wife. Not that Perdita would have been against the idea.

That was before everybody started calling him Preacher. Perdita doubted more than a handful of people around these parts even knew his name was Hiram. Sometimes she wondered if anybody knew her name. Calling her Aunt Perdie when she wasn't an aunt to the first one of them. But that was how it was in the mountains when a person got some age added on. Had to be Ma or Granny or Aunt something. Same for the menfolk. She guessed with Hiram, he got Preacher instead of Grandpa or Uncle.

He had married. A pretty little thing some younger than him since he waited a while to pick out a woman. Perdita did her best not to wish her ill, but sometimes it was a struggle. Then she had to walk through guilty hollow when poor Mary Ellen up and died, sudden like. A lung ailment. The girl didn't give it much of a struggle, but then she had always been skinny. A stiff wind could blow her sideways.

Not that such ailments didn't take plenty, poorly or not. The consumption took half her own family out in one year, and most all of them were built sturdy like Perdita. Back when they were coming up, they'd had food enough and more. Rains came down when they were supposed to and a body could keep the cupboards full.

Tansy and her horse disappeared in a swirl of snow. When the girl came back, Perdita ought to thank her for filling up the woodbox. Truth was, she was finding it harder and harder to get things done around here. She had missed Hanley coming up to help her out the last couple of days.

If it weren't for the snow, she'd hoof it down to his house to see what was the matter with him. He weren't as old as Perdita, but he weren't no spring chicken either. None of his boys must have come by for a visit or Hanley would have sent one of them up to check on her. No way could Sally, his wife, climb up to Perdita's house with the way her rheumatism gave her fits.

Perdita stepped back from the crack beside her window and rubbed her hands. Sally wasn't the only one that rheumatism bothered in this kind of weather. Just as well her cow had gone dry, so she didn't have to milk her. She fed the old girl some of Hanley's hay that morning. Could be she should have asked that Tansy throw some hay out to Daisy and look to see if one of the hens might have surprised her by laying an egg. She was down to three old hens and weren't none of them much interested in egg laying these cold days. If things didn't change, Perdita might just have to sit down and starve.

She found a rag and stuffed it in the crack. Best to keep as much of the cold outside as she could. Before she sat back down in her rocker by the fire flickering with more promise of heat now, she scrounged around in the woodbox to find the least-wet chunk of wood to put on the hearth. She'd give it a while to dry out before she fed the fire.

For a minute she considered what she might eat, but she didn't have much except an onion or two and some shriveled up potatoes in the 'tator hole under the floor. She did have a handful of cornmeal, but she was saving it in hopes Hanley would bring her a jar of milk. None of that sounded worth cooking. She wasn't all that hungry anyhow, and even if she was, it wouldn't be the first time by a lot she'd gone to bed hungry. A little fasting was good for the body now and again.

She would favor some coffee. It had been a while since she'd been blessed with a cup of that. Not since the last time she'd made a trek down to the store to see if Wallace Beatty had sold any of her carved figures to some flatlanders that didn't know any better than to waste their money on trifles. A few of her carvings sat on the mantel. If a person did some imagining, he might make out a scrawny cat from one of them.

Perdita didn't care if folks knew what they were or not. She just let her whittling knife follow what was hidden in the wood. Sometimes she didn't know herself what that was until it started shaping up in her hands. She'd been carving on wood ever since her ma let her pick up a knife.

Ma pestered her to make something that fetched a pretty eye, but Perdita never could seem to make anything fancy. She'd tried a time or two for her mother. Closest she ever came was whittling out a butterfly. Turned out to look something like a bat instead, but her ma was pleased. Said at least it was something she could name.

She'd run out of carving sticks except for one piece of chestnut she was saving for the right time. She wasn't sure what time that was but figured the wood would tell her when it was ready. She could go dig around in the snow for some bit of wood that wouldn't matter if she carved it wrong. While she was out there, she might find some chicory or sassafras roots to make something warm to drink. But like as not, she wouldn't do nothing but end up froze. Better to sit by the fire and imagine a warm cup in her hands. Her rheumatism was twisting her fingers too much to do much whittling anyhow.

Truth was, she was feeling disgruntled about everything. The weather. Her puny fire. Her hens not laying eggs. Tansy Calhoun riding up through a snowstorm like delivering books was more important than being sensible.

She picked up the book she'd grabbed out of Tansy's sack without paying it much mind except for the print. *A White Bird Flying*. What kind of title was that? Obviously this Aldrich person wasn't from the mountains. Else she'd have called her book the black bird flying or red bird flying. White birds had to be from somewhere else.

If the light hadn't been so dim, she might have tried reading a page or two to find out what kind of white bird was flying around in the story. But fire flickers didn't make enough light for her old eyes to read. Besides, if she was to strain her eyes trying to see words on a page, she'd be better off reading her Bible. Hiram Rowlett used to tell her ma that dipping into the Scriptures would calm the soul. Get rid of the worries and trembles. Might do that for her too, should she have light to see.

But she didn't necessarily need light for that. She could recall some verses that were written in her heart. *The Lord is my shepherd. I shall not want.*

Now that was foolish. She was right happy to have the Lord shepherding her, but she'd never quite got over wanting. Especially when her stomach was growling.

Prissy came over to hop in her lap. Perdita stroked the black cat head to tail. The rumble of the cat's purr did make her feel better. *He leadeth me beside the still waters. He restoreth my soul.*

She didn't bother to bend her head as she started talking to the Lord. "Thank you, Lord, for sending that Tansy to dig me some wood out of the snow. I'm right sorry that I didn't tell her thank you like as how I should've. If'n you can nudge her to come on back the way she promised she might, that would be good. Much as I hate to admit it, I could be needin' some help."

She fell silent as she stroked the cat before she started talking again. "I ain't meaning to complain, Lord, but it would have been nice to have a young'un I could have called my own. Reckon I would have had to put up with a man for that to happen, but you could have sent me a good one. One like Hiram. You sort of sent me Hiram, but I reckon my bait wasn't good enough. Now I'm jest a crotchety old woman with a cat."

She listened for what the Lord might want to say to her before she went on. "You're right. I could try not being so ornery. Tell you what. If that girl comes back like she said, I'll be sweet as pie, and if Hanley shows up, I'll even ask how his Sally is doing. But it would be nice to have somebody to sit with me by the fire. I didn't think on how lonesome a person can get when they ain't able to be out busy like they once was when I was spurning the idea of courting."

She rocked up and back while the Lord prodded her to remember her blessings. "Oh yes. I do thank you greatly for giving me Prissy, even if she does have her crotchety spells too."

As if to prove her words, the cat jerked her head up and jumped off Perdita's lap to run toward the door.

"What has got into you, Prissy?" Perdita pushed up from the rocker.

The cat yowled and scratched on the door.

"You thinkin' a mouse might come to the door wanting to be your supper? We done had all our company for the day. Ain't nobody gonna be out on this snowy night. Not even Hanley, if'n he did think about us up here with naught but a handful of meal left in the cupboard to make poor do."

She held in her breath to listen for what had the fur up on her cat's back. She did hear something. A little mouse-like knock. Nothing like how that Tansy girl a while ago had banged on the door like she figured Perdita was hard of hearing. And not like Hanley knocked either. This was a timid tapping. Could be Hanley had sent one of his grandkids up to see about her. Foolish man, if he had sent a child out in this weather. But if he had, she hoped he'd sent some vittles with the boy.

Prissy retreated to the bed, tail up and on full alert in case she needed to pounce on something.

"Ain't no worry, Prissy. Snakes ain't liable to be out in this. Ain't no tellin' what it might be. But I reckon I better open the door to see."

She considered picking up the fire poker in case it was trouble, but trouble wouldn't sound so feeble with its knocking. Then again, trouble could sneak up on a person without making a sound.

The tapping came again. Not loud, but twice in a row.

"Hold your horses," Perdita called. "I ain't as fast as I once was."

If it was somebody of Hanley's sending, she hoped he'd sent a little coffee too. He had been known to be that generous.

She pulled the door open. Snow swirled in on the wind. And there in the middle of the snow was some poor soul. About all that showed were eyes peeking out from a woolen scarf. Had to be a child. Wasn't any taller than Perdita, who had shrunk a little with the years.

"W-will you l-let me c-come in?" The child could hardly get the words out for the way her teeth chattered between the words.

Perdita reached and took hold of the child's arm. "Get on in here 'fore you let a foot of snow in. What in heaven's name are you doing out in this weather?"

She had no idea who she was bringing into her house. Could be that trouble sneaking up on her.

seven

"I-I . . ." THE CHILD STARTED TO ANSWER.

Perdita waved away her words. "Never mind. Git on over there to the fire. Explaining can come later."

A girl. Perdita could tell by her voice and the way she moved. Mincing little steps. But that could be because her feet were frozen. "Shake off that snow and shed that wet coat whilst I put another log on the fire."

Perdita placed the wood on the fire and poked the coals until the fire crackled and shot up a good flame. But when she turned around from the fireplace, the girl still stood there dripping snow water. "Don't you want to take off your coat?" Perdita asked.

"Yes'm." The girl stayed stiff as one of Perdita's carved figures, excepting for the shivers that shook her.

Perdita waited, not sure what to make of the girl. Or why she was standing in her cabin.

The girl shrugged off the poke she had slung over her shoulder and fiddled with the buttons on her coat. She looked up at Perdita. "I can't undo the buttons. I reckon my fingers are too cold."

"Well, step on over here and let me do them for you." Perdita blew out an exasperated breath and worked the buttons loose.

It wasn't an easy task with how rheumatism was paining her fingers. "Mercy's sake, child. You're nigh on as wet underneath. You need to get shed of your sweater and your long johns too. You can wrap this here quilt around you till your clothes dry." Perdita reached for the quilt that she'd left on the rocking chair. "But don't swing it toward the fire. Wouldn't want you to go from ice to flames."

"Might be for the best." The girl's voice was so quiet Perdita barely heard the words. The girl slipped off the coat and the sweater underneath. She kicked off her boots and pushed down her thick stockings with the heels of her hands.

The only light in the cabin was from the fire, but it was enough to see bruises on the girl's arms. And that she was round in the middle. She might look no bigger than a child, but she was old enough to be carrying a babe of her own. Questions toppled around in Perdita's head, but she didn't let none of them out her mouth.

"Wouldn't be better for my quilt. Took me some time to stitch that together. So pay mind to the fire and wrap it 'round you."

"Yes'm. I shouldn't have oughta said that. I wouldn't want no damage to come to your handiwork." The girl tugged the quilt tight around her. She was still shivering.

"I reckon not," Perdita said. "Now you sit there by the fire. Might be good to stick those hands up under your armpits to thaw them out."

The girl did as Perdita said. Even with the quilt around her, she hadn't unwrapped the wool scarf from around her neck.

"That scarf appears to be wet too. Let me get it for you." When Perdita started to unwind the scarf, the girl pulled in a breath and jerked back.

"I must have banged into a tree branch." She kept her eyes away from Perdita.

"Appears you must have." Perdita lifted the scarf away from

her face and gently touched her bruised cheekbone. "It ain't broke, is it?"

"I'm thinkin' not. Just sore. That'll go away after a spell." The girl pushed some wet strands of dark hair back from her face.

"Not the first time with a bruise, eh?"

"Everybody gets a bruise now and again." A tear ran down the girl's cheek.

"I reckon so. Let me fetch something to dry your hair."

Perdita tried not to think and just do, as she shuffled to the cabinet to get a towel. But she couldn't stop the thought that this girl had trouble Perdita might get mixed up in. While she had been asking the Lord for family, she hadn't expected a girl in the family way to show up on her step before she had hardly sent the prayer heavenward. Hiram used to tell her a person needed to be careful what they prayed for.

Could be she ought to send up a prayer for food now. The girl was surely hungry, and Perdita with little more than that handful of meal. Maybe the good Lord would take pity on her and keep putting meal in her crock the way he'd kept the widow of Zarephath in flour and oil. But that woman had been feeding a prophet. This girl wasn't that. Perdita wasn't right sure what she was, but once the child warmed up, she'd have to find out.

After she handed the girl the towel, she went back for another chunk of wood to put on the hearth. She dropped down in the other rocking chair that was hardly ever used. Her ma said it lacked comfort for a person's sitting-down parts, and so that made it perfect for company. Kept them from staying too long. But even if she had put the girl in this chair here instead of her own rocker, there wasn't any way she could send her away before daylight.

Prissy jumped down from the bed and came over to rub

against Perdita's legs. But then the cat went to the girl and without so much as an invitation jumped up into her lap.

"Watch out for that one. Prissy ain't always nice."

"Prissy." The girl smiled for the first time and stroked the cat from head to tail. "She seems friendly enough."

"Looks can be deceiving," Perdita warned. But the cat curled up in what lap the girl had and looked to have found the perfect laying spot right up against her baby belly.

"She's rumbling some." The girl kept her hands soft on the cat as she looked over at Perdita. "Thank you for taking me in, Aunt Perdie."

Aunt Perdie. Did that mean she was supposed to know this girl? And that "taking in" sounded some different than just letting somebody take shelter in a storm. One thing at a time, Perdita told herself. First a name. "Do I know you, child?"

"I'm thinking we met some years ago at the schoolhouse down the way when that preacher come in here from up north somewhere. I weren't but nine or ten then, but I got saved."

"That's good to know, but are you going to tell me your name or do I have to make up one to give you?" She guessed she shouldn't sound so cranky, but the girl did have her chair and her quilt.

"Oh, sorry." The girl looked contrite. "I'm Coralee Embry. From down on Hardin's Creek."

"Hardin's Creek? That's a good ways." Perdita frowned over at her. "Plenty of places between here and there a body could take shelter."

"Yes'm."

Perdita took the poker and stirred up the fire. More to give herself time to think than because it needed poking. "I remember Embrys from down that way. Maylin Embry comes to mind. Did some quilting with her once upon a time. I heard she passed on."

Coralee's mouth tightened. "Yes'm. That was my ma. She

died two years ago when I weren't but fourteen. She and the baby both." The girl lifted one hand away from Prissy to rest on her stomach. She shut her eyes a minute before she went on. "Pa ain't been the same since." Her hand went from her stomach to her jaw.

So her pa was to blame. "When's yours a-coming?"

"I'm thinking sometime in February if'n something don't happen to make him come early."

"You mean like falling and hitting agin a tree?" Perdita narrowed her eyes on the girl.

Coralee looked down at the cat. "Something like that."

Perdita blew out a breath and rocked back and forth three times. "So I'm understandin' that. But I ain't understandin' why you're here sittin' by my fire."

"I'm hoping you might let me stay a spell." She glanced up at Perdita, and then as though she didn't find any favor there, turned her gaze toward the fire. "I didn't have nowhere else to go. Pa, he put me out."

"What about your husband?" Perdita pointed at her stomach.

The girl moistened her lips and seemed to have to gather her courage to answer. "I ain't never been accused of having one of those."

"It takes two to make a baby."

Her voice was low. "I was sweet-talked into doing wrong. And then he was gone."

"Your pa could make him come back."

"More apt to shoot him."

"You might want to do the same."

"Now, maybe." Coralee's voice was sad.

"So that's that," Perdita said. "But that still don't tell why you're here. I ain't really your aunt, am I?"

"Be nice if you were, but I reckon not." She sniffed and rubbed her eyes with Perdita's towel. "Me coming here is my sister's doing. Saralynn's a year younger than me." Another sniff. "But we're the same as twins. We've got two little brothers that we've took care of since Ma died, but now she'll have to do it her own self, I suppose."

"Are you goin' to get to why you're sittin' in my rocking chair?"

"Like I said. It was Saralynn. She got saved at that same church meeting I did, but I reckon it took better with her. Anyhow, she prayed, seeing as how she was worried some about what Pa might do. I prayed too, but I didn't get no answer."

"And she did?"

"Yes'm. Said the Lord put your name in her head. Ma had one of them funny little critters you carve now and again." Coralee's eyes went to the mantel. "Saralynn, she's right attached to the thing. Thinks it looks something like a bear getting ready to growl. Ferocious like. Claims it makes her feel brave. Anyhow, she said maybe since she'd heard you were up here all alone that could be you wouldn't mind me staying with you till the baby comes. Then we'd figure out something else. My belly ain't so big yet that I can't help you with whatever you might need doing. Cleaning. Packing in wood or such."

Perdita didn't know what to think about her being the answer to somebody's prayer whilst she kept wondering if this girl was here because of her own prayer. Hiram used to say the Lord could work in mysterious ways that regular folks couldn't understand. She wished Hiram was here to help her make some sense out of this and to tell her what the Lord wanted to happen.

The girl didn't say any more as Perdita rocked back and forth. The fire crackled as it ate up the log. Outside snow blew against the door and cold crept in through cracks here and there. Come morning, she'd have to find more rags to stuff in the holes. Newsprint would work better, but she didn't have none of that.

The girl made a sniffling noise. Perdita wanted to shut out the sound. She didn't want to feel sorry for her. She'd gotten her own self in the mess. But then who hadn't been in a mess or two in their time? Poor child done been beat on and put out in the cold for doing what came natural when a person didn't shove those feelings away. Perdita didn't have no call to shame her for being young and foolish.

Another Bible story came to mind. That one about casting the first stone at that poor woman caught in doing wrong. Jesus hadn't condemned her. Just told her to go and sin no more. Whilst the Bible didn't tell the rest of the story and what happened to the poor woman then, Perdita hoped the Lord had put somebody in her path to help her sin no more. She guessed that here and now she could be the one in this poor girl's path to help her story finish up better.

She chewed on her lip and pondered what was best to say. Kind words didn't come all that easy for her lately. Then again, they never had. Maybe that was why Hiram had stopped coming to call.

The girl's sniffles got louder, even though when Perdita peeked at her out of the corner of her eyes, it was plain she was aiming to swallow them down. Prissy was rubbing her head against the girl's arm. Perdita was right proud of Prissy. If that ornery cat could attempt comfort for the girl, then Perdita surely could do the same.

"Now, now, girl. Ain't no use wasting tears." Perdita stood up. "I'm guessing as how you is probably hungry. What with eatin' for two. I have to admit that you've come to a bad place for vittles. All's I've got is enough meal for some poor-do fixings and a jar of sorghum molasses I was saving for company. And here company is."

Coralee swiped the palm of her hand across her cheeks to get rid of her tears. "Don't bother yourself."

"Not a bother. I'm feeling a bit peckish my own self." She

poked the fire and checked the kettle of hot water on a hook over the flames.

"If'n you'll look in my poke, you'll see I brung us a little. Biscuits and some fatback strips. Saralynn said she could fix more for Pa when he woke up. He'd come in with a jar of shine earlier. He ain't too pleasant to be around when he gets liquored up. The young ones just hide out and he don't never act mean to Saralynn. She looks like our ma, you know. We wrapped the biscuits up in some newspapers, so I'm hoping they didn't get wet."

Biscuits and fatback. Newsprint. The Lord was a wonder in how he could answer a prayer. Those mysterious ways, she supposed, as she pulled the bundle out of the poke and unwrapped it. "They look to have made it without much crumbling. I'll get us some plates and find that sorghum. We'll have us a feast."

Coralee pushed the cat off her lap and started to stand up. "I can get it if'n you tell me where to look."

Perdita couldn't keep back a little smile when Prissy turned and swiped at the girl's leg. That was good. She wouldn't know what to do with a sweet cat. And this poor girl would have to figure out how to make it with an old lady that lacked sweetness too.

"Settle back. You can be doing the chores tomorrow. Tonight you can be company." Perdita dropped a biscuit crumb for Prissy as she went to get plates and a spoon for the sorghum.

"Then you ain't gonna turn me out?"

"Let's take it a day at a time. Seems I remember that's what the Lord tells us to do in the Good Book. Don't be thinking on tomorrow while we're still in today." Perdita wasn't ready to embrace niceness too close. And these biscuits the girl had brought wouldn't last overlong.

Perdita sighed. Nothing for it on the morrow except to kill

one of the old hens to make some soup. She could scratch up those withered potatoes and an onion out of the 'tator hole. The good Lord might put some extras in there for her overnight. Weren't no telling what he might do, seeing as how he had delivered this girl to her door. Not family born to her but could be family born of need.

eight

TANSY WAS GLAD TO SEE THE SUN as she rode the familiar trail to Booneville. The clouds were gone. Now the brilliant blue sky, scrubbed clean by the snow, dipped down to touch snow-shrouded pines.

As the sun climbed higher over the horizon, the pink faded and sunlight set the snow aglitter like a thousand diamonds. Not that Tansy had ever seen a thousand diamonds, but she'd read about how diamonds sparkled in stories about women richer than any likely to be found in these hills.

But a person could collect riches in different ways here. In the beauty of the day. In the smell of smoke drifting through the air to bring to mind the image of a family around a hearth. That made her think of Aunt Perdie and how she had promised to go back to check on her. The old woman didn't look able to wrestle wood out of the snow.

Tansy smiled. Aunt Perdie would probably tell her to get on home. That she didn't need Tansy worrying over her. But that was another good thing about the mountains. Neighbors did look out for one another, and it was likely Aunt Perdie wouldn't chase her away until after the woodbox was full.

The ride to Booneville gave Tansy plenty of thinking time. Today with the snow-covered ground slippery in places, the going was slower than usual. Even though she'd gotten a soon start by saddling Shadrach when the first fingers of dawn were tickling away the night, she wouldn't get to the library before half the morning was gone.

She liked Tuesdays at the library when she worked on books with Mrs. Weston and the other book women. They exchanged news about the people on their routes and which books readers favored most, while they repaired the worn books for another round of reading. Most of their books were already well used when they received them and might have gone in a trash bin if not for the Packhorse Library Project started by President Roosevelt. Or Mrs. Roosevelt. The president's wife had come to visit up in Morgan County. Wouldn't that have been a thrill if she had showed up in Owsley County?

But Tansy was just glad they got their share of books and magazines for their library. If the books were falling apart, they pieced them together. When magazines were doing the same, they cut out pictures to make extra books. They copied poems to go with pretty pictures or found factual information to share about whatever was in the pictures, whether birds, trees, or cows. Sometimes they wrote out mountain legends passed down through the generations.

Tansy was good at remembering those stories. Once she'd sneaked in a story of her own, but nobody caught on. They all thought Tansy had heard it from some granny in the hills. Maybe she had. A lot of stories circled around in her head.

She remembered a few Aunt Perdie had come up with when Tansy tagged along with her father over to her place. She should ask her to tell those again so she could write them down.

Then she wanted to make Junie another book. Maybe she could come up with a story about snow on the trees, but no, that might not be best with how a tree falling wrong had deprived him of his father. Better to make it a bird or a squirrel hunting food in the snow. Those would be good words for him to learn. *Bird. Squirrel. Snow.*

Then she needed to hunt for that quilt-piecing book Ma Vesta wanted. That could be something the women on their book routes might share to let the book women make a scrapbook of quilting patterns to loan. Even Aunt Perdie might help with that. She had an artistic bent, what with the fancy quilts she pieced and those strange figures she whittled out.

Ma said a person without a family could fiddle around with things such as that, but raising children was a better use of time and what the Lord intended when he said go and be fruitful. Then she gave Tansy a direct look, as if the way Tansy played around with stories and rhymes had her started down the same kind of old maid's path. But there was nothing wrong with rhymes. Add music and you had songs. Who didn't like those?

She wondered if Caleb still played the guitar. It was good seeing him. She was surprised he hadn't found somebody to marry by now. Surely he wasn't still pining after Hilda. More probably he'd been too busy working to go courting in those different places he'd been.

At Booneville, she left Shadrach at the livery stable and happily headed to the small library with its shelves either holding books or waiting for more to come. A day chin-deep in words and stories seemed the next thing to heaven to Tansy.

Two other book women, Rayma Boggs and Imogene Wilson, were already sitting around the worktable with Mrs. Weston. With a jar of paste in front of her, Imogene shuffled through loose magazine pages to find pictures to make a new book. A pencil was stuck through the bun of dark hair on top of her head. She was so tiny that her thick woolen sweater nigh on swallowed her, but looks could be deceiving. Tansy had seen Imogene load her book bags on her horse. Nothing weak or fragile about her. A packhorse librarian had to be tough and ready for anything.

Rayma was tall and sturdy and looked ready to tackle wildcats if any got in her way. Gray traced through her brown hair, even though she was still in her twenties. Both she and Imogene were married with children, and Josephine, who worked the western route in the county, was a granny with young children to tend to as well after she took in four grandchildren when her daughter died of typhoid.

They didn't say so, but Tansy knew the others felt sorry for her with no children, while at the same time they thought she had it easier without those children to see to. Things could get complicated sometimes. Especially when a woman lost her husband to disease or to the wanderlust.

She felt guilty thinking her pa had simply wanted to be gone. He'd surely done what he thought best when he left to find work. If only he'd let them know where he was, but the weeks slid on by without a word. Could be Ma's worry that Pa had met with a bad end was right, but Tansy wasn't ready to believe that without some proof. A man died, somebody would send news of it to his family.

"Oh, Tansy." Mrs. Weston looked up to greet her. "I'm glad you made it in. I worried the snow would keep you home."

"Nothing stops a packhorse librarian." Tansy smiled over at Rayma and Imogene.

"Yesterday nigh on stopped me." Rayma shivered. "Creeks were up around my way."

"Out my way too. I had to dump water out of my boots after I got across Mad Dog Creek. Time I got to the next house, my boots were frozen to the stirrups," Tansy said.

Mrs. Weston frowned. "You did keep the books dry?" She looked a little shame-faced as she went on. "And not get frostbite."

"No worries. I made sure my books stayed out of the water." Tansy smiled at her. "And none of my toes fell off. I got to warm up some at the Barton house since I stop to read a little to them."

"Oh, those poor dears. Losing their pa like that." Imogene looked up from her pictures. "Is the boy any better? I heard nobody rightly knows why he's punying around. I'm surprised Ma Vesta hasn't come up with some kind of tonic."

"She says she hasn't figured out a tonic for bruised hearts," Tansy said.

Imogene held up a magazine picture of a boy and a dog. "This might do the trick. A pup to lick his face. But they probably already have dogs. Most folks do."

"They used to have an old hound, but I think it must have died."

"Poor boy. Losing his dog and his pa." Imogene cut around the dog-and-boy picture. "My mama cat just had kittens. Jenny Sue might let him keep a cat in the house till he gets better from whatever is ailing him."

"I'll tell Jenny Sue." Tansy hung her coat on a hook inside the door and pulled a chair up to join the others. "His uncle Caleb has come home, so that might be a better fix than a kitten or a pup."

"About time he came back," Rayma muttered.

"He had a job with the Corps. Maybe he couldn't just up and quit." Tansy took up for Caleb. "The same as none of us would want to quit our librarian jobs."

"True enough." Mrs. Weston pushed her glasses up on her nose. Her gray hair was in a twist on the back of her head. She looked like a schoolmarm and had said she considered being a teacher once. But family kept her at home. "If you get a job these days, you have to hang on to it. And I have a surprise for you all today."

Tansy looked over at her. "Oh? What's that?"

"Well, there's this works project for creative people like writers and artists that started up about the same time as our packhorse libraries. I guess hard times put all sorts out of work. So they started a Federal Writers' Project."

"What do they do?" Rayma asked. "Write books we can take out to our people?"

"No," said Mrs. Weston. "We'll have to keep relying on donations for that."

"One of my people wants some quilt patterns," Tansy said. "Do you think women on our routes would share so we could put together a book to loan out?"

"Some surely would and maybe recipes too." Imogene sounded excited.

"Great ideas, girls," Mrs. Weston said. "But back to this Federal Writers' Project. They want to do a guide highlighting places or stories that might interest strangers coming through."

"Do we have any of those kinds of places?" Rayma asked.

"The town is named Booneville because Daniel Boone camped here," Mrs. Weston said. "People would be interested in that."

"Are they picking a writer from here?" Tansy asked.

"No, they sent in a writer," Mrs. Weston said. "He got to town yesterday and will be coming by here anytime now. I told him we'd help by taking him to some people who might talk to him. Somebody like Preacher Rowlett."

"Preacher Rowlett might talk his ears off." Rayma laughed and looked at Tansy. "He lives out near you, doesn't he, Tansy?"

Tansy nodded.

"There's Miss Odella over around Cow's Creek," Imogene said. "Takes me forever to get away from her house when I stop there."

When Mrs. Weston looked toward her, Imogene shook her head. "I don't have time to take anybody up there to her house and sit listening for hours to stories I've already heard. I've got my children to see to."

They all looked at Tansy. Mrs. Weston said, "Cow's Creek isn't so far out of your area, Tansy. And I'm sure we'll come up with other people too. Maybe Perdie Sweet."

"You think she'd talk to some outlier come in here asking questions?" Rayma looked doubtful.

"Maybe if Tansy sweet-talked her into it," Mrs. Weston said. "She was cross as an old bear when I was by there yesterday, but she might enjoy a listening ear. She wouldn't admit it, but I think she needs help right now. When I ride back up there to check on her, I can ask if she'd be agreeable to talk to this writer."

"She's some related to you, isn't she?" Imogene went on without waiting for Tansy to answer. "A person has to take care of family no matter the burden. And Aunt Perdie could be a burden."

"So, it's decided." Mrs. Weston smiled at Tansy. "We can count on you to help this gentleman find people to tell him about the area."

"I suppose." Tansy didn't remember agreeing to anything, but whatever Mrs. Weston wanted her to do, she'd give it a try.

"You won't have to stay with him all the time. Just make introductions so they'll be easier talking with him," Mrs. Weston said. "Folks can be suspicious of strangers showing up on their front stoops. You know how it's been a challenge to get some of the people to take our books."

"That's getting better," Rayma said. "Old Mrs. Johnson told me yesterday that book she finally agreed I could leave with her was right entertaining. And could she have another one?"

"That's wonderful." Mrs. Weston beamed. "I knew once we got started, people would love the books. They will probably like telling this writer about things too."

"Have you met him?" Tansy asked.

"Not yet. I got a note from the head of the project here in Kentucky asking us to help last week and then a note from the writer this morning saying he'd come meet with us today."

"What's his name?" Rayma asked.

"Damien Felding."

"That sounds like a name right out of a book," Imogene said.

"It does, doesn't it? Damien Felding." Tansy let the name roll off her tongue just as the door pushed open to let in a burst of cold air and a stranger.

"Did I just hear my name?" the man said. "If so, here I am. Damien Felding in the flesh. Or rather, wrapped up in a wool coat. It's nearly as frigid as New York out there."

He snatched off his felt hat to give a little bow. He was tall and slim with dark hair smoothed back, without a strand out of place. His name not only sounded like it belonged in a book, he looked as if he could have just stepped out of one of the stories Tansy read. A man with aplomb.

nine

THE WALLS OF THEIR LIBRARY ROOM seemed to expand to hold the energy Damien Felding brought in with him. Without invitation, he shrugged off his coat to come sit at the table.

"Not many books for a library," he said.

Rayma looked ready to fight those wildcats as she gave the man a hard look.

"Nothing like you might have in the cities, I'm sure." Mrs. Weston didn't sound bothered by his remark. "We depend on donations to fill our shelves. And those donations are divided up between all the packhorse libraries here in Eastern Kentucky."

"I didn't mean any criticism. Just noting a fact," he said. "Perhaps you should have a book drive."

"Good idea," Rayma muttered. "Wonder we hadn't thought of that."

The man looked at Rayma, seemingly unaware of her irritation. "That's why you need fresh eyes examining a situation sometimes."

That was more than Rayma could stand. She looked straight at him across the table. "Your fresh eyes keep on examining situations around here like that, the only thing you're going to find is people pointing you out of town."

The man looked taken aback by her words. "Do forgive me, madam. I fear I've gotten off on the wrong foot with you. Are you not receptive to new ideas?"

Before Rayma could respond, Mrs. Weston lightly touched her arm. "Why don't you look through our books to see if you can find some to interest the people on your route, Rayma? Mr. Felding is simply trying to help."

"I was. For certain." He looked at Rayma as she stood up from the table. "Rayma is a beautiful name."

Without bothering to answer, Rayma turned toward the bookshelves.

Damien shrugged one shoulder and looked at Imogene and Tansy. "So am I going to find out your lovely names?"

Tansy's heart beat a little faster when he smiled directly at her. His light brown eyes held as much smile as his lips, but she wasn't sure the smile was for them or at them. She wanted to reach up and push back some unruly strands of her hair, but she kept her hands on the table.

Mrs. Weston made the introductions. "This is Imogene Wilson and Tansy Calhoun. Rayma Boggs is choosing books for her readers. Are you a reader as well as a writer, Mr. Felding?"

"One can hardly become cognizant with words without being widely read."

"What kind of writing do you do?" Tansy asked.

"General interest articles for a magazine that went broke in this economic downturn. So here I am. But I aim to write a novel someday." He looked over at the shelves again. "And see it on shelves like these. Perhaps the mountains will give me inspiration for that book."

"Did you choose to come here with the Writers' Project?" Mrs. Weston asked.

"I won't say it was my first choice. Somewhere in the south

would have been nice during these winter months." He flashed his smile at Mrs. Weston. "But I was glad to get an assignment with the project wherever they sent me."

"What are you expected to write?" Imogene spoke up for the first time. She had continued to paste pictures on her book pages while they talked.

"Guides to the area. I am to describe how people can get to Booneville and then what they might want to see here. But also, especially in rural areas where there aren't a lot of tourist attractions, I'm to record legends. People's life experiences. That sort of thing."

"And will you then put that in a book?" Imogene asked.

"Not me. I'll turn over what I write to a central location here in Kentucky where others will look over all the submissions and determine what to publish. I'm covering several counties. Not just Owsley."

"Have you already written guides for other places then?" Tansy asked.

"I've been to Bloody Breathitt. Found out about the feuds there. Do you have a history of feuds here in Owsley County too?"

Tansy and Imogene looked at Mrs. Weston. Feuds were part of the history in nearly every town in Eastern Kentucky. Some Booneville families were part of the Clay County feud in the 1800s.

"Some," Mrs. Weston said easily. "But as you say, that's history. Nothing like that in our modern day."

"But the history of the feuds is fascinating," Damien said.

"Inspiration for that book you want to write?" Rayma turned to say.

"Novels need plenty of action. I'm sure you've found that to be true with the books your readers like best." Damien didn't seem to notice the challenge in her voice. "But I'm here to write for the Federal Writers' Project. So you needn't worry.

The project is not trying to expose bad things about a place but rather highlight the good. Open up an area to the rest of the country. For instance, you here in Eastern Kentucky might be interested in knowing more about New York City, and should you decide to visit there, a guide would be helpful."

The thought popped into Tansy's head of walking those big city streets with a tour guide like Damien Felding. She stared down at her hands to hide the blush climbing into her cheeks. Whatever had made her imagine such a thing? Better to think about how she was going to find people for him to interview and then keep him from getting off on the wrong foot the way he had with Rayma. That might not be an easy task.

"Do you have a horse?" Mrs. Weston asked.

He had a puzzled frown. "I have a car."

Rayma snorted but didn't say anything.

"That will be fine if you only want to talk to people here in Booneville," Mrs. Weston said. "But if you want to see more of the county, you'll need a horse. You might rent one from Mr. Hardesty at the livery stable."

"Rent a horse?" He looked surprised at the idea.

Tansy glanced up to ask, "Can you ride?"

"I'm sure I can." He waved his hand as though riding was of little concern. "How hard can it be to sit on a horse and point it to wherever you want to go?"

"Somebody is liable to find out how hard the ground is," Rayma muttered.

Mrs. Weston pretended not to hear her. "It's not always quite that easy. You might be wise to inform Mr. Hardesty you aren't an experienced rider when you see if he has a horse you can use. Or a mule."

"A mule?" Damien shook his head. "My friends would never stop laughing if they hear I'm riding a mule."

"I use a mule to make my rounds," Imogene spoke up. "Mules are perfect for our mountain trails."

"You don't say? Still, I think I'd prefer a horse. Do you think I could get riding lessons?"

"Tansy will be taking you to meet and talk with some local people. She might give you some pointers." Mrs. Weston looked from Damien to Tansy. "But you need to keep in mind that our hilly terrain can give even experienced riders trouble at times."

"I'm a fast learner." Damien's smile was back as he looked at Tansy. "So you're going to be my teacher and my guide. Wonderful. When do we start?"

"I . . . I don't know." Tansy looked at Mrs. Weston.

"How about now? You're here. I'm here. No time like the present."

Before Tansy could think what to say, Mrs. Weston spoke again. "While Tansy is being kind enough to volunteer her time to help you, she still has her duties as a packhorse librarian." Her voice wasn't quite as friendly. "Besides, you will need to obtain the use of a horse. Or a mule. Until Tansy is available to go, you can interview people here in Booneville. I'm sure Judge Milner or others at the courthouse will be glad to talk to you about our county's history."

"Maybe some of them will know about the feuds."

"That could be. Any feuds here happened so long ago, what you might hear now could be nothing more than rumors." Her voice sounded even cooler.

"Plenty of people were ready to talk about feuds in Breathitt County."

"Then perhaps you should have stayed in Breathitt County." Mrs. Weston's smile didn't involve much of her face.

"I'll get back there eventually. First, I want to see the place where Daniel Boone did some surveying and camping. I'm told the old explorer even carved his initials in a rock somewhere around here."

"Judge Milner can tell you all about that. The site of Boone's encampment is only a little south of town." Mrs. Weston sounded doubly polite.

"Well then." The man stood up. "I'll leave you ladies to your work and go see if I can obtain a suitable mount to head up into the hills." He looked over at Tansy. "Whenever you can make time for me, I would be most appreciative."

"If you get a horse, I could take you to Preacher Rowlett's house on my way out of town this afternoon. Then you will know how to get there to follow up with him tomorrow. I could meet you at his house to take you around to other areas on Saturday." Tansy didn't smile. She wouldn't want him to think she was too eager, even if the thought of riding out of town with him did have her heart bumping around in her chest. "If that suits you."

"It seems I'm at your mercy. So—"

Mrs. Weston interrupted him. "Not at all, Mr. Felding. I'm sure you could find others here in town willing to guide you."

"No, no. I'm very happy with a packhorse librarian guide." He smiled at Tansy. "Especially such a pretty one. This will work out perfectly."

When heat warmed Tansy's cheeks yet again, the man laughed. "Good day, Mrs. Weston, Miss Imogene, and Miss Rayma. And Tansy, I will see you later."

He put on his coat and hat and was out the door before any of them could say so much as goodbye.

Tansy blinked. Had she heard him right? Had he actually said she was pretty? Sitting there with stray hairs escaping her braid, wearing wool britches and a black sweater that had seen better days. Nobody had told her she was pretty since Jeremy first came calling before a prettier girl enticed him away. Pretty didn't matter anyway. What was in the heart and mind, that was what mattered.

Rayma frowned toward the door. "Was he for real?"

Tansy was wondering the same thing, but not exactly the way Rayma was.

Mrs. Weston shook her head a little. "Flatlanders are different."

"You mean rude and highfalutin?" Rayma said.

"Well." Mrs. Weston seemed at a loss for a response.

"I suppose you think I could have held my tongue better, and you're right." Rayma sighed. "I've never been good at that. Has gotten me in trouble more than once." She looked at Tansy. "But I can guard my tongue and be the one to show him around instead of Tansy. He wouldn't be trying to sweet-talk me."

Mrs. Weston gave Tansy a long look. "He was a bit forward acting. I wouldn't want to put you in an awkward position, Tansy." She looked worried. "Or compromise your reputation. Perhaps we can find a man to show him around. Preacher Rowlett might enjoy helping him."

Tansy couldn't argue against the wisdom of that, even though she didn't want to give up the chance to be around Damien Felding, someone from a big city with that aplomb. The idea of talking to a real writer, someone who actually intended to write a novel, was exciting. Tansy had wondered about writing something more than the odd poem or story. An actual book. But it seemed a dream too big for a mountain girl like her.

Instead of saying any of that out loud, she said, "I can ask Preacher Rowlett if I take Mr. Felding up to talk to him this afternoon. If he finds a horse."

"And can ride it," Rayma put in.

"Very well. That's settled." Mrs. Weston looked around the table. "Now we have library work to do. If we all make a picture book today, that will add to our collection. Plus, we need to repair any books that came back in bad condition."

Imogene pointed toward a box on the table. "The judge's wife gave me her old greeting cards and postcards. We can make bookmarks from them for folks to use so they won't dog-ear the pages."

Mrs. Weston picked up a card out of the box. "Perfect. We have ribbons and string over there on our hodgepodge shelf we can use for tassels."

Tansy started looking through the cards.

"You'd better gather up the books for your rounds before you start on this," Mrs. Weston said. "So you can be ready if Mr. Felding comes back any time soon. Loaning you to him for part of one day won't hurt our library work that much. Josephine might get in this afternoon to help here."

"All right." Tansy took her saddlebags over to the shelves to unload her books. Picking different books to take out to those on her rounds was one of her favorite tasks. Each book she chose made her think of a different person who might like it. They had such few books, they were only able to leave one book per reader and three per family. She had broken that rule at Jenny Sue's house, but then that was two households if she counted Ma Vesta and Caleb as a different family. She hadn't run out of books. The books and magazines she picked up from one house, she could then loan to the next.

After she had her saddlebags packed again, she went back to the table. The others were talking books now. Which they'd read. Which magazines had been the most wanted. Gentle talk wrapped around them as they worked. And all the while Tansy kept listening for the door to open and Damien Felding to blow in again.

ten

WHEN DAMIEN FELDING CAME BACK that afternoon, he brought an armload of magazines and newspapers to spill out on the table.

He smiled at their surprised faces. "I thought I'd start off your book drive. Well, not books, but magazines and papers I had in the car. I don't need them, so I'm donating them to your library." He looked straight at Rayma. "I should have been less direct in my comments this morning. I do hope you will forgive me." He turned to Mrs. Weston, then Imogene and Tansy. "That you will all give me a second chance to be a friend and not an adversary."

"We never thought of you as an adversary," Mrs. Weston said. "Directness is not necessarily a fault."

"It is if it causes ruffled feathers." His smile got broader. "Ruffled feathers. Does that sound like country speak, Miss Rayma?"

"I'm guessing feathers can be ruffled up country or city, and I shouldn't have let my feathers get ruffled so easy." Rayma gave him a little smile. "I have as much to learn about getting along with those brought in as you brought-in people need to learn about us."

"Brought-in?" he asked.

"People from other places," Tansy explained.

"Sorry. Mountain talk," Rayma said.

"Don't be sorry. I love it." Damien looked from Rayma to Tansy. "As long as I have an interpreter. I did get that horse. No mules available. But Mr. Hardesty assures me I'll the same as think I'm in a rocking chair aboard Belle. And who can't ride a rocking chair?"

Tansy looked at Mrs. Weston, who nodded. "Then I can get my horse and show you the way to Preacher Rowlett's."

Damien walked with her to the livery stable, leading his horse. "No sense making her tired before we start."

"Have you ridden her yet?" She draped her saddlebags over Belle's saddle since the writer didn't seem in any hurry to mount up.

"No. Seemed easier to lead her along so she could get to know me before I climbed aboard."

"That won't be a problem." Tansy stroked the mare's neck. "Belle will be used to different riders."

"Yes, well, maybe I need to get used to her." He gave the mare an uneasy look.

As they walked on, Damien asked, "This man you're taking me to see. Did you say he's a preacher?"

"Sort of. He doesn't have a church, but he steps in when no regular preachers are around."

"Don't you have regular preachers? Churches?"

"There are churches here, but none up where I live. Sometimes a preacher comes in to preach at the schoolhouse or wherever they can gather the people to listen."

"Brought-in preachers." He shot a smile over at Tansy.

"I guess so." Tansy had to laugh. "But as for Preacher Rowlett, folks just gave him that name because he knows Bible stories."

"I hear they call you packhorse librarians 'book women' because you know books. Do you like being a book woman?"

"Oh, yes. I love it. To get paid for sharing books with people.

What could be better than that?" She turned the question back on him. "How about you? Do you like writing?"

"Most of the time." He looked over at her and then away down the snowy street.

She waited for him to say more, but when he didn't, she said, "But not all the time?"

"I don't think anyone can like what they do all the time. You surely have days when you aren't as eager to ride out with a load of books."

"No, I want to take the books out. People expect me."

"Admirable. At the same time, I have to ask how long you've been a packhorse librarian?" He raised his eyebrows at her.

"A few months," Tansy said.

"Then it could be the new hasn't worn off and later on you'll get weary of loading up books."

"I don't think so."

"Maybe not, but I admit to sometimes being weary of writing. I do like to write. I want to write, but at times, I wish I could write stories of my choosing instead of the choosing of others. That's when motivation to come up with the best words lags."

"Is that how you feel about being here in Owsley County? Something you're only doing because of the Project paying you?"

"Getting paid for writing is always good. Plus, I have the bonus of experiencing new things. Like riding a horse and meeting new people. One especially delightful packhorse librarian willing to escort me up into the hills."

His smile sent tingles through her all the way to her toes. Tansy tamped down the feeling. The man was simply being nice. He'd smiled the same way at Rayma and Imogene. She needed to get her imagination under control. While he might look like he stepped out of a novel, she definitely wasn't a character in a storybook.

Damien's mare was every bit as safe as promised. The writer

sat stiff and straight in the saddle, gingerly holding the reins with one hand while keeping a death grip on the saddle pommel with his other hand. Belle plodded along and paid no heed to the man's jerky movements.

As they rode up into the hills, the horses' hooves in the snow and the wind whistling down through the pines made it hard to talk. Besides, Damien appeared to need all his attention to stay on his rocking-chair horse as the terrain roughened. But he managed to keep his seat and not find out how hard the ground was. A smile sneaked out on Tansy's face as she thought that might disappoint Rayma.

She doubted the writer would get off on the wrong foot with Preacher Rowlett. She'd never known anybody to have a harsh word to say about Preacher. Except Aunt Perdie. But then Aunt Perdie had harsh words about everybody. That was just her way. If Tansy took the writer to talk to Aunt Perdie, she'd have a chore smoothing down that introduction.

Tansy looked at the sun. Even as slow as they were going, she should still have time to go on to Aunt Perdie's from Preacher's house. She could fill up her woodbox again and look to see if the woman had any meal fixings. Tansy should have thought to take her a poke of food. Maybe Preacher would have something she could carry up to her. He generally had a full cupboard.

Aunt Perdie and Preacher Rowlett were about the same age, both in their sixties. Pa said once that he thought Aunt Perdie was sweet on Preacher when she was younger, but if so, Aunt Perdie had been too ornery to catch his eye. Too ornery to catch any man's eye. Then Pa had narrowed his eyes on Tansy and told her she should take notice of what orneriness could bring on a person. That was after Jeremy married Jolene.

Tansy had bit her lip to keep from saying she knew plenty of ornery married people. Then again, maybe they weren't ornery until after they stood before a preacher to promise forever.

Preacher Rowlett never married people. He said he could pray over them when they were sick, point them at the right Scriptures to help them through troubling times, and speak words at a burying, but he wasn't fit to join two people in matrimony. He claimed he'd never actually felt a calling to preach. Other folks saying he had a calling didn't matter. That had to come from the Lord, and so far the Lord hadn't called him.

Damien Felding interrupted her woolgathering. "Are we about there? It's cold on this horse."

"Not far now," Tansy said. "Just down the creek a ways."

She guided Shadrach into Mad Dog Creek, which was flowing easy now. When the snow melted in a few days, the water might rise again. She looked back at the writer, who yanked on the reins to stop Belle in her tracks.

"Do you have to ride out into the creek to give your horse a drink?" he asked.

"It's one way. But the creek is our trail for a little ways here."

"The creek?" He looked more than a little uneasy.

"Creeks make the easiest trails. Better than fighting our way through the low-hanging pines here." She pointed toward the trees on both sides of the creek. "Just give Belle her head."

"How do I do that?"

"Loosen the reins."

"But I might fall off."

"Probably not." She managed not to smile at his worried look. "Remember, just like riding a rocking chair. All you have to do is keep your feet in the stirrups."

"A challenge a minute for this greenhorn." When she did have to smile then, he went on. "Trust me. I'd be smiling too if my face wasn't frozen."

"You fall off in the creek, you'll freeze for sure."

After Belle stepped down into the water, Tansy flicked Shadrach's rein to move on. She heard the mare following and hoped she wouldn't hear a big splash, although that would be sure to make Rayma laugh.

When they came in sight of Preacher Rowlett's house, his bluetick hound let out a volley of barks. Preacher stepped out on the porch to put his hand on Rube to stop his noise. Then he smiled out at them. "Company. A surprising blessing on a snowy day like this. Especially if you brought books, Tansy Faith."

"You aren't on the route until Friday, Preacher, but you can go ahead and get something new today. I've got last month's issue of *National Geographic*."

"Last month's? I didn't think you ever had one that current."

"A donation from our visitor, Damien Felding." She gestured toward the writer. "He's here with the Federal Writers' Project to gather information about our county for a guidebook."

"Well, sir, welcome to my mountain home." Preacher Rowlett was a short man, not much taller than Tansy, and appeared to have never gone without food. He had an ample waistline and a plump face that showed only smile wrinkles. "Tie your horses up and come on in to warm by the fire and let me know what I can do for you."

Tansy dismounted, but Damien stayed on Belle. When she looked up at him, he said, "I'm not sure getting off this animal is exactly like standing up from a rocking chair."

"I guess not." Tansy had to laugh. "Wait a minute. I'll hold her for you."

After tying Shadrach's reins to the porch railing, she grasped Belle's bridle. "All right. I have her."

"That's wonderful, my dear girl," Damien said. "But then what?"

"Then you get off the horse."

"Easy for you to say from down there on the ground." He looked from her to Preacher Rowlett. "I fear I'm totally inexperienced with horses. I'm more comfortable behind the wheel of a car."

"No cars up this way. I'll handle this, Tansy." Preacher Rowlett stepped over to the edge of the porch. "All right, sir. Just slip your feet out of the stirrups. Now hold onto the saddle or grab a handful of mane while you lift your leg over the horse's haunches. The back of the horse. Then drop down to the ground. Bend your knees a little to land easy."

The writer looked awkward doing it, but he landed on his feet with only a little stumble.

Preacher Rowlett smiled. "Practice will make it easier."

"Practice. To be frank I don't care if I ever ride a horse again." Damien rubbed his hands down his backside.

"It would be a long walk back to town shank's mare," Preacher said.

"That's on foot," Tansy interpreted for him. She looked up at the sun with a frown. "You won't have long to talk if you want to retrace our route back to town before dark. Do you think you can find the way if night falls?"

"By myself? I'm not sure I could find the way in daylight. Aren't you going to escort me?"

"I need to go on home."

"Worry not, sir," Preacher Rowlett said. "You can spend the night, and in the morning, I'll point the way. I'd go with you, but I seem to have picked up a cough that wouldn't be improved by a trek to town. Rube and me, we need to stay right here in my warm cabin."

"Rube?" Damien looked around as he went up the steps.

"My faithful hound." Preacher Rowlett touched the dog's head again, then opened the door to let Rube go in first.

Tansy didn't follow him up the steps. "If you're staying the night, I'd better take care of your horse."

Preacher Rowlett turned around. "A man takes care of his own horse."

"Of course," Damien said. "If said man knows how to do that."

"Come on. I'll show you." Tansy took the reins of both horses and started toward the barn.

"I'll make some tea while you see to the horses," Preacher Rowlett said. "Don't you head home without leaving that magazine, Tansy Faith."

"Tea." Damien caught up with Tansy. "No coffee?"

"The tea will be warm." Tansy opened the barn door.

Preacher's other horses looked out of their stalls as she led Shadrach and Belle in. Shadrach neighed and she turned him loose to go nuzzle the neck of another horse. Preacher knew horses. His Morgans were prized in these parts.

Damien followed her in. "Warm would be nice. So what do we do to horses after we ride them? Take off their saddles, I presume."

"That and brush them down." She pointed toward the cinch strap. "But saddle first. Just unbuckle that strap there. Did you saddle Belle at the stable?"

"No, the man there did it for me. Said he didn't want to have my demise on his conscience if I did it wrong and the creature threw me off into a ravine."

"Then you'd better get Preacher Rowlett to help you saddle her in the morning. To make sure you get things right."

"Important to get things right." Damien unbuckled the cinch and lifted off the saddle. He placed it on a plank bench and came back to watch Tansy rubbing down the horse. "Will I be safe spending the night here?"

"Safe?" She frowned at him over the mare's back. "You'll be fine. You'll have plenty of time to hear his stories."

"Don't you want to stay and hear them too?"

"That would be good, but I need to go check on a relative a little ways from here before I head home. Ma would worry if I didn't come in."

"Still mama's little girl?" He had a teasing voice. "You don't look that young."

"Young or not, I don't want to worry my mother."

"Uh-oh. There I go ruffling feathers again." Damien made a face.

"I don't have feathers to ruffle." Tansy picked the mare's hoofs, then led her to an empty stall. She might not have ruffled feathers, but just being there alone with him in the barn was knocking her off-kilter.

She didn't remember ever feeling like this with a man. Unsure of how to act. The day before in Caleb's barn, talking with him was easy, as natural as buttering a biscuit. But then Caleb was an old friend. This man watching her with curious eyes was a stranger. She didn't know him. He didn't know her.

"If you did, those feathers would be bright and beautiful." He stepped closer to her when she turned away from the stall.

Tansy stood still a few seconds while her heart did a jig before she remembered Mrs. Weston's concerns about Tansy's reputation and stepped away. "That sounds like a line out of a book. Maybe something Mr. Darcy might say."

"A Jane Austen reader, I see." He laughed. "I doubt Mr. Darcy ever said that, but maybe I should jot the line down to save for my own book someday."

"Maybe you should." She lifted the saddlebags off Shadrach to give his back a rest. No need unsaddling him. They'd have to leave as soon as she took Preacher Rowlett the magazine. She pulled it out of the saddlebags and glanced over at Damien. "Preacher Rowlett will wonder where we are."

"Sometimes I wonder where I am." He looked around before he followed her out of the barn. "Far from home. Certainly that."

eleven

PERDITA SWEET HAD DEALT with bad times aplenty in her years, but she wasn't sure she'd ever had a day quite as bad as this one. Not even when her ma died, because Ma was more than ready to step over to the other side by the time she breathed her last.

But this day had gone wrong from the moment she'd opened her eyes that morning and saw the girl in the bed beside her. Hadn't none of it been a dream. Coralee Embry had her eyes closed, but she wasn't sleeping. Perdita knew that right away. The girl was too stiff. Something was hurting her. Guess that was probably natural with a load of baby in her belly.

Prissy was settled down between them, looking content as a flea on a shaggy dog. Perdita had hated disturbing the cat, but a woman her age couldn't put off using the slop jar just to placate a cat.

She hadn't begrudged the girl jumping up as soon as she realized Perdita was awake to get to the slop jar first. She had a lot pressing down on her bladder, and anyway Perdita needed to sit on the side of the bed a while to get her knees working. Rheumatism was not a good sleeping companion.

The girl did carry the slop jar out to empty it later. So that was good. She'd built up the fire and offered to share the last bit of food she'd brought from her house. But Perdita pretended she wasn't hungry. The girl was eating for two. She needed the biscuit more than an old woman who never had the pleasure of eating for two or the joy of making a baby.

Not that she would have ever got carried away like this girl must have done. Perdita would have made sure somebody stood before a preacher with her first. But weren't no use bashing the girl about that now. What was done was done. Her pa had bashed her enough already. Like he'd never done nothing wrong. Men didn't get caught from those kinds of wrongs if'n they were willing to own up to being the seed planter.

She hadn't minded skipping breakfast, seeing as how she was thinking on that chicken soup she was going to make out of poor Clementine. The old hen wouldn't be the best eating, but it would be food on the table for a couple days at least. By then maybe Hanley would show up with some vittles. Or that Tansy might come back the way she said, and she could impose on her for some fixings, what with this girl in the family way here. It wouldn't be like she was begging for herself.

But when she did finally get on her coat and boots and head out to the henhouse with her axe, all she found were feathers. Clementine's feathers. Some old fox must have been hungry too. The other two hens had scattered to who knew where. The fox could have found them out in the woods and had a feast, but whether it did or not, Perdita wasn't likely to find them.

The girl hadn't come out to the henhouse with her. Perdita left her inside heating water for the feather plucking once Perdita did the killing. But naught but feathers were left of their chicken soup. She peeked in the nests just in case the hens had laid an egg before the fox came calling. Nothing there but the old doorknob left to trick the hens into thinking they were sitting on an egg with the need to add more to the clutch.

The cow had wandered off too. Looking for grass where the snow wasn't so deep, she supposed. Not that she was thinking on butchering it. She couldn't have killed the fatted calf even if she had one. Too old for that. Too old for most everything.

But not too old to be hungry. She pretended she wasn't when she gave Coralee the last of the meal to make her some poor fixings of mush. Supper would have to be watery soup out of those withered potatoes in the 'tator hole unless one of the hens came back. Or Hanley showed up.

Could be the Lord would somehow provide. Wasn't that what preachers were always saying? If a person was faithful and lived right, the Lord would provide. Then again, there was that problem with living right to earn the Lord's favor. Perdita often as not was prickly as a cocklebur. That kind of crankiness was surely a sin.

She tried to remember if Hiram had ever said that about the Lord providing, but she couldn't recall it if he had. He was big on praying, and hadn't she had answers to her prayers just the night before, with the girl showing up to want to be the same as family for a while and then having a little poke of food besides? Maybe she hadn't done enough praying when she got up, because the day kept going from bad to worse.

So bad that when Prissy brought in a bird she must have snuck up on in the snow, Perdita considered snatching it and roasting it over the fire. Hunger could make a body think some weird things.

Make them do some foolish things too. Like build the fire up too high in the old fireplace. It just hadn't seemed right to be hungry and cold both. But she knowed better. Some of the chinking had fell out of the old chimney, so it was a bad thing to let the flames shoot up too high. But oh no, a sweet little fire hadn't been good enough for her.

Above the mantel caught first, the newspaper covering the wall curling black. Perdita stared at it a minute, not wanting to believe what she was seeing. Coralee grabbed the water bucket and threw it on the wall, but that didn't do nothing but splash water back on the poor girl.

She looked at Perdita with big round eyes and said something, but Perdita couldn't take in her words. Instead, she watched the flowery drawing around a poem she knew by heart from reading it a hundred times on that spot of the wall disappear into black ash.

Coralee grabbed Perdita's carved figures off the mantel and the quilts off the bed and piled them in the best rocking chair to tug out the door. The air swooshed in through the door to give the fire a boost.

Perdita gave the girl little notice even when she came back to yank on her arm. Funny how a body could be standing in the middle of a whirlwind of noise and hear nothing but silence. Still, deep silence.

The flames licked out onto the ceiling. The smoke rolled down around her and Prissy clawed up her skirt. Perdita blew out a breath and let the noise come into her ears again. She wouldn't want Prissy's tail to catch fire. She picked up the white bird book and her mother's sewing basket and let the girl pull her out the door into the snow. Coralee grabbed their coats and boots on the way out.

The girl draped the coat around Perdita's shoulders and ran back inside. She came out with a stack of pans and a butter churn. Pans with nothing to cook in them, although Perdita supposed there was fire enough. She almost laughed but stopped herself. She didn't want the girl to think she'd gone into hysterics.

Perdita looked around. No Hanley coming up the hill to see what was burning. The man must have died. That was all that explained it. He'd about worried her to death coming when she didn't need him to, and now he was nowhere to be seen. People did die unexpected at times.

She might just sit down in the snow and die herself. Seemed

the easiest thing. Everything gone. While the cabin hadn't been much, it was all she had.

Coralee started back for another load of who knew what, but Perdita grabbed her arm. "'Twouldn't be good for the baby. All that smoke."

Tears coursed down through the black ash on the girl's cheeks. Poor child. Thrown from her home and thinking she'd found refuge with Perdita. Refuge Perdita had been willing to give, such as she could. Empty pots and a house with flames licking out on the roof. No refuge now.

"What are we going to do, Aunt Perdie?" She was holding her stomach and breathing hard.

Aunt Perdie. She weren't her aunt. No kin at all, so far as she knew, but somehow she liked how the girl said "we." Of course, the girl could have been meaning her and the baby she was carrying, but Perdita didn't think so. She thought she was in that we.

"You take a calming breath first off, child. You ain't wanting to drop that baby before your time."

Coralee pulled in a shaky breath.

Perdita put a cold hand on her cheek. "I don't rightly know what we're gonna do, but the Lord does. He'll let us know in his good time."

"Could be we'll freeze first."

"No way to freeze. Leastways till the fire goes out." Again Perdita had to clamp down on the urge to laugh. Weren't nothing about none of it funny. She reckoned she should take a calming breath too. "Right now, me and you, we're gonna pray for the Lord's provision. You claimed to be a believer last night, didn't you?"

"Yes, ma'am."

"So you know how to pray."

"Ma taught me soon's I could talk." More tears spilled out of her eyes. "But you think the Lord will pay any mind to my prayers, what with how I didn't pay mind to some of his rules?"

"The Lord ain't one to hold our sins against us if we ask forgiveness and aim to do better."

Perdita couldn't believe she was standing here preaching a sermon while her house burned down. She was the one in need of a sermon to keep the hopelessness from overcoming her. Empty pans and no roof. She wanted to ask what the Lord was up to, and after she'd taken in the girl. An answer to prayer, she reminded herself. Even if it hadn't been the kind of answer she was expecting.

She didn't rightly care what kind of answer he was ready to give now, so long as it led out of the snow. Her feet were cold. The girl was shivering too, her teeth chattering. Maybe from the cold. Maybe from being scared about the fire and all. She was a right pretty thing. Something Perdita hadn't ever been able to claim. But she hadn't been too hard on the eyes when she was marrying age. Not exactly pretty like Coralee here, but not ugly either.

That hadn't been Perdita's problem with the men here in these hills. She'd just been too choosy. No need thinking about that. Any thought of marrying had washed on down the creek long ago. Praying for help was what needed doing now.

"Wrap one of those quilts around you, child. You're gonna shake that baby out if you don't stop the shuddering."

"Yes'm."

At least the girl was willing to do what Perdita told her. Once Coralee had the quilt around her shoulders, Perdita held out an arm. "Step closer and we'll huddle here to offer up that prayer."

"Will you say it for us?" Coralee sniffed. "The Lord maybe knows you better."

"The Lord knows us all and our prayers even when we don't speak them out loud."

"But will you speak it out loud? Hearing your prayer words will be a comfort."

"Then I reckon I can." Perdita paused a moment before she started talking to the Lord. "Lord, me and this child here are in an awful spot. No food. No roof. Just a rocking chair and a sewing basket with nothing to mend. I'd be mighty beholden were you to send us some help. If'n you do, I'll aim to be less contrary to them around me and this one here, this girl will work on being a good mother to this gift of a baby you've blessed her with."

Coralee started sobbing then and Perdita pulled her closer. So close she felt the baby inside the girl's tummy kick against her. The wonder of it made Perdita smile as she said, "Amen."

"Amen." The girl choked out the word, then relaxed against Perdita as she stopped crying. "Thankee for that, Aunt Perdie."

When Perdita didn't say anything more, Coralee asked, "And now what do we do?"

"We hold on and wait for what the Lord answers."

"Will he answer?"

"I'm thinking he will. If we give him time." Perdita tried to sound sure, but if the smoke from the fire didn't draw somebody to them, she'd send the girl down the hill to Hanley's house. Even if Hanley had up and died, his wife would be there. A roof for the girl would be there. And she and Prissy could sit here in the rocker the girl had pulled out of the house and wait for whatever came next.

She kept her arm tight around the girl and kept praying in her head. When a thank-you prayer for this gift of family rose up in her thoughts, she wondered at it and prayed it might be for more than this one day.

"'Wait on the Lord,'" Perdita whispered. "'Be of good courage, and he shall strengthen thine heart. Wait, I say, on the Lord.'"

"That's Scripture, ain't it?" The girl kept her voice at a whisper too.

"Out of Psalms, best I recollect." Perdita wished she'd thought to grab her ma's Bible. She looked up at the sky where the sun was shining bright as ever. Never seemed right for the sun to be so bright when things were going bad, but it weren't the sun's fault she built the fire up too hot. "We're waiting, Lord."

twelve

TANSY SMELLED THE SMOKE while climbing through the pines toward Aunt Perdie's cabin. Good. She hadn't run out of wood for her fire. Then they were out of the pines and a cloud of smoke darkened the sky. Too much for chimney smoke. Panic raced through Tansy as she kicked Shadrach's flanks. Please let it be the barn and not the cabin.

But the old barn leaned a bit to the side as always, while flames licked across the cabin's roof and smoke puffed out its open door. Tansy went weak with relief at the sight of Aunt Perdie wrapped in a quilt. A rocking chair sat in the snow and pans were scattered around her.

Tansy slid off Shadrach but kept a firm grip on the reins. The crackling of the fire made the horse anxious. She yanked off her scarf and wrapped it over Shadrach's eyes. That calmed him enough that she could tie him to a tree before heading over to Aunt Perdie.

"Afternoon, Tansy." The woman must have seen her riding up.

"Are you all right, Aunt Perdie?"

"I ain't quit breathing just yet, if that's what you mean. But I can't exactly say I'm all right, what with my home going up in smoke and my feet turning to ice." She glanced around at Tansy. "But I do admit I was some glad to see you coming up the hill."

Another head poked out of the quilt. A young girl barely taller than Aunt Perdie was sheltered under there. "I reckon I'm feeling glad too, Miss Tansy."

"Coralee! What are you doing here?" Tansy was beginning to think she was in the middle of a story she was making up to tell Livvy at bedtime.

"Freezing her feet same as me," Aunt Perdie said. "This girl on your book route?"

"No. I used to help her with her reading back when I was in school," Tansy said.

"You did. You were as good a teacher as Mr. Raymond, but I'm surprised you knowed me, that was so long ago."

"You haven't changed that much."

"She's done some changing since her little-girl days. And fixing to do more." Aunt Perdie peered over at the girl and then back to Tansy. "But if you're wanting to know why she's here, she needed a roof and I had one. At least, I did have one."

Tansy looked toward the fire. "I'm sorry, Aunt Perdie."

"Don't you worry none about that book you left me. I brung it out. It's right over there in Ma's sewing basket."

"I'm not worried about a book. I'm worried about you. Both of you." Tansy looked at Coralee. "You still live over on Hardin's Creek, Coralee? That's a good piece from here."

"It is for a fact," Aunt Perdie said. "Poor girl walked the whole way. She was nigh on froze when she got here and she's 'bout as froze now. Storytelling can wait till we get somewhere warm." Aunt Perdie looked past Tansy down the hill. "I'm reckoning Hanley must have died since he ain't up here trying to throw snow on the cabin."

"You think that would work, Aunt Perdie?" The girl started to move away from the old woman. "I could try it."

Aunt Perdie held her where she was. "Ain't nothing gonna help this old place now." She looked over at Tansy. "You hear anything about Hanley?"

"I did. I was by Preacher Rowlett's and somebody brought him news." Tansy hesitated to heap more bad news on top of what was already happening.

"He's dead, ain't he?" Aunt Perdie didn't sound surprised.

"I'm afraid so. His son told Preacher his father had been feeling low for a few days, but they hadn't expected him to pass on. They figure his heart gave out. Said his sister had taken their mother home with her, but they wanted Preacher to come say a few words to the family, even though they'll have to wait for the burying until the ground thaws some."

Aunt Perdie blew out a long breath. "I knowed it. What I can't figure is what the Lord must have been thinking. Taking a good man like Hanley with more years to count yet and leaving an old woman like me standing here in the snow."

"Don't say that, Aunt Perdie," Coralee spoke up. "The Lord knew I needed you to help me."

"I ain't gonna be much help with no house. No nothing. Don't even have no hens. Fox done got them." Aunt Perdie pressed her lips into a thin line and shook her head.

"Maybe we should find a way for you to go home, Coralee," Tansy said.

The girl's face turned white as the snow around them, except for the purple bruise on her cheek. "I can't do that, Miss Tansy."

Aunt Perdie put her arm around Coralee and pulled her close. "She ain't going back there just yet. She's my family right now. Leastways till something better comes along." Aunt Perdie narrowed her eyes on Tansy. "You is family too."

"I am. So you'd both better come home with me until that better happens." No other choice about that. "But getting you and these other things across the hill to our house might be a quandary."

Shadrach was already loaded down with her saddlebags of books.

"Ain't nothing here worth worrying over. 'Cept my mother's sewing basket and that book you lent me." Aunt Perdie waved her hand, dismissing the rest of the plunder.

"Your quilts and carved-out pieces," Coralee put in.

"Weather won't hurt them," Aunt Perdie said. "We can gather them up later. I'm some surprised folks haven't seen the smoke and come to investigate already."

Tansy was surprised at that too, but Aunt Perdie didn't have any near neighbors other than Hanley and his wife. Still, the smoke had to be visible a good way.

She no more than thought that when a man on a mule rode into sight. "Looks like somebody is coming now."

Aunt Perdie squinted her eyes toward the man. "That ain't your pa, is it?" she asked Coralee.

The girl shook her head. "He won't come after me. You don't have to worry none about that."

"Looks like Caleb Barton," Tansy said. He sat easy on his mule. Nothing like how Damien Felding had perched on Belle like an uneasy crow ready to take flight at the first bump.

"So he's back on the mountain." Aunt Perdie scowled at the approaching rider. "About time."

Tansy said, "He came home when he could."

"Not soon enough," Aunt Perdie growled.

"Coming home wouldn't change that tree falling on Reuben," Tansy said.

Aunt Perdie peered at Tansy. "You must be sweet on the boy the way you're taking up for him."

"He's a friend. And now he's here just when we need help getting you and Coralee somewhere warm." Tansy waved at Caleb, who stopped his mule a little away from the fire. "Right on time."

"I remember him," Coralee said. "A right handsome fellow."

"He ain't the one, is he?" Aunt Perdie said.

"No, no. I done told you the daddy of my baby is long gone."

"Baby?" Tansy gave Coralee a better look. The quilt didn't completely hide her baby bump.

Aunt Perdie noted the direction of Tansy's look. "She's getting ready to have my grandbaby. Family connection or not. Think your ma will have trouble with it?"

"She'll probably fight over which of you gets to be called Granny," Tansy said.

"A baby can have more than one granny," Coralee said.

"True enough," Aunt Perdie said. "So let's get going. Ain't a bit of use standing here watching every last thing I ever owned turn to ash. Ain't like the fire will spread, what with all this snow."

"A spark could fly over to the barn," Tansy said.

"I reckon it could, but what would we do about it if it did 'cept stand here and watch it burn too?" Aunt Perdie raised her eyebrows at Tansy, then bent to get the book to hand to her. "You best carry your book away from here instead of me. I'll have enough trouble with Prissy." She picked up Prissy and stroked her back. "You have any idee how a cat takes to horseback riding?"

Prissy fixed a green-eyed stare on Tansy, as if daring her to speak against her coming along. Tansy had no doubt Aunt Perdie and Coralee would be welcomed by her mother. Even in Coralee's expectant condition. Maybe especially in Coralee's condition. But the cat? Ma liked cats well enough as long as they stayed out in the barn to tend to the business of catching mice.

One thing sure, no way would Aunt Perdie want to leave Prissy here to fend for herself. But how to get the cat from here to there? Tansy looked back at Caleb. Maybe he would know what to do.

"I saw the smoke," Caleb said after he tied up his mule that didn't seem as nervous about the fire as Shadrach. "Got to say

I'm relieved to see you standing out here, Aunt Perdie. And looks like you've already gathered some help." He nodded toward Tansy.

"Some help's better than none, I reckon," Aunt Perdie grumbled. "But Tansy appears ready to let us stand here in the snow till our toes freeze off."

Tansy looked heavenward for patience. She might not be doing Ma any favors bringing Aunt Perdie home. The woman could be prickly. So much so, Tansy had to wonder why Coralee had come to Aunt Perdie's. She might have found help at other houses nearer to where she lived. But then, Aunt Perdie was claiming her as family. Kin was kin and neighbors were neighbors. Prickly or not.

"I was trying to figure out how to take you and Coralee and that cat all on Shadrach." Tansy looked from Prissy to Caleb. "You know how to carry a cat on a horse?"

"Not sure I do," Caleb said as he eyed the cat.

Prissy was being sweet as pie while Aunt Perdie stroked her, but Tansy didn't expect that to last. The cat wriggled out of Aunt Perdie's grasp and streaked across the snow to disappear into the barn.

"I ain't leaving here without Prissy," Aunt Perdie said.

"Even if your toes freeze off?" Caleb asked.

Aunt Perdie narrowed her eyes at Caleb. "You ain't got no call to be sassy. But froze toes or not, I ain't leaving Prissy here for the coyotes to get."

"Then I guess we better go catch that cat." Caleb didn't sound at all upset with Aunt Perdie's contrariness. "Or maybe if you call it, the cat will come to you."

"Prissy is a mite peculiar about doing as she wants," Aunt Perdie said.

"Why does that not surprise me?" Caleb held up his hands to stop Aunt Perdie saying anything. "I'm not being sassy. Just observant."

"Did you learn that off wherever you were when you oughta been at home?" Aunt Perdie frowned at him.

"I guess that could be." He smiled at Aunt Perdie. A good smile full of kindness.

Whether his smile worked on the old woman or not, it made Tansy feel better. She was more than glad Caleb had seen the smoke. "Caleb is trying to help, Aunt Perdie."

"Hmph." Aunt Perdie didn't appear to want to give up her huffiness.

Tansy gave Caleb a quick look, but he didn't seem bothered.

Caleb looked over at the quilts on the rocking chair. "Maybe we can wrap the cat up in a quilt to carry to Tansy's house. If that's where you're aiming to go."

"Ain't got nowhere else, don't seem like. Less'n I just sit out here in this rocker all night and wait for the sunrise."

"We'd get awful cold out here." Coralee's teeth were chattering. "Might not be good for my baby."

Aunt Perdie let out a long breath. "Well, if you won't go without me . . ."

"I won't," Coralee said.

"Me either," Tansy added.

"Same here," Caleb said.

"I ain't got but the one rocking chair for sitting," Aunt Perdie said. "So I reckon I'll have to let you folks have your way. But I ain't wanting to leave Prissy." Tears popped into her eyes.

Coralee hugged her. "We can catch her."

"Might get scratched in the bargain," Aunt Perdie said.

"I've got gloves," Caleb said.

"Prissy don't like menfolk."

"She won't like anybody corraling her in a quilt. You better let me do the catching so she won't hold that indignity against you." Caleb grabbed a quilt off the rocking chair and headed toward the barn. He looked back at Aunt Perdie. "Call her out into the open, so we can do what has to be done before those toes freeze."

There was that smile again that had Tansy breathing easier. Somehow they would get things done.

thirteen

CAT CATCHING wasn't something Caleb had ever tried. At least not since he was a kid chasing kittens in the barn loft. But once down in the Smokies, he helped round up a wild hog that was terrorizing their camp. Surely a cat couldn't be worse than that. Even if it was as crotchety as its owner.

Caleb had to smile thinking about Perdie Sweet. Not much sweet about the old lady, but he was sort of glad she hadn't changed while he was off the mountain. He liked some things being the same. Of course, she might be some changed if she was taking in that girl who appeared to be lacking the proper kind of family.

Then again, maybe Aunt Perdie was her actual aunt, even though Caleb had never heard about her having family around here. Poor old thing was hobbling along behind him now. Maybe her toes were frozen.

Tansy stepped up to help her on one side and the girl, whoever she was, on the other. He was surprised he didn't recognize the girl, but she didn't look more than sixteen. She would have been a little girl when he left the mountains. Whatever her story was, it appeared to be a sad one.

Not his worry. Other than getting them all somewhere

warm and safe. Including this cat that would probably screech like a banshee when he trapped it in the quilt.

His worry was figuring out how to let Tansy know he had never been interested in Hilda. Always only her. He was still sending up thankful prayers she hadn't found somebody to marry before he got back home. After Tansy was by their house yesterday, Ma had filled him in about Tansy's broken romance. Ma said Tansy was liable to end up like Perdie Sweet if she didn't watch herself. That wasn't likely to happen if he had anything to do with it.

He peeked over his shoulder at Tansy before he tugged the barn door open. Just the sight of her made his heart do a sideways beat or two, and when she smiled at him, his feet wanted to dance. While right now wasn't the best time to let Tansy know he aimed to come courting, he would tell her. He'd have a fine excuse to show up at her house after he gathered up Aunt Perdie's plunder from the snow. He would ask her to go riding with him Sunday afternoon. Ma could do her preaching at him in the morning hours. She was still pushing Jenny Sue at him as his duty. Another thing that wasn't liable to happen.

But first the cat. He pulled on his gloves. In spite of the glow of the fire slipping through the cracks between the logs, a black cat might still be hard to spot inside the barn.

Aunt Perdie stepped through the door and stopped, not making the first attempt to call the cat. When he looked at her, she frowned. "I told you Prissy ain't one to come less'n she wants to."

"How do you know she doesn't want to?" Caleb asked.

"She'll know we're up to something. A cat can read a body's mind if it takes a notion." Aunt Perdie's shoulders drooped and she sighed. "I ain't aiming to be contrary, though I know you're thinking that."

"No, ma'am. I wasn't thinking that about you at all." It wasn't much of a falsehood. "About the cat, maybe. Do you see her?"

"She's over there." Aunt Perdie nodded toward the corner.

"It wouldn't hurt to try calling her, Aunt Perdie," Tansy said. "I would, but Prissy doesn't like me."

"She likes me," Coralee spoke up.

"She does?" Tansy sounded surprised.

"Surprised me too," Aunt Perdie said. "But Prissy went right and sat in her lap, purring up a storm last night."

"She was probably smelling this scrap of bacon I just found in my pocket. I forgot I put it in there to munch on while I was walking here yesterday." Coralee held out her hand. "Let me see if that'll get her to come."

"It's worth a try." Caleb held up the quilt.

"I don't think you oughta do that with the quilt till I have hold of her," Coralee said. "Could be you'uns need to go wait outside. If I catch Prissy, I'll bring her out. Aunt Perdie can stay if'n she wants."

"I think we've been kicked out." Caleb smiled and followed Tansy out the barn door.

Behind them, Coralee called the cat. "Here, kitty, kitty. Good Prissy."

"I hope this works." Tansy looked at the fire and then at the sun sinking below the tree line. "If it doesn't, we'll have to hog-tie Aunt Perdie to carry her away from here without that cat."

"That's a chore I don't think either one of us would want to undertake." Silence fell between them as they listened to sounds from the barn. After a minute, Caleb asked, "Who's the girl? Kinfolk of Aunt Perdie?"

"Coralee Embry. Not actually kin, but Aunt Perdie is ready to claim her anyhow. Aunt Perdie says she showed up needing a place to stay."

"You wouldn't have thought she'd pick Perdie Sweet."

"That's for sure, but seems to be working for her." She looked back toward the fire. "Except for the cabin burning down."

"She run away from her husband?"

"From what she and Aunt Perdie were saying, I'm guessing she doesn't have one. And no ma to help her either. Her mother died a while back. She appears scared to go home."

"I see." And he did see. Poor girl. Finding a place and then losing it so soon. But she might find another place with Tansy's family. "How come you came back today?"

"I was worried about Aunt Perdie being able to get wood in out of the snow. Then when I heard Hanley Scroggins died, I had to come for sure. Hanley took it on himself to see to her." Tansy pressed her lips together and shook her head. "Sad about Hanley. Do you remember him?"

"I do. Ma always said he was more preacher than even Preacher Rowlett. That Preacher Rowlett did the talking but Hanley did the walking."

"I don't know that Preacher would like her saying that." Tansy frowned a little.

"Probably not, and he is a good man too. I haven't seen Preacher since I got back. He has to be getting on up in years. Something like Aunt Perdie here. How's he doing?"

"Same as always. Busy with his horses and loving these library books I take him. Says he was getting too old to trek over to Jackson to the library there. He wants to be first in line for any science books or newspapers we get in. Especially if they have to do with the chestnut trees."

"The chestnuts." Caleb looked away from Tansy toward the tree line. "I remember we used to talk about the trees. How it pained us to see them dying."

"Were there any down in those other mountains where you were working?"

"Plenty. They say they grow all the way down into Georgia. But they were dying in the Smokies, and a forest man said the blight was spreading south."

"What started it?"

"Forest people say it came over when people brought in

Chinese chestnuts. Doesn't hurt those trees. Only the ones native here." Caleb looked off toward the woods, imagining how they used to look in the spring when the chestnut blooms would turn the ground white with their drifting petals. "Nobody seems to know any way to stop it. They tried burning a break up north, but it didn't help."

"Preacher says it started in New York City. That seems a long way from here." Tansy was looking off at the trees too. "Have you been there?"

"Nope. I guess the Corps thought a mountain boy like me would be better working in the woods."

Tansy's gaze came back to him. "Were you sorry about that?"

"Not at all. I liked the Smokies. You should see them, Tansy." Caleb smiled at her. "It's a fine thing, standing high on one of those mountains."

"I'd love to do that."

"It's not so far away. Maybe you can someday." What he really wanted to say was maybe they could both go someday to tour the new park. Together. He did dare to add, "You could hike up some of the trails I helped carve out."

A yowl came from inside the barn as the door opened and Coralee carried out a poke full of cat. Blood was running down her hand.

"Are you all right?" Tansy asked.

"Just scratched a mite."

"Prissy didn't take to the poke I found in the barn," Aunt Perdie said. "Let's get going afore she tears her way out of it."

Caleb took the sack from Coralee. "I should have given you my gloves."

"It ain't bad." She picked up a handful of snow and washed off the blood. Then she leaned down close to the cat fighting to get out of the sack. "I'm real sorry, Prissy. I know it ain't no fun in there, but it'll just be for a little while."

Aunt Perdie stepped over to add, "You listen to her, Prissy."

Caleb was surprised when the cat went still.

Coralee must have noticed his look. "She's probably done tuckered out. That's all." Then the girl turned to Tansy. "Are you sure I can find a place at your house? Your ma won't turn me away?"

"Ma will be glad to have you, Coralee." Tansy looked at the cat sack. "A wild cat on the other hand . . ."

"She's a good mouser," Aunt Perdie said. "She'll earn her keep. But if'n we just keep standin' 'round talking till the moon comes up, we won't never get there. Go fetch my ma's sewing basket to carry with us. Can't worry about none of the rest of that stuff Coralee brung out."

"I'll come back and gather it up for you," Caleb said.

Aunt Perdie gave the scattered household plunder a look. "Ain't nothing to be overly concerned about. 'Cepting maybe the rocker. It's a fine rocker. But if'n you do come back, you can take my cow back to your place and see if my two old hens came home to roost. They'd do a heap better in a sack to carry away from here than Prissy."

Aunt Perdie rode with Tansy on Shadrach. Coralee rode behind Caleb on Ebenezer. He held on to the cat that made a fitful struggle now and again. The girl had been smart to fold a quilt to put under the cat to keep any claws from making it through the sack to stab Ebenezer, although a sharp jab or two might have made the mule move a little faster. The cat's yowls did nothing to speed up Ebenezer, but he did flick his ears. Not a good sign. If he pulled a stubborn and stopped dead in his tracks on a hillside, that might pitch Caleb, along with the cat and the girl, off. Caleb and the cat might be fine, but the girl didn't need bumps. Not in her condition.

"Maybe you better talk the cat down again," Caleb said.

So the girl kept up a singsong, nice-kitty talk, which did calm the cat. Made Ebenezer walk smoother too. The girl had a way with animals.

Up ahead, Tansy and Aunt Perdie moved along a little faster than Ebenezer. The way they had decided to ride was the only way to work it. He couldn't expect Aunt Perdie and Coralee to ride on their own, not with Aunt Perdie so wobbly and Coralee full of baby. But right or not, Caleb did wish Tansy, instead of Coralee, was the one riding behind him and holding on to his waist.

fourteen

NIGHT HAD EDGED DAY OUT OF THE WAY by the time Tansy shouted a halloo to warn her mother she was home. Ma was used to Tansy coming in late, but she wouldn't be expecting her to bring a houseful of people with her. Not to mention a cat.

Explaining it all to her mother before they carried in the cat might be best. But she couldn't very well leave Aunt Perdie and Coralee outside while she tried to convince Ma to let Prissy stay. At least the cat had quit yowling, but poor Coralee would be hoarse from singing to it all the way.

Aunt Perdie didn't seem at all worried about Prissy once they were packing the cat along with them. The poor woman was probably too exhausted to worry about anything except not falling off Shadrach. Tansy kept a grip on the woman's arms wrapped around her waist while somehow managing to hold the reins and Aunt Perdie's sewing basket.

At least Tansy wasn't the one holding the cat sack. Caleb had undertaken that task. She wouldn't have blamed him for dropping it, but he wouldn't. Caleb was someone you could depend on to do the right thing. The best thing.

Ma opened the door but didn't come outside. Her bad hip

made even that first step down onto the porch hard. "Tansy, are you bringing company this time of day?"

Her voice was even, but Tansy could hear the bother under the words. Livvy squeezed past Ma out on the porch. She jumped up and down, as excited to see visitors as Ma wasn't.

"It's Aunt Perdie, Ma," Tansy said. "Her house burned down."

"There's more than Perdie, but come on in afore you all catch your death of pneumonia. And Livvy, get back in here. You ain't got on no shoes."

Tansy slipped off Shadrach. Poor fellow. He'd had a long day. But she couldn't take care of him until she got Aunt Perdie inside and let the cat out of the bag. Once the door was closed.

"Let me help you down, Aunt Perdie." Tansy practically lifted the woman off Shadrach and set her on her feet.

When she started to help her up on the porch, Aunt Perdie knocked her hand away. "I kin walk without you hanging on to me." The old woman looked around at Caleb. "You still got Prissy?"

"She seems in fine fettle." Caleb held up the sack. "Could be you should worry more about Coralee here." Caleb reached up his free hand to help the girl slide down to the ground.

"I knowed you wouldn't lose the girl. Wasn't that sure about Prissy." Aunt Perdie held a hand out to Coralee. "Come on, girl. Let's go see if Eugenia will let us stay. All three of us."

"Nigh on four." Coralee stepped over to Aunt Perdie, who didn't make the first complaint when Coralee helped her up the steps.

Tansy hung back beside Caleb.

He whispered, "Your poor mother is going to need the patience of a saint."

The cat let out a yowl from inside the sack, now that Coralee wasn't doing her singsong chant. Tansy had to laugh. Sometimes that was all a person could do.

Aunt Perdie frowned back at her. "Ain't the first thing funny, Tansy Calhoun. Poor Prissy."

"You're right, Aunt Perdie. I'm sorry." Tansy twisted her mouth to the side to hide her smile.

Caleb didn't worry about hiding his. "Prissy's fine. It's the rest of us you'd better worry about when we turn her loose."

"What in the world?" Ma stepped back to let them inside. "Did I hear Tansy right? Your house burned down? Seems like nothing but trouble all around."

Ma didn't give Aunt Perdie time to answer. "Arlie, take their coats, and Livvy, stay out from underfoot. Come on over to the fire, Perdie. You have to be froze. Riding out in this weather. And who's this you brung with you?" Her gaze went from Aunt Perdie to Coralee and then to Caleb as he came in behind Tansy. "Caleb Barton. Tansy told me you was home. I was mighty glad to hear that."

"Good to be home," Caleb said.

Livvy scooted past Aunt Perdie and Coralee to grab Tansy's leg and peer up at Caleb with big eyes.

Caleb smiled down at her. "Well, if it isn't baby Livvy practically grown up. And look at Arlie and Josh. A fellow goes away for a spell and he can't hardly recognize anybody when he comes back. You married yet, Josh?"

Josh grinned and ducked his head. "Not yet, Caleb."

"Better watch out or Arlie here will beat you to it," Caleb said.

Josh was tall and lanky and sometimes looked as though he simply wanted to curl up out of sight when, like now, he sank down on the steps that went up to the loft. Arlie, on the other hand, was never bothered by anything anybody said. That could be the difference between being eight and eighteen.

"No, sir. Not me." Arlie took Coralee's coat and then Aunt Perdie's. "Ain't wanting no extra girl around here telling me what to do all the time." He shot a look at Tansy. "Got enough of them here already."

"I do know what you mean there." Caleb winked at Tansy.

Aunt Perdie scowled at Caleb. "Could be if you'uns would give a person a minute to get a word in edgewise, I could tell Eugenia what she's wantin' to know." She looked back at Ma with what seemed to be an attempt at a smile. Not something Tansy had seen that often on Aunt Perdie's face.

"Sit down first," Ma said. "You look ready to fall."

"The way you're limping, you don't appear in that much better shape than me, Eugenia." She sat in the chair Ma pointed out.

"You could be right there. So I'm sitting right along with you." Ma's smile wasn't a bit forced as she settled in the other chair by the fireplace. "So who's this with you?"

Coralee had crept over beside Aunt Perdie as if afraid to get more than a few inches away from her.

"This here is Coralee Embry. She's needing a place to be. The both of us are," Aunt Perdie said. "Leastways for a while. My old cabin went up in smoke."

"Arlie, fetch the girl a chair." The fact that she was carrying a baby hadn't escaped Ma's eyes. "You're looking as tuckered out as Perdie, honey."

Coralee scooted the chair Arlie brought close to Aunt Perdie's rocker before she sat down. "I reckon you might want to know how come I'm with Aunt Perdie." She hurried out the words as though she wanted to get them all said before she lost the courage to say them. "My pa turned me out. My sister and me, we hoped Aunt Perdie would take me in. Let me do chores and such for my keep. But then the chimney caught fire."

"My fault. I knowed better than to build up the fire that high but weren't nothing to do once the ceiling went to burning." Aunt Perdie looked directly at Ma. "There is this one more thing, Eugenia."

"What other thing?" A little frown wrinkled the skin between Ma's eyebrows.

"My cat, Prissy. Over there in that sack."

"Your cat?" Ma's frown got deeper.

"She ain't a bad cat nor a sweet one neither. But she's all I've had to keep me company for some time now. She's a fine mouser, but she does expect to abide with me. Inside and not out in no barn. I'll see to her."

"I'll help," Coralee said. "Clean up any mess she makes and keep her from troubling you, should you let us stay a while." She paused. When Ma didn't say anything, she added, "Aunt Perdie does think a heap of Prissy."

"Well." Ma looked over at the sack Caleb held.

Tansy couldn't tell if she was at a loss for words or if she was trying to keep from saying what she was thinking. The room was silent except for a pitiful mewing that came from the sack. No yowls now.

Finally Ma said, "I reckon you'd best let her out."

"If you're sure you're ready." Caleb held the sack in front of him. The pitiful mewing continued.

"Poor kitty." Livvy stepped toward Caleb, but Tansy pulled her back.

"Let me do it." Coralee got up. "She's done drawed blood on me once. Maybe she won't see the need to do so again."

Caleb looked hesitant to give her the bag.

"Don't worry yourself, Mr. Barton. After how I sung to her the whole way over, she knows I mean her no harm." Coralee reached for the sack.

"You could bring her over here to me," Aunt Perdie said.

"Give us a minute first, Aunt Perdie," Coralee said. "I'll turn her loose over here in the corner so she can have some peace to pull herself together."

"This is a cat you're talkin' about," Ma said.

Nobody answered her as Coralee took the sack and gently laid it down in the corner where the least light reached.

Tansy didn't know if it was right to pray for a cat with so

many other needs around, but she sent up a prayer anyhow. It was as much for Aunt Perdie as for Prissy. Again the room got extra quiet. Even Prissy quit her mewling and the only sound was the fire popping.

"Sweet Prissy." Coralee knelt down and kept her voice low. "Sweet, sweet kitty."

She worked the knot loose on the sack and sat back on her heels. "Here, kitty, kitty."

Tansy pulled in a breath, expecting the cat to explode out of the sack and attack Coralee and anybody else in her way. But nothing happened.

"You must have kilt her dead." Aunt Perdie glared over at Caleb.

"A dead cat can't mewl like it was a minute ago," Caleb said.

"Prissy don't never carry on like that neither." Aunt Perdie's lip was trembling. "I reckon I should've left her there at the barn where she knowed where she was."

"Don't fret, Aunt Perdie," Coralee said softly. "I'm thinkin' she's fine. Come on, Prissy."

Ma stood up to get a better look, and Tansy let Livvy inch a bit closer, although she kept a hold on her shoulder. Arlie stepped up behind Coralee, and even Josh uncurled from the steps to peer over at the sack. Only Aunt Perdie stayed still in the rocker, staring down at her hands.

At last Prissy pushed free her black head to look at Coralee before she slinked out of the bag, her stomach dragging the floor. Then to Tansy's surprise, the cat sat up and began licking her paw to wash her face, as though the indignities she'd suffered weren't worth acknowledging.

"Pretty kitty," Livvy said.

Coralee joined her voice to Livvy's. "Pretty kitty."

The cat finished washing her face before standing up with her tail high in the air. The tip of it twitched a bit. She gave

Coralee another look but didn't go to her. Instead, she walked regally straight toward Livvy to rub against her legs.

Tansy caught Livvy's hands to keep her from reaching down toward the cat. "Sometimes that's just a tempting move so she can swipe at you," Tansy warned.

The cat stared up at Tansy as if unable to believe her words before she smoothed her side against Livvy's legs again.

"She's purring," Livvy said.

Tansy wasn't sure she'd even once heard Prissy purr when she stopped by Aunt Perdie's house. But she was making a loud rumble now, as if she was the sweetest cat ever born. After a second, she sauntered over to Aunt Perdie, where she jumped up into her lap, turned around twice, and settled down for a nap.

Aunt Perdie deserved credit for honesty when she spoke up. "I ain't aiming to fool you, Eugenia. Prissy ain't always this nice. Don't know if she's putting on a show or if that ride over bounced her head around so that she's forgot who she is."

"Well." Again Ma seemed to be searching for the right words.

"Just be out with whatever it is you's wantin' to say," Aunt Perdie said. "It's your house. I'm the one come begging."

"I don't rightly know what to say," Ma said. "But I wouldn't ever think on turnin' you and the girl out, Perdie. Seems the cat comes with you, so I reckon I won't turn it out either. Long as the cat doesn't scratch Livvy." She looked at Livvy edging closer to the cat. "Without cause anyhow. You leave the thing alone, Livvy. I expect it needs to rest a mite without being bothered."

"I've got to take care of Shadrach." Tansy looked over at Arlie and Josh. "Josh, you come help me. Arlie, you get whatever Ma says for Aunt Perdie and Coralee to eat."

"I can help him." When Coralee struggled to get up off her knees, Caleb reached a hand down to help her. "Thankee. I can't get used to not being able to move like I used to."

Ma gave her a long look, but she must have decided more

of Coralee's story could wait. Instead, she asked Caleb, "You want to stay to supper, Caleb? We done eat, but there's plenty of beans. Not fancy fixins, but filling."

"Thank you anyway, but Ma will have my supper waiting."

"She's liable to be worried some about you being out past sundown with what she's gone through. I sure was sorry about your brother. He was the nicest boy." Ma shook her head. "How's little Reuben doing?"

"We're doing our best for him."

"Course you are. Vesta will find a way to bring him back to health."

"Yes, ma'am." Caleb eased toward the door.

"Don't be talking his ears off, Eugenia. Let the boy go on home." Aunt Perdie looked over at Caleb. "I reckon I owe you some thanks."

"You don't owe me a thing, Aunt Perdie. I'll gather up your stuff and bring it around soon."

"Ain't no hurry. Eugenia might not want none of the plunder anyhow."

"I took in your cat. I reckon I can take in whatever else you saved from the fire." Ma's voice had a little edge to it.

"And I'm right grateful to you." Aunt Perdie stroked Prissy a couple of times. "I reckon I forgot to say that I weren't always the sweetest old lady, same as Prissy weren't always the sweetest cat."

That made Ma laugh. A good sound that followed Tansy out the door behind Caleb. Josh moved past her to untie Shadrach and lead him toward the barn.

"Maybe your ma won't want to kick her out before morning," Caleb said.

"Coralee will pull enough sympathy to keep things settled." Tansy smiled. "And like Aunt Perdie, I owe you thanks. It would have been a long trek on foot leading Shadrach carrying the two of them."

"And the cat. Don't forget the cat."

"I don't think anybody is going to forget Prissy." Tansy laughed.

"When I bring Aunt Perdie's things, maybe we can sit a while and do some catching up." He sounded almost shy.

"That would be nice. I'd like to hear about where you've been."

The snow bounced up enough light that she could see his smile before he got on his mule and rode off.

She watched him out of sight. It would be good to have somebody to talk books and trees with again. She wondered if Preacher Rowlett had told Damien Felding about the chestnuts. A smile slipped out on her face as she headed toward the barn and thought about how the writer had been nervous about sleeping at Preacher's house. Unless Tansy missed her guess, the poor man might not get much chance to sleep. Preacher Rowlett got started telling stories, he was apt to talk the night away.

Saturday might be interesting when she took Damien around to meet more of the mountain people. Maybe she'd bring him up here to talk to Aunt Perdie. No telling what stories she could tell.

fifteen

THE WEEK PASSED AS AUNT PERDIE and Coralee settled in. Ma seemed happy to have them there and put up with Prissy. But things were a little crowded. So when Saturday dawned clear and cold, Tansy was glad to escape the house to the barn to see about the animals. Josh was out before her.

"It's busy in there," he said as he glanced back in the direction of the house.

"That's the truth."

"You ain't had the full brunt of it, having the chance to ride off with your books every morning." He handed her the milk bucket. "You milk the cow. That's women's work."

"I'm thinking you drink the milk and eat the butter." But Tansy took the bucket. She didn't mind milking, but she couldn't let him think she'd do it during the week. "A man's hands can milk a cow the same as a woman's, and you can't be weaseling out of milking during the week. Ma can't do it anymore. Not with her hip the way it is. And Livvy's too little."

"She's four. I was helping Pa time I was five."

"But not milking."

"No, you or Hilda did the milking back then. Women's work." He climbed up to the loft to throw down some hay.

"Women's work or not, you can't get out of milking." Tansy looked up at Josh in the loft while she warmed her hands under her arms. Cold milking hands could make Bonnie kick.

Josh jumped back down from the loft and picked up the pitchfork. "Coralee offered to do the milking, but Ma said no."

"Well, I guess so." Tansy frowned at him. "Not in her condition. Bonnie can be ornery at times." She patted the cow's side as she sat down on the stool and positioned the bucket.

"Don't I know it," Josh said. "She's nigh on as mean as Aunt Perdie's cat."

"Has Prissy been giving you trouble?" The first squirts of milk made a pinging noise in the bucket.

"Ma ain't happy with the cat, but it ain't been bad. Coralee keeps it out from underfoot, and the cat did catch a mouse in the woodbox."

When Josh leaned on the pitchfork and didn't make any move to go feed the cow outside, Tansy glanced over at him. He looked uneasy about something.

Tansy tried to wait him out, but he was still standing there when she finished stripping the milk out of Bonnie's bag, so she asked, "Something bothering you, Josh?"

"Naw. Not really."

Tansy let Bonnie out of the stanchion and smacked her haunches to send her out to join the other cow. Josh would eventually get around to throwing hay out to them. She picked up the milk pail. She needed to get ready to ride down to Preacher Rowlett's to meet Damien Felding. Josh could do the rest of the chores, but he did look bothered.

"All right. Out with it. I don't have all day." She wanted time to fix her hair a little nicer before she left.

Not that it mattered how she looked, but on the other hand it wasn't so cold that her ears would freeze off if she didn't wear her hat. She could catch her hair back in a ribbon instead of a plait. And maybe she could wear that nice sweater that used to be Hilda's. Not quite blue. Not quite green. But the color did make Tansy's eyes look caught between the colors and appear a little mysterious. Like a character out of a Jane Austen book.

She was ready to laugh at that. Jane Austen's heroines wore fancy dresses and sat in parlors or walked across meadows. None she'd ever read resembled a woman in work pants holding a bucket of milk while waiting for her brother to voice whatever had his cheeks going red from more than the chill of the morning. No Jane Austen scene here.

Josh stirred the hay with the pitchfork. He didn't look at Tansy. "I was wondering some about Coralee."

"What about her?"

"Well, about whoever it was who didn't marry her like he should've."

"I wouldn't know anything about that." Tansy hadn't foreseen this. That Josh would be attracted to Coralee. But she was pretty in a sweet way. Josh was certainly old enough to be ready to look at a girl with favor. But not a girl on the way to being a mother.

"Yeah." Josh kept his eyes on the hay he was stirring about. "But do you think he, whoever he is, might show up here or anything?"

"I don't know that either. But from what I gathered from Coralee, it doesn't sound likely."

"But what if he did?"

"I suppose that would be something Coralee would have to deal with."

"You think she'd be glad to see him?" Josh shot a look over at Tansy, then stared down at the hay again. Tansy could barely make out his next mumbled words. "That she still loves him?"

Tansy paused to consider her words. The barn sounds kept on. Shadrach moved in his stall. A bird flew into the loft. The cows mooed outside, impatient for their hay. Josh didn't look up at Tansy as he waited for whatever she might say. Tansy wasn't sure if she should be irritated with him or sorry for him. She supposed patience was best. Things hadn't been easy for him since Pa left.

She kept her voice even, hiding the worry that was scratching at her. Best not to let Josh know what he'd asked bothered her, even if it did. "I can't know that without asking Coralee, but I'm guessing she did love him not so long ago or thought she did."

Josh's face went a little redder, but he still didn't look up at Tansy.

Tansy went on. "I don't know if him leaving her high and dry changed that. I'd think it would have, but sometimes we can be foolish with our feelings. You read about that happening all the time in storybooks."

"This ain't no story in a book," Josh said. "It's Coralee who's right back there in our house."

"That's so. And she's not that far from being a mother."

"What's that got to do with anything?" Finally, Josh stared up at Tansy. "That don't make who she is any different."

"Being a mother might make her a little different." Tansy kept her voice gentle. "I can see you like her and you're too old for me to tell you what to do. But if you were to ask, I'd advise you to just be her friend a while. Things are hard enough for her without you adding complications."

"I ain't doing nothing to complicate matters for her, and I never said I wanted to be any more than a friend." Josh stood up straight and lifted his chin a little, obviously taking umbrage at her words, even when she'd tried her best not to upset him. "I just wanted to know should I get the gun and run the scoundrel off if'n he were to come knocking on our door."

"We don't even know who he is."

"Coralee does."

"Then if he were to show up at our door, which I don't think he will, it would be up to Coralee to send him on his way if that's what she wanted to do."

"But what if he wouldn't go?"

"None of that is about to happen, but if it does, we'll figure out what to do then. But don't you be pointing a gun at nobody. You hear me?" Tansy stepped closer to him and stared straight at him. "We don't aim to start any trouble."

Josh didn't back down. He stretched up a little taller and peered down at Tansy. He was a good five inches taller than she was. "I hear you. So tell you what. I'll just give the gun to Aunt Perdie."

Tansy couldn't do anything but laugh. "Don't do that. No telling what might happen then."

Josh laughed too. "Might get interesting."

"That's for sure." Tansy put her free hand on his arm. She still had the milk pail in the other hand. "Look, Josh. Ease back a little and give it some time to see what comes."

"I reckon that's a baby."

"You're right there." Tansy had to smile again at that. "Now you need to feed the cows and I need to get this milk in to Ma so I can head down to Preacher Rowlett's to meet that writer guy I told you about."

On the way to the house, she thought the advice she'd given Josh might be good for her to think on herself. She was getting too worked up by the idea of being with Damien Felding. Thinking about wearing her best sweater. Still, a girl had to wear something.

The house hummed with activity when she carried the milk inside. Ma sat at the table rolling out biscuit dough. Livvy, on her knees in a chair beside her, cut out the biscuits with a jar top. Arlie dumped a load of wood in the woodbox inside the kitchen door. Aunt Perdie rocked back and forth in the chair by the fire. Prissy wound in and out between Coralee's legs.

Coralee paid the cat no mind as she turned from stirring something on the stove to take the milk from Tansy. "I'll strain it up. You won't mind if'n I give Prissy the foam off the top, will you?"

Livvy looked up from her biscuit cutting. "Can I cut a little corner biscuit for Prissy, Ma? Please?"

Ma blew out a breath. "You'uns are spoiling that cat rotten. No wonder she's so ornery."

"But if we're nice to her, shouldn't that make her nice?" Livvy asked.

"Should." Aunt Perdie spoke up from her chair. "Should ain't always the truth of it. But I do think you is making a difference in my rascal cat, young Livvy. She ain't hardly scratched at me for a day and a half."

Livvy giggled and cut a half-sized biscuit.

"How are you feeling, Aunt Perdie?" Tansy asked. The old lady had been coughing some at night. Probably from getting chilled standing in the snow watching her house burn down.

"With my hands same as always," she said. "Same as you, I expect."

"I never knew you were such a jokester," Tansy said.

"No doubt a lot you don't know about me. Reckon me here sitting right down in the way of the bunch of you will change that some. Me and Coralee done crowded you up a bit."

They had fixed a cot for Coralee over in the corner of the sitting room. No way could she share the loft with Tansy, Livvy, and the boys. Aunt Perdie was taking Pa's place in Ma's bed. Tansy thought Ma might be having more trouble living with Aunt Perdie than with Aunt Perdie's cat. But Coralee had found a spot right in Ma's heart from the first. And now after her talk with Josh, it could be Ma's wasn't the only heart the girl had won. Tansy shook that worry aside. Maybe Josh would pay mind to what she'd told him out in the barn.

"I thought it was funny, Aunt Perdie. That about you feeling with your hands." Arlie laughed as he carried some wood over to the fireplace.

"You think everything's funny," Tansy told him.

"Ain't a bad way to be. Me and you both could think on cheering up a little ourselves." Aunt Perdie gave Tansy a look that didn't have a bit of cheer in it.

Arlie laid the big chunk of wood in the flames, then held out a small piece to Aunt Perdie. "You told me to be on the lookout for a piece of wood that wasn't too big. You think this one will do?"

Aunt Perdie took the wood and turned it this way and that in the light from the fire. "Can't know for sure until I think on it for a while."

"Reckon a dragon might be hiding out in there?" Arlie stepped closer to peer at the wood.

"I'm thinking it's a mite small for a dragon, but could be a two-headed old man." Aunt Perdie ran her fingers along the grain of the wood.

"I think I druther see a two-headed dragon."

"Fact is, I might druther see that too." Aunt Perdie laughed. Not something Tansy had ever heard her do, but then the laugh turned to a cough.

"Your cough doesn't sound so good," Tansy said.

Aunt Perdie pulled a handkerchief out of the end of her sleeve to wipe her mouth. "My cough ain't worth thinkin' about. Old women just get to coughing now and again. Once the snow goes off, I'll find some gingerroot to make me a tonic."

"I can find it for you if'n you tell me what to keep an eye out for," Arlie said.

"Well maybe you kin go with me when I go root huntin' and learn that way," Aunt Perdie said. "But you done found me this fine piece of wood. I thank you for that. Once I sharpen up my whittlin' knife, I'll see if it might not have a dragon hid in it after all."

"Does it take a special knife?" Arlie asked.

"Got to be one sharp and sturdy enough to carve away the part that ain't whatever's hiding in the wood."

Coralee looked over at Aunt Perdie. "I wish you'd told me where your carving knife was. I could have brung it out of the fire."

"No worry about that, Coralee." Aunt Perdie patted her pocket. "My knife is right here in my pocket. Ain't used it for a while, but if Arlie hunts up one of his pa's whetstones later on, maybe I'll try some whittling if'n my fingers will work right. Rheumatism keeps them all knotted up."

Aunt Perdie sounded practically congenial. Maybe all she had needed was company and somebody to do the chores so she wouldn't have to get out of her rocking chair. The smell of biscuits baking followed Tansy up to the loft.

She shrugged off her old sweater and pulled on the blue-green one. It didn't do any good to save something and never wear it. Could be while she was saving it for good, moths would get into it. Better to get some use out of it. That didn't have a thing to do with Damien Felding. Not a thing. Besides, the sweater would be covered up by her old jacket most of the day while they were riding around the hills hunting down stories.

When she went back down the stairs, Aunt Perdie gave her a look. "That's a nice-looking sweater you got on. You must be expecting that Barton boy to show up. You know he done brung my pans and things on Wednesday. He appeared some disappointed you weren't here. Asked about you."

"Stop it, Aunt Perdie. Caleb was just being a good neighbor."

"Appeared to be a right friendly neighbor. Wanting to be friendlier to some than to others." Aunt Perdie raised her eyebrows at Tansy.

"Well, if he comes today, I won't be here. I'm taking that writer man around to gather some stories for the guidebook the government has hired him to do." Tansy shook her head at the woman. "I'm supposed to meet him at Preacher Rowlett's. Do you think you might like to talk to him about some of the things you remember happening around about here?"

"Hiram Rowlett might be a better one for stories."

"He's already talked to Preacher. He spent the night there on Tuesday."

"For mercy's sake, there weren't be nothin' more to tell then."

Outside, Josh's dog, Hunter, started barking. Arlie ran to the window. "Somebody's coming. Two of them."

Tansy looked out over top of Arlie. She knew who it was before Arlie said, "Looks to be Preacher Rowlett and some stranger."

sixteen

MA PEERED TOWARD THE DOOR. "Oh my, I should have made more biscuits."

"I can add some milk to the gravy," Coralee offered.

"Gravy without biscuits is poor fixings," Ma said.

"They aren't here for breakfast, Ma," Tansy said. "I was supposed to meet Mr. Felding at Preacher's house, but I guess he was in a hurry to get a soon start. But they won't be coming to eat."

Aunt Perdie spoke up from the fire. "I ain't never knowed that Hiram Rowlett not to be ready to eat. But if'n you're feared you don't have enough, Eugenia, I ain't all that hungry. Hiram can eat my biscuit."

"That's neighborly of you, Perdie," Ma said.

"I reckon it is." Aunt Perdie frowned. "Don't 'ary a one of you let Hiram know I said that. He'll think I'm getting senile for sure."

Livvy and Arlie were both peeking out the window now.

"Get away from the window, you two. You'll see them plenty fine enough once they're in the house. We could try to act civilized." Tansy shook her head and went toward the door.

Behind her, Aunt Perdie said, "Don't you worry, Tansy. I'll

do my best not to spit on the floor while that city feller is here. He is a city feller, ain't he?"

"He's from New York City."

"All the way from New York City." Ma sounded amazed. She looked up from setting plates on the table. "Now what did you say he was down here for?"

"He's part of a government program for writers. Something like the packhorse librarians. Except he's writing about places around here for a visitors' guide."

"Never cared all that much whether visitors came in here or not. Generally just bring trouble with them," Aunt Perdie grumbled. When Tansy looked around at her, she raised her hands up in the air. "I done told you I'll act proper as can be. But if he's here to write about us, then he don't need no fancied-up version."

When Tansy stepped out on the porch, Josh was leaned against a post, watching the men. "Early for company," he said.

"It's Preacher Rowlett and that writer man I told you about. You need to call off Hunter." The dog ran toward the horses, still barking.

"He won't bite them."

"But he could make their horses nervous. Mr. Felding isn't much of a horseman."

"I noted that when they first came into sight. Guess he has one of them motorcars to take him places. But not this place." Josh clapped his hands.

Hunter came back up on the porch to lie down with a disappointed huff.

"I wish that would work on Aunt Perdie's cat," Tansy said.

Josh shook his head. "No training a cat."

"Hello there, Tansy Faith," Preacher Rowlett called when he got closer. "Josh."

"I was ready to head down to your house." Tansy looked from Preacher to Damien. "And here you are."

"Damien come up last night, and seeing as how I was feeling some perkier this morning, I decided to ride up here with him. I heard about Perdie's cabin burning. She still here with you'uns?"

"She ain't going nowhere. She likes the rocking chair by our fireplace," Josh said.

Tansy frowned at him, but he just shrugged. "I didn't say nothing that wasn't true."

"Don't pay mind to Josh. We're glad to have her here," Tansy told Preacher and then smiled at Damien. "How are you, Mr. Felding? Are you getting more used to riding?"

"Belle's a fine horse," Damien said. "Good as a rocking chair the way the man at the stable said. So your Perdie and I can have something in common. Liking those rocking chairs. And look." He laughed as he dismounted. "I'm getting better at getting on and off."

When Tansy gave Josh a warning glance, he rubbed his hand across his mouth to wipe away his smile before he said, "If'n your horses need tending, I can see to them after breakfast. Come on in. Ma always cooks plenty."

"That's kind of you, but we already ate," Damien said.

Preacher slid down off his horse and tied the reins to the porch railing. "Don't be hasty, son. A man would be foolish to pass on Eugenia's biscuits. Especially if she has some blackberry jam to go with them."

Josh led the way into the house. Preacher followed him in, shouting greetings out to Ma and Aunt Perdie as he went through the door.

Damien hung back. "Hello, Tansy. I've looked forward to you being my mountain guide today." His smile looked as if it was for her and her alone.

Tansy hated the burst of heat that shot up into her cheeks, but some things a person couldn't help. He noticed the blush. She could tell by the look in his eyes. "Oh? I thought maybe since Preacher Rowlett was with you that he had decided to take you around today."

"Just to here. As he said, he wanted to see your aunt who had the misfortune of losing her house."

"She's a relative, but not my aunt." Tansy shook her head a little. "Not that it matters. We do call her Aunt."

"Right. The way you call Mr. Rowlett Preacher. All those nicknames, Miss Book Woman."

Tansy clamped down on her runaway feelings and stepped through the door in front of him. It didn't feel exactly proper the way Damien kept looking at her, but then maybe city men were simply a little easier around strangers than people here in the hills. And while she might know Damien's name, he was a stranger. An outlier.

As though he knew he was making her uneasy, he laughed softly before he followed her in. He stopped right inside the door and looked around. Her unease grew as she saw the crowded room the way he might be seeing it. Coats with patched sleeves hanging on hooks just inside the door. Mud-encrusted boots on the floor under them. Newsprint covering the walls. A wash pan and water bucket on a table in the kitchen part of the room, with a towel hanging on a nail beside it. At least Arlie or Coralee had already carried out the slop jar.

She imagined how his house back in New York City might look. He probably couldn't even smell his breakfast bacon frying when he was in his sitting room. Might even have a maid doing the cooking. There would be rooms aplenty, with no need for a cot in the corner or for the eating table to protrude out into the sitting area of the room.

She blinked her eyes to erase the images of houses she'd read about in books. Same as she was no storybook heroine, their cabin wasn't city shined either. But it was home.

Preacher Rowlett had a voice that took over a place when he was talking. "Perdita, I was sorry to hear about your cabin catching fire. Something all of us up here in the hills have to think about with our fireplaces." He reached down to take one of Aunt Perdie's hands.

"'Specially if a body's chimney ain't chinked proper." Aunt Perdie pulled her hand away from him. "How about you, Hiram? I heared you was some under the weather."

"A cough. Not much more than an aggravation."

"I know some about such as that."

The two exchanged coughs as if to prove it.

"Preacher, we were just about to sit down to breakfast. Won't you join us? You and your friend too." Ma looked over at Damien.

"I apologize, Eugenia." Preacher straightened up from talking to Aunt Perdie. "Tansy or I should have made introductions." He looked over at Damien still standing by the door, as though unsure of whether to come on in or escape outside.

Tansy supposed she wasn't much better, standing beside him and wanting to rewrite the scene in front of him. The thought shamed her. She had a good family and their cabin might not be big, but it served their need for a roof over their heads.

She stepped on into the room. "This is Damien Felding, the writer I told you about." She motioned toward Damien, who smiled friendly enough but with a little hesitation mixed in. Could be he wasn't as confident as he acted.

She went on. "My mother, Eugenia Calhoun. My cousin, Perdie Sweet. Brothers, Arlie and Josh, and little sister Livvy peeking out from behind Ma." Tansy had to smile at Livvy's rounded eyes. Coralee scooted away from the stove and had one hand on the kitchen door, as though ready to make a run to the outhouse. She shook her head at Tansy, but Tansy couldn't pretend she wasn't there. "And this is our friend Coralee." She left off her last name.

But Preacher didn't let that little bit of information slide.

"Coralee Embry. What are you doing way over here? Come on out to where I can see you. I haven't been over your way in forever. Not since your mother died. God rest her soul." Preacher Rowlett's face went somber for a moment before he started toward Coralee with outstretched arms.

Ma moved in front of Coralee. "Sit on down to breakfast before everything gets cold, Preacher. You can catch us up on all the news."

Preacher appeared a little surprised but let Ma turn him toward a chair at the table. "Well, I guess I could eat a biscuit if you have some to spare." He peered around Ma toward Coralee. "But I wanted to say hello to Coralee."

Aunt Perdie was up and coming over to the table now. "And I heared you do that very thing already. The girl don't need you overpestering her."

That seemed to astound Preacher even more. Tansy hoped if she ever needed defending against somebody, Ma and Aunt Perdie would be on her side.

Coralee let out an audible sigh as she came to the table. "Don't bother yourself, Aunt Perdie. I wasn't hiding back there in the shadows because of Preacher Rowlett. It's that other one that has me worried." She shot a look over at Damien.

"Me? I can't imagine what you mean, miss," Damien said.

"Tansy says you're writing about folks in these parts." Coralee narrowed her eyes at him. "I ain't wanting nothing written about me nowhere no time."

Damien held up his hands as though to ward off her words. "I wouldn't write about anybody who didn't want me to. Besides, I'm not exactly writing about the people here. More life stories or tales passed down through the years. That and describing interesting places to see."

"Like where Daniel Boone camped back in the settlers' days." Preacher reached toward Coralee. "Come here, child. I am so glad to see you. I'd heard things weren't going too well for your father and worried that meant things might not be going well for you children either."

"I guess they weren't for me." Coralee blinked back tears and nodded as she let Preacher take her hands. "But Aunt Perdie took me in."

Aunt Perdie spoke up. "The Lord sent her my way, Hiram, and I didn't think I should be arguing against what the Lord aimed to happen. Even after my house burned down. Could be he was behind that too, seeing as how I had naught but empty pots, with poor Hanley gone on to his just reward. He was a fine neighbor. But the fire brung Tansy, and Tansy brung us here."

"You could have come and asked me for help, Perdie," Preacher said.

"If I rightly remember, it ain't no short trek over to your place. I can't walk like I used to, sorry to say." Aunt Perdie sat down at the table.

"You're right as rain. Forgive me for that. And forgive me for forgetting to come see you for too long."

"Are you saying you done forgot all about me, Hiram?" She peered up at him with a look that was practically flirty.

"Not at all. But I never intend to let so many weeks go by without checking on my friends." He shook his head.

"I ain't faulting you about that. The years do slide past."

"It's fine to see you now and Coralee too." He still held Coralee's hands. "Let me go on and pray for this fine breakfast. Be a downright shame to let Eugenia's biscuits get cold."

Preacher looked up toward the ceiling. "Dear Lord in heaven, bless this your child as she walks down this hard trail and bless all the others here in this house with your kindness and love. Forgive us when we don't do what you would rightly have us do. Keep us safe in your mighty hands. And thank you for this food and the good hands that prepared it. May it be for the nourishment of our bodies to enable us to do the work you lay out for us. In your almighty name, amen."

Aunt Perdie echoed Preacher's amen and then so did Ma.

"About time," Arlie grumbled. "A feller could starve waiting for food around here."

"Enough of that, Arlie, or you can wait for supper." Ma gave him a stern look before she smiled over at Damien. "Come join us, sir. Tansy didn't get to the store for coffee last week when she went to town, but I think we still have a spot of tea in the cupboard if'n that's to your liking."

"No, ma'am. Don't go to any trouble. I already had breakfast." Damien leaned against the wall.

"You can have a chair. The boys can sit on the bench here." Ma worried with her apron.

"No, really. I've been sitting on that horse. It feels good to stand a while."

"Sit down, Eugenia," Aunt Perdie said. "The man wants to stand up, let him stand up. Maybe that's how they do in the city. Stay on foot in case they need to run off somewhere fast." She gave him a sour look and took that biscuit she said she wasn't hungry for earlier.

Ma sat down, but she didn't stop fiddling with her apron. Coralee looked near tears. That kept Josh frowning. Livvy stayed stuck to Ma's side. Arlie, Preacher Rowlett, and Aunt Perdie were the only ones who didn't seem bothered by Damien's presence.

"Best come get something to eat, Tansy, before these boys eat it all," Aunt Perdie said between bites.

"I'm not hungry right now." Tansy looked at Damien and wished he'd sat down somewhere. Him standing there was worrying Ma. "Tell you what, Mr. Felding. If you aren't interested in breakfast, we should go see to the horses. Then maybe Aunt Perdie can share some stories with you."

"I can probably come up with something," Aunt Perdie said. "If'n Hiram hasn't already told him every tale I know."

Preacher laughed. "I couldn't even begin to touch the tales you could tell, Perdita."

Prissy emerged from wherever the cat had been hiding and sauntered toward Damien.

"What a pretty cat," he said.

He leaned over to stroke the cat, but Arlie spoke up. "I wouldn't do that, Mister Sir. Prissy has a way of tauntin' a feller so she can sink her claws in your hand once you get close."

"That ain't always so," Aunt Perdie said without looking around.

"It's so often enough." When Tansy pushed Prissy away with her foot, the cat swiped at her. "I've got the scars to prove it."

"She don't never scratch me," Livvy said.

"You must have cat magic," Damien said. "You'll have to tell me how I can get it sometime."

That made Livvy giggle and hide her face against Ma's arm.

Coralee called Prissy, and the cat actually listened and sauntered over to rub against the girl's legs.

"She's the one with the cat magic," Josh said.

"No magic," Coralee said. "I just like her."

"Is that what does it?" Josh asked.

Tansy could tell Josh was hoping Coralee would like him for as simple a reason.

Damien put his coat back on. "It seems I'm going to get another lesson on the necessity of a man taking care of his own horse."

"A good lesson to learn," Preacher Rowlett said. "And thank you, Tansy, for seeing to Samson." He looked over at Livvy. "You know the story about Samson in the Bible, Livvy?"

She peeked around at him.

"The Lord made Samson extra strong, and not cutting his hair was the secret to his strength. But old Samson let a woman trick him into getting his hair shorn. We can get in trouble when we don't listen to what the Lord wants, and Samson found trouble sure enough. So I named my horse

Samson because he has to be strong to carry around a fellow like me who likes to eat his biscuits. But I don't trim his mane none."

"Appears you like eating Eugenia's biscuits too." Aunt Perdie gave him a look.

"I do indeed, Perdita. These are fine biscuits. You must have the biscuit-making magic, Eugenia."

The sound of his laughter followed Tansy and Damien out on the porch.

"I've had a great time with Preacher Rowlett," Damien said after the door shut behind them. "He's been a wealth of information."

"Aunt Perdie said he'd probably wear out your ears." For some reason, Tansy felt easier with the man outside away from Ma and the others. She buttoned up her coat to protect her sweater before she untied Preacher's horse and led him away from the house.

Damien followed with Belle. "Miss Perdie appears to be quite a character. What do you think I should ask her?"

"Whatever you want, but you can be sure she won't sugarcoat anything."

"I wouldn't want her to." When Tansy led Samson on past the barn, he said, "I thought we were taking them to the barn."

"First to the creek so they can drink. The water's running, so it won't be iced over." Tansy looked up at the sky. "Doesn't feel as cold today anyway. The sun feels good." She gave Samson his head to step down to the creek.

Damien did the same with Belle. "What happens if they run away?"

"They won't unless something spooks them."

"Good to know." He looked around. "This is a pretty place. The creek running through the snow under the trees. The blue sky with clouds almost close enough to touch."

In the house, she'd worried about what he might think

about her home, but out here his words made her appreciate things she sometimes forgot to notice.

His gaze came back to Tansy. "You have a nice family."

"I do." Sometimes a person could imagine a person's thoughts all wrong.

seventeen

CALEB HADN'T PLANNED to ride over to Tansy's house until Sunday, but he was too eager to have her in front of his eyes again to wait. Sadly, she hadn't made it home from her book route on Wednesday when he took Aunt Perdie her rocking chair and other plunder.

His mother feared something was amiss about him heading back to the Calhoun house so soon, but her thinking was what was amiss. Before he left, she frowned at him. "Now, don't you go getting yourself in a situation over there, son. You can't think on rescuing that young Embry girl from a mess she stepped into on her own. If you're hankering to give away your name, you need to consider your brother's family first."

Ma wouldn't let loose of that idea no matter how often he told her it wasn't happening.

"I'm not giving my name away to anybody, Ma," he told her. "I just have it in mind to be neighborly by offering them some help."

"That oldest boy of Eugenia's is plenty big enough to do a man's work."

"But he's young and his pa is gone."

"So's Junie's."

"Yes, and I'm right here every day to do for him."

His mother sighed. "I wish it appeared to be helping more. The poor boy keeps punying around."

"Could be if Jenny Sue would pull herself together, things might change."

"I'm not saying that's not so. Whilst the girl does lack some in backbone, we should give her some slack. She did lose her husband. I can attest to how hard that is from walking down the same sorrowful path, but I didn't lose your pa until you were nigh on grown." She looked down at the dishes she was clearing off the table. "It's some different for Jenny Sue."

"I guess so." Caleb tried to come up with something to appease Ma. "Looks to be a nice winter day coming. How about I take Junie out for a ride this afternoon?"

That brought new worry to Ma's face. "I don't know that he should be out in the chilly air. Might give him a lung ailment."

"Maybe you coddle him too much. Makes him believe he's sick."

"He is sick." She dropped down in a chair by the table. "Heartsick."

"I know, Ma." Caleb went over and put his hand on her shoulder. "But getting outside might give Junie a change."

"I think he's feared to go outside. For himself that something bad might happen and feared for us too whenever we're out of his sight. That we won't come back." She reached up and covered his hand with her own. Her hands were rough from years of hard work.

"But we have come back. Time and time again."

"So did his pa. Until he didn't."

Once on his horse headed toward Robins Ridge, Caleb shook away thoughts of his mother and Junie. Instead he soaked in the fine morning. The air had a winter chill, but the sun on his shoulders had the promise of a warm-up. He wasn't on Ebenezer. No wood hauling today unless Tansy's brothers needed help bringing some in. If that was true, the Calhouns had a mule.

Ebenezer deserved a day off. Caleb deserved a day off too. He was on Phoebe, his sweet riding mare he'd left with Reuben when he went off to the Smokies to work with the Civilian Conservation Corps. Those mountains were more breathtaking than these hills, but it was still good to be back where his feet connected with home ground.

Other men he'd met in the Corps didn't feel the same about where they came from. Some claimed to have no plans to ever head home. They aimed to move on to cities as soon as they had a little money in their pocket. That took awhile since part of their pay was sent back to their families. The Corps was a way of keeping young men working while helping their families. Now Caleb would need to find another way to take care of Ma and Jenny Sue and the kids. And maybe his own family soon.

Without thinking about it, he started whistling "Happy Days Are Here Again." President Roosevelt's campaign song. Caleb had happy days in view with Tansy right over the hill. Unmarried. Available.

Guilt trickled through him from feeling that happiness when Reuben would never again know earthly happiness. But Reuben wouldn't want him to stop living. He wouldn't want his family to stop living either. More guilt poked Caleb as he couldn't block away those last words his brother had shared with Ma. She hadn't made them up. She wasn't one to twist things so they would sit more comfortably in a person's thoughts.

No, he had no doubt Reuben had said them. Even more, that he'd meant them. He was used to Caleb cleaning up things behind him. But not this time. At least not that way. Jenny Sue was young. She would marry again. That could be what was worrying Ma. That Jenny Sue would marry, and Ma would have to watch some man she might not like stepping into Reuben's place as pa to his children.

Caleb's sad thoughts drowned out his whistling. Life could get too messed up at times to figure out what was next. Reuben used to tell him he fretted about things too much. A man might not be able to change the course of a river, but he could put his raft in it and enjoy the ride wherever he might end up.

"You're right, brother," Caleb said aloud, then felt silly to be talking to a brother who could no longer hear him. Better to talk to the Lord. He looked up at the sky. The sun leaked down through the bare limbs of the oaks and maples along this stretch of the trail.

The sun's rays bounced off a dead chestnut. Folks probably used it as a marker tree for the trail now. He stopped Phoebe beside it and touched the trunk. He remembered this tree. It once had a thick umbrella of branches and dropped bushels of chestnuts every year. Now nothing was left but this gray ghost of a trunk.

Caleb sighed. Some things a man could do nothing about no matter how much he wished he could. He looked up toward the sky the color of bluebirds' wings. No bluebirds in the trees now until the spring. In the Smokies last summer, he'd think he was fine, not homesick at all, but then one of those bluebirds would flit by to put a longing for home in his heart.

He pulled in a breath to still his mind. He said nothing aloud. The Lord knew his thoughts. Caleb was better able to lift up prayers out in the woods than in a churchhouse. That didn't necessarily seem a bad thing. Plenty of the Scriptures were inspired by the wonder of God's world. He had a psalm marked in his Bible about how the Lord made grass to grow upon the mountains. He made the trees grow too and filled the woods with his creatures. Caleb was one of his creatures that belonged where he was.

Phoebe shook her head and danced to the side. She wanted

to be on the move. He did too. To where his gaze could rest on Tansy Calhoun.

With a smile, he flicked the reins to head on across the familiar trail he'd ridden many times. He guided Phoebe down into the creek that made the easiest trail for a ways. Sparkling water flowed past the mare's hooves.

As they climbed out of the creek and through the trees to Robins Ridge, he thought about how he could explain showing up. But maybe he didn't need words for that. A neighbor should be able to call on a neighbor without stating a reason. Especially when that neighbor wasn't quite ready to state his real reason.

He came out into the open area around the Calhoun house for the third time in a week. The first time night had already fallen, and then Wednesday when he'd brought Aunt Perdie's plunder, twilight was gathering. He hadn't really given the place a good look either time, but he did today. The sun flooded down on the cabin in a sweet little nook of the hill with trees clustered around it.

The log barn sat not far from the cabin. His pa had helped Joshua Calhoun raise those logs into place before Caleb was born, but the barn looked as sturdy as ever. A mule and a couple of cows were eating hay pitched out on the snow. A rail on the fence around the lot looked ready to fall. Something Caleb could offer to help fix. A whinny from inside the barn made him smile. Tansy was home.

Another whinny came from the barn. A different horse. Not Shadrach. Caleb pulled up Phoebe and dismounted. Voices drifted out to him, and then Tansy and a man came out the barn door. Tansy was laughing.

The man kept his eyes on Tansy as they walked toward the house. A flatlander. His coat was too fine for a mountain man. Caleb didn't know the man. Had never laid eyes on him before now, but he didn't like him on sight.

eighteen

LOOKS LIKE YOU have more company," Damien said. "Your fellow, maybe? Or perhaps the husband of that girl inside who doesn't want me to know her story." Damien raised his eyebrows at Tansy.

Tansy had been so wrapped up in his talk about New York City she hadn't noticed Caleb. Weeks had gone by with no one showing up to visit and suddenly their cabin was a hive of activity. Ma was liable to take to her bed with all the strain of entertaining.

"Caleb?" Tansy shook her head. "He's a neighbor. That's all."

That felt wrong when she said it. Caleb was more than a neighbor. He was a longtime friend, but she didn't change her words.

"I like how neighbors matter up here in the hills," Damien said.

"Don't you have neighbors in New York?"

"Oh yes. Some of them right on top of you, living in the same building, while up in these hills, neighbors are spread out." He looked around at the trees, then back at Tansy. "Here you have so much space. In the city, people are always in your face, whether you want them to be or not."

"I guess it can be that way here sometimes too," Tansy said.

"Are you saying you're not glad to see this guy coming?" Damien smiled the way that made Tansy uneasy.

"I didn't say that. I just wasn't expecting him, but then I wasn't expecting you and Preacher to show up here either." Even to her ears her voice sounded stiff.

"But we did have a date," he said.

"A plan anyway. So you could get some stories." The word *date* made her even more uncomfortable. She moved toward Caleb, who had dismounted to wait for them. "Caleb, what brings you out so early?"

"I came by to see if Josh might need help getting up wood if you were running low." He gave Damien a look before he went on. "I don't believe I've met you, mister."

Damien stepped forward, his hand outstretched. "Damien Felding here."

"Good to meet you." Caleb took Damien's hand with a smile that seemed to have a question in it that he wanted to ask but wouldn't. "Caleb Barton."

The two men looked made out of different cloth. Caleb was tall with wide shoulders. His big hand engulfed Damien's. Damien was nearly as tall as Caleb, but he seemed much smaller because of the look of never having swung an axe. He was all smooth edges in his fancy riding coat with the splits in the back to sit a horse better. She couldn't imagine where he'd found that coat in Booneville, since he'd had nothing that looked like riding clothes when he rode with her to Preacher's house last week.

"I'm sorry." Tansy's cheeks warmed. She'd blushed more around Damien Felding in two meetings than she had for months. "I should have introduced you."

Damien had that smile back on his face. "We took care of it for you."

"Yes." Caleb had no smile at all as he looked from Damien to Tansy with questions lurking in his eyes.

She opened her mouth to explain why Damien was there, but Damien spoke up first.

"I'm not a revenuer, if that's what has you looking so dour." Damien looked amused by his own words. "I'm part of a works program to write a guide to note interesting places in the area a tourist might want to visit and also to record residents' recollections."

"A program like the packhorse library, only for out-of-work writers," Tansy put in.

"For artists too. All sorts of projects going on." Damien's smile didn't waver. "So in spite of me being a kind of government man, your moonshine stills are safe. At least from me."

"That's good to know." Caleb's face looked chiseled in stone. "Then could be nobody will shoot you before you get whatever you aim to get written."

Damien's smile got bigger. "That's where sweet Tansy comes in. She's volunteered to smooth the path for me to talk to people up here. She's already introduced me to Preacher Rowlett."

Caleb's eyes tightened on Damien. Tansy wasn't sure how she felt about Damien sounding so familiar. The "sweet Tansy" was unexpected, for sure. Something like Damien saying earlier that they had a date. But he wasn't from around here. He couldn't know mountain ways, and Tansy didn't know city ways. Her heart was pounding in her ears, but she wasn't sure if it was because of Damien acting so familiar or because of how Caleb looked ready to knock the man flat.

"Mrs. Weston asked me to show Mr. Felding around, and who better than Preacher to tell about our county." Tansy didn't know why she felt the need to explain, but she did.

"Mr. Felding?" Damien laughed. "Please, not so formal, Tansy."

Tansy flashed a look at him before she kept talking to Caleb. "Preacher is in the house having breakfast. We had to see to the horses before we get Aunt Perdie to tell some stories."

"She might come up with plenty. How true they might be is anybody's guess." Caleb smiled.

Tansy breathed easier as she smiled too. "I planned to warn Mr. Felding that sometimes Aunt Perdie remembers things the way she wanted them to be instead of how they really were."

Damien laughed a little when she said "Mr. Felding" again.

Caleb ignored him. "She and Coralee settling in?"

"Ma's making it work." Tansy wanted to ask Caleb to talk to Josh, man to man, about stepping in water over his head with Coralee, but Damien beside them kept the words unsaid.

Another awkward silence swelled up between them. Damien looked between them as though taking notes for that book he intended to write someday.

Caleb's mare tossed her head to jingle her bridle and break the silence. Caleb grabbed her bridle. "Easy, Phoebe."

Tansy stroked the mare's soft nose. "I remember Phoebe from when you used to come over." She glanced over at Damien. "Caleb was sweet on my sister when they were in school together."

"You don't say." Damien looked amused. "Maybe he picked the wrong sister."

Caleb gave Damien a narrowed-eye look that was anything but friendly. When he turned back to Tansy, his smile looked forced. "Guess I'll be on my way to let the two of you get on with your story gathering."

"You can't leave without coming in. Josh would be mad if I let you leave when he might need help with something. I'm gone so much I don't always know what needs doing around here." Tansy wanted that easy feeling to be back between her and Caleb the way it had been the day they'd helped Aunt Perdie. "Besides, you haven't seen your favorite cat."

"Prissy?" He did smile then. A real smile. "She'll probably attack me when I walk in the door."

"The devil cat?" Damien said.

"So you've met her," Caleb said.

"I was warned away before damage was done. I'm not much of a cat or dog person anyway. No place for them where I live in the city. But at Preacher Rowlett's house, I wouldn't have been surprised if he'd set a plate on the table for his dog."

"That sounds like Preacher. Did he tell you some dog stories?" Tansy breathed easier with the two men friendlier now.

"No, does he have some good ones?" Damien looked at her and then Caleb.

"I'm sure he does," Caleb said. "He has stories for everything, but he likes his Bible stories best."

"I found that out." Damien laughed. "You think Jesus had a dog when he was a little boy? Or a horse?"

"I'd wish a dog or horse for any boy," Caleb said. "But Bible stories have Jesus going by boat or shank's mare."

"By foot." Damien looked proud to remember that. "Same as hoofing it."

"Right." Tansy smiled at Damien and then at Caleb. "Please come in, Caleb."

"Yeah, no need to rush off, Barton," Damien said. "The more people telling stories, the better for me. Unless you're like that girl in there who doesn't want to tell me anything."

"Coralee?" Caleb said.

"That was her name, but I don't think she wanted me to know that either. But with or without her, let's get the party started. We'll have a grand old time."

When Tansy stepped back from Phoebe, Damien slid his arm around her shoulders to turn her toward the house. Caleb's frown came back, fiercer than ever, but he had no reason to look so bothered. Even if Damien was her fellow. Which he wasn't. But Caleb wasn't either.

Damien did seem to want to get familiar. Maybe too familiar. Not that it felt bad to have a man's arm around her. Even so, she sidestepped away from him. No need to let her emotions rush her down a river of trouble. A girl had to consider how things looked as much as how they felt.

Damien let his arm fall down to his side, with that smile again.

Tansy pretended not to see it as she glanced over her shoulder at Caleb. "I'll let Ma know you're here."

He muttered something she couldn't make out as he took Phoebe toward the corral.

"I get the feeling your friend doesn't like me all that much." Damien kept his voice low. "Guess I got off on the wrong foot with him the way I did with Rayma."

Tansy kept walking. "You do seem to like stirring things up when you get the chance."

"A story needs a little conflict."

"We aren't in a storybook here."

"No, but sometimes a little stirring, as you say, makes things more interesting." He grinned at her. "Tell me, sweet Tansy, do you like it when things are stirred up?"

"I'm not your sweet Tansy."

"I apologize if I'm being too familiar." He looked ready to laugh. "Would you prefer I call you Miss Calhoun instead? You are a miss, aren't you?"

"I'm not married, if that's what you mean, but Tansy will do."

"I'll agree on one condition," he said. "If you call me Damien. Mr. Felding sounds like my father's name, and trust me, you don't want to be riding all over these hills with my father."

"Don't you like your father?"

"I like him well enough when he's in New York and I'm here in the wilds of Kentucky. How about you? Do you like your father?"

"Of course. He's my father." That seemed an odd question, but then she had known some fathers who might be hard to like.

"Am I going to meet him?" Damien asked.

"He's gone right now. Off to find work."

"A man does need work. Except my father has never considered writing work. He believes in muscle power, not brain power."

"Oh." Tansy didn't know what to say to that. Her own father might think the same.

"I could tell you some stories about my dad, but I'm here to listen to stories. Not tell them. I want to hear your stories."

"I don't have any stories to tell."

He grinned over at her. "No stories about delivering those books to people? Never been challenged by a bear, frightened by a rattler, hurried on your way by a gun-toting granny?"

Tansy shook her head. "Not so far. Rattlers are wintering under rocks somewhere, and we haven't seen a bear in forever. No gun-toting grannies either. People are happy to see me coming with my books. They wouldn't shoot me."

"You sound sure of that." Damien looked behind him. "I'm just relieved your guy back there isn't packing a gun."

"He probably has one." Tansy could hear Caleb coming up behind them. "Most men up here want to be prepared. You know, in case they run into that bear or rattlesnake. Or see a squirrel to shoot for supper."

"Or a city guy where he doesn't belong," Damien said.

"I wouldn't worry too much." Tansy kept her voice light. "There's a government limit on how many city fellows we can shoot. One per year, I think. Caleb will probably want to save his one chance in case somebody comes along who needs shooting worse than you. So I think you're safe from him. Now Aunt Perdie? I'd be worried if she had a gun in her pocket. She doesn't pay much attention to government rules."

"You're right about that." Caleb stepped up beside them. He managed a worried look in spite of the twinkle in his eyes. "You did hide her gun, didn't you? If not, she's liable to shoot first and ask questions later." He glanced over at Damien.

Damien smiled. "All right, guys. I wasn't born yesterday. I know when somebody is pulling my leg." But under his smile, he looked a little uneasy.

Tansy and Caleb both laughed. "We are, Damien." Tansy shook her head. "Sorry about that, but outsiders are always ready to think we're going to shoot them. You'll be fine."

Tansy led the way up the porch steps. Just before she reached to open the door, Caleb said, "But all the joking aside, you did hide Aunt Perdie's gun, didn't you?"

Tansy played along. She looked back without the least hint of a smile. "I hope Ma did."

She had to give Damien credit. He only flinched a little. "Everybody's got to die sometime, I suppose. I hope I get a good story first."

nineteen

"WHAT YEAR WERE YOU BORN, Miss Sweet?"

Perdita gave the man a look. He sat in a straight-back chair directly in front of her rocking chair that Tansy had scooted away from the fire to face him. Perdita started to ask if the poor man was hard of hearing and had to lip read. He appeared to hear fine when Tansy said something, but seemed he couldn't hear Perdita's words if they came sideways to his ears.

He held his pencil over a notebook of some sort where he must intend to write down her answer. He'd done written her name. Had her spell it out. As if anybody didn't know how to spell Sweet. She wasn't so sure he hadn't asked her the spelling just to see if she knew her letters. And now asking when she was born, as if that was any of his business or that of the rest of those with big ears lurking around behind them.

Not that the others in the room didn't already know how old she was or at least close to it. That didn't mean she was ready to claim all her years right in front of them. It was bad enough being four years older than Hiram without speaking it out loud. But she couldn't not answer something. They'd think she didn't know and had gone senile.

"1876." That was close to right. Just borrowing three years

from time and still admitting to being older than Hiram. She wondered what stories he'd told this man and if any of them were true.

True or not, she figured they'd been stretched a little. Hiram always could make a story some better in the telling. Less'n he was telling something from the Bible. Then he stuck to the way it was written. Never had David using a long rifle on Goliath instead of that slingshot, whether the people listening to the story could read it for themselves or not. Besides, the slingshot was better than the rifle. Anybody ought to be able to take down somebody with a rifle no matter how big he was. But young David would have had some troubles if his rock hadn't found its mark.

She didn't rattle out all that talk but sat quiet and waited for this city fellow's next question. He wasn't hunting Bible stories. Said he wanted stories about mountain folk. Folks like her.

"After the War Between the States. So you wouldn't know how that conflict may have divided people here in the mountains." The man sounded disappointed.

"I knowed the South fought agin the North and folks up here took sides and got all crossways with their neighbors. My pa, he fought for the Union. Said he didn't want no divided states of America."

"I've heard that some of the feuds that happened after the Civil War were ignited by the hard feelings between men who fought for different sides. Was there truth in that?"

"Coulda been. But I'm thinkin' most of them coulda started over a woman or a grievance over a spot of land. Who knows what? Some idjit shot some other idjit and then that idjit's people had to shoot somebody to make things even. 'Cept things never did seem to even out."

Now the man looked interested. He leaned forward in the chair toward her. She rocked back and chocked her feet to stay there.

"So you remember the feud they had around here?"

"Don't remember a thing about it. Before my time and didn't have nary a thing to do with my people. We were too busy trying to find a way to live ourselves to worry about stealing anybody else's breathing chances. Didn't do so good at that."

The man looked like he didn't believe her. The others around them waited along with her to see what he'd say next.

Perdita had heard the tales they told about the feud that spilled through this county and beyond, but they were tales. Could be true. Could be lies, for all she knew. She did know pretty sure for certain that one of the main players in it all had gone off up in the hills to ponder things after some killing years and come back a preacher. Perdita supposed the Lord could change anybody for the better. Wasn't she herself doing her best to not be so contrary here in Eugenia's house?

After all, she had prayed that prayer back when her house was burning in front of her eyes and she and Coralee needed help. Promised to work on the contrariness. But this writer fellow was poking around at her ornery bone with his nosy questions.

Even so, she supposed she owed it to Tansy to let him ask what he wanted to since she seemed so set on the man writing something down. She kept looking at him as though he won the prize at the fair just because he had that pencil and paper. The girl ought to have more sense than to be letting a city fellow stir her interest. Be a heap better for her to look at that Barton boy.

Perdita had given him a hard time about being gone from the mountain when his ma needed him, but he had come back. A good mountain boy. He'd beat a city man as somebody to hang your future on, hands down. She bet this Damien, or whatever his name was, wouldn't even know how to skin a rabbit. Probably starve to death if he couldn't buy store fixings.

He was talking on about the feud. Things he'd heard from folks in Booneville. Probably old Judge Milner who never had known how to keep his mouth shut. The writer man appeared to want her to say she'd heard the same, but she just smiled and nodded a little. Not enough to show she was agreeing with anything he said. Just enough to show politeness.

Most of the others seemed to be doing some the same. Excepting Tansy, who kept looking at him as though she thought he'd been the one to write every one of them books she carried out to folks.

Perdita peered past the man over toward Coralee off in the far corner of the kitchen. She was making out like she was simply holding Prissy to keep the cat from making trouble, but that wasn't it. She was hiding back there. Didn't want nobody mentioning her to nobody.

That was why Perdita had agreed to answer the writer man's questions. So he wouldn't pay any more notice to the girl. And he hadn't. She had to give him that. Coralee had told him to leave her alone and he'd stuck to it. Too busy trying to impress the rest of them. Seemed to be working on Hiram, who looked extra proud to be the one to bring the writer to Eugenia's cabin. But then, Hiram had always been as crazy about books as Tansy. He ought to have written one himself. Maybe he should write about that feud this city fellow wanted to know about. Might be a bestseller.

Finally, the man sat back in his chair and tapped the pencil on the paper as though nothing else she might say would be of that much interest. "All right, so you don't have any stories about the feuds right now. But if you think of some later, I'd like to hear them. We'll move on." He put his pencil back in working position.

She couldn't see if he'd written anything down other than her name and that lie she'd told about when she was born.

Putting it on paper made the lie some worse, but she didn't admit to it. She'd have to beg the Lord's forgiveness later.

"So, you say you had some hard times. Was this when you were young or perhaps after you married?"

That had the whole bunch of them watching, waiting for her to tell the man what for, but it was an honest enough question. "I've been a Sweet all my life. Never married to take on any other name. Mind I said I was a Sweet and not that I was sweet." She chuckled a little at that and was pleased when Hiram and that younger boy of Eugenia's laughed too. Some people just naturally knew when a person said something aimed to be funny.

The writer man smiled and waited for her to say more. At least he wasn't poking her with questions. That made her consider answering him proper.

"But those hard times I was meaning were during my young years. Lost three brothers and one sister to the consumption before they got grown. Never knew why I escaped it. Ma and me both did. Pa got it some, but it didn't kill him the way it did the young ones. Pa always said I was spared 'cause somebody had to take care of things. Ma never was very spritely. At least in my memory. I was born fourth. The oldest two brothers had done headed out for who knows where before the young ones took sick. Never heared from them again. So maybe they went off and died down in the flatlands somewheres."

Perdita rocked forward and back. She wished Prissy was in her lap to give her hands something to do. She hadn't thought about these hard memories for some time now. But there weren't no reason for her to be shedding tears about what happened fifty years before. She hadn't shed that many tears even then. Not after her sister Miriam passed. She had cried a river for that sweet girl who followed her around like a shadow. Prettiest child ever. She got all the looks Perdita didn't. She weren't but five when the Lord took her.

"That must have been very difficult for you. How old were you?"

"When the first one, my baby sister, died, I was thirteen. Old enough to help with laying her out. She was a little thing. Looked like a doll after we put on her best dress. Ma made it for her after Miriam started coughing. Cried some while stitching it, as if she knew it would be her burying dress."

"So was it tuberculosis that took her?"

Perdita liked how he said took her instead of killing her dead. "I did hear some call it such. Others named it the white plague. The brother next up from me died after Miriam and then the two younger brothers. Not all at once. Spread out over some time, but once they went to coughing and spitting up blood, they just wasted away. Ma sort of wasted away with them."

"Didn't you bring in a doctor to treat them?" The man frowned over the question.

"No doctors much around back then. No doctors much around now. We had a granny healer who made them a potion. Didn't help them one whit, but she tried." Perdita pulled in a long breath. "Ma seemed to get some stronger after they all passed. I reckon once the worst had happened, she figured she might as well accept it and try to keep going."

Perdita looked over at Eugenia, who had settled in the other rocking chair by the fire. A tear was sliding down her cheek. "Eugenia over there knows about that, don't you, Eugenia? Once the Lord takes them, there ain't nothing to do but give them over to heaven and keep doing what has to be done."

"That's the truth," Eugenia said.

Perdita went on with her story. "Pa, he made it on a few more years, but one morning he didn't get up. Died in his sleep. Ma got up to fix breakfast, thinking he was just sleeping sound. 'Bout did her in when she went to wake him up and couldn't. Don't think she ever got over that."

Perdita stopped talking and looked down at her hands. That morning changed everything. Sometimes things went that way. Before that, she'd thought of going off to school somewhere. Maybe learn enough to talk right and be a teacher. But weren't no way she could leave her mother after that.

She'd made the best of it, but she couldn't deny a little bitterness had taken root. Not at her pa, who couldn't help dying. Not even at Ma, who couldn't help grieving. She'd known it was wrong to look askance at the Lord, but she had some back then. The good Lord had miracles in his pocket, but he hadn't sent none down for her. At least that's how she felt at the time.

Took a while before she came around to thinking on how her escaping the consumption might have been one of those miracles the Lord had thrown down. Nigh on another miracle was how the Lord never stopped loving her in spite of her doubting ways. But even after she embraced that truth, she stayed some contrary when other folks came around making out like she weren't doing the best she could. Maybe they were right.

She waited for the man to ask something else, but he kept staring down at his paper like what she'd said wasn't what he wanted to hear. If that was so, guess he shouldn't have asked.

Eugenia broke the silence that had fallen over the room. "No wonder about that. Your ma was blessed to have you to help her." Her gaze slid over to Tansy.

Perdita had to bite her lip to keep from telling Tansy to run and get away from that expectation right now. But with the way the girl was watching the writer fellow, she would be liable to run the wrong direction. Besides, Eugenia had other children. Little Livvy would grow up to be a good daughter. The boys might marry helpful girls. Her situation was not at all like Perdita's. Not at all.

Weren't nobody else to take over the care of her mother, but she could have looked around for somebody to marry. That wouldn't have kept her from doing for her mother. She was still young then. In her twenties. Could have had a houseful of children. But the men able and single weren't any she wanted to invite into her house.

She let her glance slide over toward Hiram. The one man she'd been ready to consider had never considered her. She looked quickly back toward the fire, but not before that writer man's sharp eyes caught the direction of her look, even if nobody else in the room paid the least bit of attention. If he had the nerve to make mention of it, she'd show him where the door was and not care whether he had time to grab his coat before she pushed him out of it.

She braced herself for the man's next question, but he at least had enough sense not to pretend like he knew more than he did just because she'd looked around the room. He asked, "How did you and your mother make out without a man to do the plowing and planting or whatever needed doing on your farm? Your father did farm, didn't he?"

"Mostly. He worked in the mines a while before he married Ma, and then he did some lumbering too. But mostly we got by with what we raised. I was able to keep raising us a sass patch." When the writer frowned, she added, "That's a bean and potato patch to you, I reckon."

He smiled. "I've never had any kind of garden at all."

"Figures," Perdita said. "Never could understand how a person could get by without some connection with the dirt. But I'm supposing your dirt up there in New York City is all covered up with something or other."

"You're right there, Miss Sweet. But we do have parks where we can walk and commune with nature. And some people with a little land around their houses sometimes raise a tomato or two. I've never been fortunate enough for that, but I'm hoping to still be here in the mountains when winter ends to see some things growing and flowering."

That softened Perdita's heart some toward him. At least he did know he needed dirt.

"Tell you what, mister," Arlie spoke up. "If'n you're still around come summer, I'll let you do my part of the weeding in the sass patch."

"Arlie," Eugenia called him down.

"That's all right, Mrs. Calhoun." The writer looked over at Eugenia. "I might like the experience, but I'm not so sure I could tell the difference between a weed and a bean plant."

"It ain't as hard as all that," Arlie said. "I could teach you."

That made everybody laugh and the air felt easier in the room. Perdita looked around at them. Tansy was smiling like she'd just got a new load of books and everybody else was looking easy too, excepting the Barton boy. He hadn't smiled since he'd come in that Perdita had seen. He kept looking at the writer man as if he'd like for him to get on his horse and ride away. Far away.

The writer man started talking again. "So, you were able to grow everything you needed?"

"We had chickens. Everybody has hens up here in the hills. And we had a cow. Managed to raise enough corn to feed the chickens, and neighbors kept the cow in hay in the wintertime and brung a bull up for her when it was needed. They shared from their hog killings too. Bacon and such. Neighbors know how to watch after one another up here."

Perdita rocked back and forth. The writer man didn't push extra questions at her but gave her time to think. Maybe he wasn't all bad. She was some surprised that Hiram had managed to stay quiet so long and let her be the talker.

"We did need a thing or two at the store now and again. After Pa died, we couldn't raise enough to sell anything. I was able to gather some chestnuts to sell. Mighty sad when those trees went to dying. Then I've always been handy with my needle. I made some quilts others took a liking to. Those and my carvings brought us in enough to buy some extras like coffee and flour."

"Carvings? What kind of carvings?" he asked.

"Nonsense things. Nothing that amounted to much. Never could make anything a person might recognize." Perdita reached in the pocket of her dress and fingered her folded carving knife. "Naught but trifles."

"That ain't so." Coralee spoke up from the kitchen. "They're works of art. Arlie, get one of them down off the mantel to show Mr. Felding."

Perdita didn't know who was the most surprised to hear her. The writer man or her. She said, "The girl kept them from the fire that got our cabin last week." Arlie looked at her and she nodded. "Go ahead and get one of them down to show him if'n he wants to see it."

She'd never cared a lot whether anybody liked what she whittled out of the wood or not. At least not since she'd tried to make that butterfly for her ma that turned out to look more like a bat. Guess that carving burned up in the fire since it was put back in a chest with her ma's things. All gone now. Funny how a person hung on to things that weren't of no use at all, but then those were the things that pierced the soul when a body lost them.

Arlie picked up the odd figure that a person might imagine was somewhat the shape of a man with shoulders rounded and arms hanging low. Markings in the wood brought to mind teardrops.

That wouldn't have been the one she'd have picked to show the man, but she hadn't told Arlie which one. It wasn't a bad carving. She'd rubbed it down until the wood was nigh on as smooth as silk. She'd considered taking it down to see if Wallace Beatty could sell it to some unsuspecting soul who might not see the tears in the piece, but she'd kept the sad thing instead.

She wondered what the writer man might see in it.

He ran his fingers over it, top to bottom, and then held it to the light of the fire. "What is it, Miss Sweet?"

"Whatever you think it is," she said.

"Fair enough, but I'm not sure." He looked around the room. "Do any of you know what it is?"

Hiram spoke up first. He'd been quiet about as long as Hiram ever was quiet, so Perdita wasn't surprised he put in his two cents' worth before anybody else. "I think it looks like somebody praying."

Of course Hiram would say that, but Perdita didn't say he was wrong. She'd already said it was whatever anybody wanted to say it was.

"That's a good thought, Preacher." The writer man looked over at Perdita. "But you don't look like you agree, Miss Sweet. Tell me, what did you intend it to be when you started carving it?"

"I never intend nothing when I start out. I let the piece of wood tell me."

"But it must mean something or represent something to you when you're finished," the man persisted.

She guessed he had to badger people for an answer if he wanted to get stories. Not that the piece of wood he held was any kind of story for him. "Most things mean something."

Arlie gave his idea. "I think it looks like a boy in trouble with his ma."

"That could be." Eugenia laughed.

Josh spoke up then. "It just looks like a piece of wood to me. All smooth and nice but not like anything in particular. A pretty to sit on a mantel." The boy took a quick look over at Coralee, as if afraid he'd said something she wouldn't like. When she looked up at him, he added, "Not that it ain't good having something pretty to look at."

Perdita had seen him eyeing Coralee, but she didn't think

the girl had noticed it yet. She was too turned inward on that baby she was carrying. But she did smile at the boy now. Made his face light up and must have given him the courage to say more.

"What do you think it is, Coralee?" he asked.

She looked down at Prissy still in her lap. The girl had a magic touch with that cat. "I don't know. That's for Aunt Perdie to say."

"But she's already said she's not saying," Josh said.

The writer man was sitting back, obviously curious about what the others might claim to see in the piece he held. Perdita was a little curious herself. She'd never had more than one person at a time giving an opinion about one of her pieces.

"Then I guess we won't know." Coralee's voice wasn't much over a whisper.

The writer man looked over at Tansy. "What do you think, Tansy?" He smiled a little, as if giving her some kind of test.

Color pushed up in her cheeks. Perdita was ready to speak up to get the girl off the spot, but the Barton boy did first.

"It's sorrow." He didn't look at Tansy or the figure in the writer man's hand but at Perdita. "If you look, you can see the tears."

"Really?" The writer man held the piece of wood up for a closer look, then peered over it at Perdita. "Is he right, Miss Sweet?"

"I done told you there weren't no right or wrong thinkin' about the thing. It's naught but a trifle. Something to keep my hands busy."

She reached to take it from him. It felt warm in her hand. Not from his hand but from the memory of carving it.

She knew one thing. She wasn't telling the writer man any more this day. "I done talked myself out for a spell," she told him. "But you come back some other time and I'll think up some mountain story with some fun in it if'n Hiram hasn't already told them all to you."

She nodded a little at the Barton boy. Guess he did know some about sorrow after losing his brother. For a fact, she'd carved out this piece after her ma died, not mourning her ma so much as the lost years of her own life.

twenty

TANSY WAS GLAD ENOUGH that Aunt Perdie was through talking so she could head to the barn to saddle Shadrach and take off toward Cox's Creek with Damien. She loved her family. Truly she did. But sometimes it was good to leave them behind for a while. Of course, this day, her house was running over with more than her family.

She supposed Aunt Perdie was family. Cousins, even some removed, were still kin. And Coralee needed a place too. Josh was ready to make her family, even if he hadn't known her but a week, not counting the time the two were in school together. But they were kids then. Too young to be thinking romance.

Could be, Josh was too young to be thinking romance now, and Coralee definitely wasn't thinking it. Poor girl was still wounded by whoever had led her on with sweet words, loved her, and then deserted her. Perhaps the love had only been on her side and not his. If so, things must be changed in her mind now. From the way she shrank away from Damien's notice, she didn't seem to want anybody to know where she could be found.

Still, Tansy hoped to have time after taking Damien to Miss Odella's to swing by Hardin's Creek so she could let Coralee's sister know where she was. Then again, with the way the mountain grapevine spread news, the girl might already know Aunt Perdie had found a place with them after her house burned and would hope Coralee was with her.

Tansy blinked back a few tears thinking about Aunt Perdie's story. Little wonder the woman rarely seemed happy. And that carving. Sorrow. Sometimes you thought you knew a person and then found out you didn't know half enough. She'd had wrong thoughts about Aunt Perdie. How she'd taken in Coralee proved that.

Tansy pulled on her coat while Damien said his goodbyes, and Caleb offered to help Josh get up firewood. Ma would be relieved to see them out of the house. Except for Preacher, who had settled down by the fire and made himself at home. Ma wouldn't mind him staying a while since he promised to tell Livvy about Jonah and the whale. Preacher knew how to make a Bible story entertaining for young and old alike. Coralee might even creep out of the shadows to give him a listen.

Caleb walked with Tansy and Damien toward the barn while Josh and Arlie headed out to get the mule. He still wasn't smiling. When he used to come visiting, he was always ready to smile, but that had been a few years ago. A man could change.

On her other side, Damien was smiling too much. That smile that made her feel like he found everything about them backward and amusing. She supposed they did seem backward to somebody from a big city like New York. She had a hard time even imagining what it would be like there. She'd seen pictures, but that wouldn't be the same as standing in the middle of all those buildings with the ground covered by cement. No mud anywhere. That might not be so bad. They were apt to find plenty of mud on the trails today with the way the sun was warming the air.

"The snow is melting." A silly thing to say, but the silence needed breaking. She glanced over at Damien. "Do you get a lot of snow in New York?"

"Do we ever. Slows down everything until the plows scrape it out of the way," Damien said. "Snow is a real nuisance there. Here I haven't minded it so much. Guess that's the difference in going by horse rather than motorcar. No cars here. I'll wager plenty of people up here have never ridden in one. Like Miss Sweet." He peered over at Tansy and Caleb. "Or maybe you two."

"We had trucks in the Corps," Caleb said. "They might go faster, but a person is always having to tinker with the things to keep them running."

"Sounds like you were ready to trade your truck for a horse," Damien said.

"Ready enough." Caleb still wasn't smiling.

Damien turned to Tansy. "Next time you come to Booneville, Tansy, if I'm still around, I'll take you for a ride."

Caleb spoke before Tansy could say anything. "Better hope it's not snowing or you're liable to end up in a ditch and need a mule to pull you out."

Damien laughed. "I don't doubt that I might get in that fix sometime before the winter is gone, but if the company is good, being stuck in a ditch might not be so rough."

Tansy could feel both men looking at her, but she kept her eyes straight ahead and kept walking. Sometimes nothing was the best thing to say.

Before they got to the barn, Caleb touched her arm. "Got a minute to talk, Tansy?" He shot a look over toward Damien. "Alone."

Damien shrugged. "Guess I'll go pack up my notebook."

"Did you need something, Caleb?" Tansy asked after Damien went on to the barn. "A new book for Junie? I'm not supposed to give out more books to a family until I get the ones you have back, but I could give him another book this one time."

"I don't want you to break the book rules. Junie and Cindy

Sue don't mind hearing that Doctor Dolittle book over and over. And Junie has just about memorized the one you made for him to help him with his reading. Thank you for that."

"Is he better?"

"Hard to say. Maybe. I aim to take him riding later on today."

"He should like that. It promises to be a nice day." Tansy couldn't imagine what Caleb wanted if it wasn't about books.

"Tomorrow might promise a good day too." Caleb looked up at the blue sky, then back at Tansy. "It being Sunday, I wondered if you might have time for a ride with me. We could go back over to Aunt Perdie's to see if we can find anything to salvage."

When she hesitated, he rushed on. "If you aren't too busy with other things." He looked straight at her then, his blue eyes full of life the way she remembered them when he came to visit Hilda.

She tried not to wonder if Damien might ask her to take him around to more people for stories. Or even to come to town for that car ride. But she couldn't turn down an old friend for a maybe. "We should check on Aunt Perdie's place. So come on over if you want to."

"No worries about that. I'll want to." A smile lit up his eyes. He started to turn back toward where Phoebe was tied to the fence, but then he hesitated as he glanced toward Damien waiting in the barn door. Caleb lowered his voice. "I worked with city men in the Corps. It would be good for you to keep in mind that city people don't necessarily think about things the same as those of us up here in the hills."

"It's just part of my job."

Caleb raised his eyebrows a little. "Are you getting paid extra for working on Saturday?"

"Well, no."

"Then I guess it's a favor and not a job." His smile was gone.

"Just pay mind to what I said. He seems a nice enough fellow, but the two of you will be on lonesome trails together."

"Weren't you just asking me to ride along some of those same trails with you?" Tansy frowned.

"But you know me. We're neighbors and I'd never do anything to hurt you. I can't be as sure about him."

"I'm not a little girl anymore, Caleb. I can take care of myself."

"Of course you can. I'm counting on it." He gave her another smile before he headed over to where Arlie waited with the mule while Josh went to the woodshed to get the axe and saw.

Tansy should have taken the opportunity to ask Caleb to give Josh some advice about Coralee since he seemed to have everybody's well-being in mind. Whether that person wanted him to or not. That might have been how Josh had felt that morning when she was giving him unasked-for advice.

She blew out a breath and headed on to the barn. She had simply been trying to keep Josh from getting hurt. Perhaps Caleb was doing the same for her. Simply trying to help, but because he'd known untrustworthy men in the Corps was no reason to think Damien was the same.

"Everything all right?" Damien asked.

"Fine." Tansy wasn't going to talk about Caleb to Damien. Or Damien to Caleb. "I'll get Shadrach saddled and we can take off."

"Sounds good," he said, still with that smile. Maybe Caleb was right about him. But then she was right too. She could take care of herself. She was out on the trails every day. People were getting used to seeing her with her books. Plus, whether she was getting paid extra or not, Mrs. Weston had asked her to take Damien around.

They didn't talk much as they rode away from the barn. Tansy led the way, only glancing back now and again to be sure she didn't get too far ahead. Belle plodded along as slow as a plow horse.

Damien looked less nervous on the mare than he had earlier, riding to Preacher's house. She wondered what stories Preacher might have told him. While Bible stories were Preacher's favorites, he did tell a Jack story now and then. Those mountain tales always made her smile. That Jack was forever getting into messes but coming out good at the end.

She led the way into a creek where the going was easier and waited for Damien to catch up beside her.

"I can't get used to wet roadways." Damien looked over at her. "You're looking happy. Want to let me in on the joke?"

"No joke." Tansy let Shadrach move on down the creek after Belle fell in beside her. "I was wondering if Preacher might have told you a Jack story."

Damien laughed. "Fact is, he did. But when I asked for more, Preacher said I'd have to find someone else to tell them. Shepherd boy David stories from the Bible served him better."

"Preacher's good at Bible stories."

"But the Jack stories are more what I need for my project. Do you think Miss Sweet knows any? She seemed agreeable to talk to me again sometime."

"I really don't know what stories Aunt Perdie knows, but I wouldn't doubt she could tell some entertaining ones." Tansy listened to Shadrach's hooves hitting the rocky bottom of the creek. "Then again, they might be sad."

"She did have a sad story, didn't she. I guess you might have heard it a dozen times before."

"No," Tansy admitted. "Really haven't been around Aunt Perdie that much. Just went with my father now and again to make sure she didn't need anything. She didn't encourage visits. Always seemed to want to be alone."

"Didn't sound so much like that when she was talking, did it?" Damien's smile was gone. "Sounded as if she'd had a lonesome life. Guess that's why she carved out something that meant sorrow to her." He shook his head a little. "Not that I could see sorrow or anything but a polished piece of wood myself. What did you think it was?"

"I didn't have much imagination about it either."

"That sounds bad when you put it that way. A writer needs plenty of imagination."

"Maybe not if he's writing the truth."

"Ah, the truth. But what is the truth?" Damien looked over at her. "Take those Jane Austen books you like. Do they sound like something that could really happen?"

"Maybe across the sea." Tansy looked around at the trees pushing in on the creek. "Not something that would happen here."

"But people are people wherever. Think about Elizabeth in *Pride and Prejudice.* She was a girl much like you. Loved books. Walked where she needed to go. You're riding, but you'd walk if you didn't have a horse. And then she meets a man who lives on a grand estate with pockets stuffed with money."

"That's not much like me. Plus, she doesn't know about the estate at first."

"True. If I'm remembering the story right, she doesn't much like him to begin with. Or he her."

"Oh, I think he loved her all along. From his first sight of her."

"A true romantic." Damien laughed. "I like that in a girl. Somebody who can see past the surface to the possibilities. The underlying story."

"But I didn't see under the surface of Aunt Perdie's carving."

"Perhaps that's because, like me, you need words to take you into a story." Damien was silent for a moment before he went on. "Your friend Caleb nailed it though."

"He did. Perhaps because he lost his brother in an accident a few months ago."

"That could be it. I wasn't expecting him to be a deep thinker."

"You think we can't think about things up here in the hills?" Tansy kept her voice light.

But he must have heard irritation in her words. "Uh-oh. Here I go again. Ruffling feathers. No wonder you people had so many feuds up here. Always ready to take offense."

You people. She let out a slow, quiet breath. She might be ready to say "you people" about him and anybody from the city too. No need in getting her back up over his words. "No offense here. The feuds were a long time ago. We work on getting along these days."

"You're right. I'm impressed with how you people up here take care of one another. Like your mother taking in Miss Sweet, and not only her, but that girl too. And I thought it neighborly your Barton fellow offering to help your brothers." Damien gave her a sideways look. "Then again, it could be he did that to make points with the brothers' sister."

Tansy didn't even try to not sound irritated this time. "He's a neighbor. That's what neighbors do. Help one another."

Damien had a smug look. "I think he wants to help you a little more than your neighbor down the way. But if we did get stuck in that ditch sometime, I'll hope he'd want to be a helping neighbor to get my car back on the road."

"I'm sure he would."

"I'm sure he would too. Especially if you were in the car with me. He wouldn't want us to get stuck in the same place together for too long." He had that smile again.

"Here we are together today hunting up people to talk to you."

"So we are, and I wouldn't be surprised to look behind us to see him following along to keep an eye on us. Or rather, on me."

"He wouldn't do that."

"Maybe not, but he did warn you about me, didn't he?" When she looked over at him, he waved away his question. "Don't answer that. None of my business what he said. But you can tell him the next time you see him that I promise to be a perfect gentleman."

"That's good to know," Tansy said.

Very good to know. But she didn't plan to tell Caleb anything about Damien when they went riding tomorrow. Damien was off-kilter to think Caleb was sweet on her. What she needed to do was stop letting Damien Felding make her feel off-kilter.

twenty-one

CALEB SHOULDN'T HAVE SAID anything to Tansy about Felding, but he couldn't help himself. He saw the way the man kept smiling at Tansy, like he knew she thought he was better than anybody up here in the hills. Just because he was a writer from a big city.

With a hard swing of the axe, Caleb cut into the tree trunk. Then he hit it again, taking out some of his frustrations of Tansy taking up for the man. Comparing being with Felding as the same as being with him. He struck the tree again and then stood up to breathe in and out to tamp down his temper.

"Want me to take a turn?" Josh reached for the axe.

"No, I've almost got it notched. Then we can saw it through. Did your pa teach you how to notch a tree to make it fall the way you want?"

"He generally did the notching, but I've done it some since he took off. I pick littler trees that are easier to saw down on my own."

"Smart thinking," Caleb said. "Don't want to take any chances."

Every time Caleb felled a tree, Reuben came to mind. That made him glad he was here to help Josh. The boy was taller than Caleb, but he was a slim model. He lacked some getting his muscle growth. Taking on a man's load for his family was weighing him down. Not that the boy would admit it. He put up a good front as he led the way to a dead chestnut. While it wasn't one of the monster trees, its trunk was as big around as a good-sized man.

Caleb ran his hand up and down the trunk. No doubt it was dead, but he did wish it wasn't.

"I've been eyeing this one," Josh said. "But was waiting. Hoping Arlie would get big enough to help with the saw."

"I could handle my end," Arlie said.

Caleb turned his head so Arlie couldn't see the smile that slipped across his face. The kid was at that age where everything seemed possible. He wasn't much older than Junie. If only Junie thought he could do everything. Or anything.

"Sure you could," Josh said with a big brother's disdain. "It ain't as easy as you think."

Caleb wanted to laugh, but he remembered being the little brother. Of course, Josh was more than twice as old as Arlie, while Reuben only had a couple years on Caleb. By the time Caleb was a teenager, he could generally outwork Reuben no matter the chore. That didn't mean Reuben didn't act the oldest.

"How old are you, Arlie?" Caleb asked.

"Going on nine, but Josh there don't think I can do anything but feed the chickens and pack in wood." Arlie picked up a stick and snapped it in two.

"Both things that need doing," Caleb said.

"Girls' work." Arlie screwed up his mouth in disgust.

"And what girl you think is gonna do it at our house with Ma not able, Livvy no bigger than a minute, and Tansy gone till who knows when." Josh scowled at his brother.

"She go out on her book routes every day?" Caleb asked.

"Every day excepting Saturday and Sunday and then today she's off with that other feller."

Caleb took off a glove and ran his finger along the axe blade as if checking how sharp it was. "She been out with Mr. Felding before?" He tried to sound casual, as though it didn't mean anything to him if she had or hadn't.

"Naw. Think she met him this week when she went to Booneville. The day Aunt Perdie's cabin burned and the two of you brung her and Coralee over to our house. Tansy took the writer feller to Preacher's house before she went to Aunt Perdie's."

"Then I guess she told you all about him." Caleb pulled his glove back on.

"Not more'n a mention. We were too busy figuring out what to do with two extra women in the house."

Arlie spoke up. "And that cat."

"I guess that did take some thinking." Caleb smiled. "Especially the cat."

"I don't know," Arlie said. "The cat might be easier for Ma to get used to than Aunt Perdie."

"Ain't good to talk bad about kin." Josh frowned over at his brother.

"She ain't much kin. But I didn't mean it bad. I sort of like her. I want her to show me how to whittle something." He picked up a stick and studied it. "But something happy. Not sad."

"I guess you could give it a try," Caleb said. "All it takes is a whittling knife."

"And calling your whittled wood whatever you want to call it," Josh said.

"I think I'm the one that named it," Caleb said. "Not Aunt Perdie."

"I reckon I didn't have as good an eye for it as you did." Josh shrugged. "I don't think that city feller had any idea what she had carved out either." Josh peered over at Caleb. "What did you think about him?"

"Don't know what to think." Caleb kept his voice even. No need letting Josh know his real thoughts and how he'd like to punch the man in the nose and see the last of him in this part of the country. "Only just met him."

"I didn't like him much," Josh said. "Seemed a mite unfriendly turning down Ma's breakfast."

"That suited me," Arlie said. "Left more biscuits on the plate. If a person could beat Preacher to them. You'd think Preacher hadn't seen a biscuit for a year."

"I remember your ma's biscuits from when I used to visit before I went off the mountain," Caleb said. "Not something anybody with any sense would turn down. But Felding hadn't ever had your ma's biscuits, so guess we shouldn't fault him for not knowing what he was missing out on."

"If you say so," Josh said. "Tansy looked to think he was fine. Skipped breakfast too when he turned down eating. Course he did say he'd had breakfast before they came our way, but Tansy went hungry, I reckon."

That made Caleb want to pound his axe into the tree again, but instead he gripped the handle tighter while keeping a grip on his temper too. "Guess we better get this tree down. You stay back out of the way, Arlie. Once it comes down, you can chop off some of the branches and maybe find you a good whittling piece and one for Aunt Perdie too."

"How will I know if it's a good one?"

"That I don't know. But Aunt Perdie might tell you to eye them and see if they speak to you."

Arlie pitched aside the piece of wood he held. "That one ain't saying a thing."

Caleb laughed as he swung the axe again. Maybe he should bring Junie over to see Arlie. But not tomorrow. Tomorrow he was coming to see Tansy. Only Tansy. He looked up at the sky through the trees. He hoped the Lord would send down more sunshine tomorrow. And that Felding would have other things to do in Booneville or, better yet, back in New York City.

Josh held up his end of the sawing, and after some steady work, the tree crashed down. Caleb always paused a minute after downing a tree. Seemed a time to be thankful for God's provision, and now after Reuben died, a time to be glad nothing went wrong.

"I'm always kind of sorry when I take down a tree." Josh stepped up beside Caleb. "Pa laughed when I told him that. He thought I was being silly. I reckon you probably think so too."

"Nope. When a man has to get his living from the land, it's good to appreciate what the ground gives us. That includes trees. But this one was already standing dead. No sense letting it go to waste." Caleb looked over at Arlie. "Take your hatchet and trim off some of the little branches while Josh and I take a breather."

"Don't chop off your fingers," Josh said. "Ma would be a mite perturbed at me if'n you did."

Arlie started hacking at the branches with all the enthusiasm of a boy too young to be of much real help.

Caleb watched the boy for a minute to be sure he knew enough about using the hatchet to keep from chopping off those fingers Josh had warned him about. Then he said, "Things hard for you since your pa left?"

"We get by," Josh said. "Better now that Tansy has the book woman job. Ma was fearing we might have to go on the dole to keep food on the table after Pa left. Tansy going to see about getting help from the county was how come them to consider her for the packhorse librarian job. They give them government works jobs to folks to keep them off the dole, you know."

"Right," Caleb said.

"We were glad enough Tansy got the job. We didn't want to have to take a handout." He looked over at Caleb. "I reckon

I could try to sign on with the Corps like you did, but I don't know what they'd do at home if'n I was gone. Was you glad you went?"

"I was. Got to see places I might not have otherwise. But things were different for me. My brother was still there then to take care of things around home. And the money the Corps sent Ma for my work was a big help."

"Yeah, it was bad about Reuben. We were all some sorry when we heard that. You never think something like that will happen." Josh stared at the tree. "Ma don't think Pa will ever be back. That he must have drowned in the river or something."

"Well, don't give up on him yet," Caleb said. "He might still be hunting work."

"If that's so, he might as well of stayed here, but I reckon he'll either come back or he won't. Naught I can do about it."

"That's a good way to think, Josh. Just consider the things you can do something about."

"I wasn't meaning to complain. Ma don't push me to do nothing I can't. Tansy, she can be bossy, but we'd be up a creek without her book job. Probably wouldn't have much but shucky beans to eat this time of the year, and I sure do get tired of bean soup. Leastways with Tansy pulling in money we can buy a few store fixings."

"That's good."

"You don't think Tansy will run off with that city feller, do you?"

Josh's words made Caleb's insides freeze up. He had to force out a breath before he could answer. "She's got more sense than that."

"I don't know. Pa said her reading those books had her thinking about romance like it don't ever happen up here in the hills."

"People fall in love up here same as anywhere," Caleb said.

"Well, yeah. Have a pile of babies most of the time too. But

I tried to read one of those stories Tansy likes once. Those book people weren't nothing like us."

"Tell Tansy to bring you a Mark Twain book. You'll like that better."

"He have love and stuff in his stories?"

"Some, I guess. But more adventure and fun."

Out in front of them, Arlie wasn't making much progress as he attacked the branches. They needed to help him if they aimed to get this tree back to the house in time for Caleb to get home and take Junie for that ride, but he hesitated to pick up the axe. He had the feeling Josh had more to say.

Josh shuffled his feet and didn't look at Caleb. "You've been around a while, Caleb, but you ain't married or nothing."

Caleb had to smile. "With a little luck, I might still find somebody willing to marry this old man."

"I didn't mean that. But, well, you know. Plenty your age are already married. Some my age."

"You have somebody in mind?" Caleb did have somebody in mind, but he wasn't confessing that to her little brother.

"Sort of. Not exactly." Josh looked up at him then. "You ever been in love, Caleb? I mean really in love with a girl?"

Caleb hesitated, wondering how to answer without revealing too much. But the boy so sincerely needed some kind of answer. "I have." He could have said he was in love right now.

"What happened? She throw you over?"

"Nothing's happened yet. It's kind of wait and see right now."

"Oh. Well, I guess you being gone from the mountains so long might have put a kink in any romance you had around here."

"That's so."

"But how do you know you're really in love?"

"It's just something you do know. Maybe by the way you feel

when you think about the girl. Does the very sight of her curl your toes up in your shoes? Or make your heart jump around in your chest like it's trying to find a way out? Just thinking about her can make you smile even when there's nothing much to smile about."

"But can it happen fast? Like maybe overnight?"

"I'm not sure about that. I've heard people say so, but if that's how you're feeling about somebody, then it might be good to give it a little time to see if it lasts. Do you have your eye on a neighbor girl?"

"You might say that." Josh stared down at his boots again. "A little closer than neighbors right now."

Of course. "I see. Coralee."

Josh's gaze flew up to Caleb's face, his cheeks fire red. "I always did like her back in school, but that was different from now. I just want to help her."

"Love is more than feeling sorry for somebody."

That brought a spark of anger to Josh's eyes. "I'm not feeling sorry for her, but I'd give whoever did this to her a beating if I could."

"I don't think she was forced."

That turned the boy's cheeks even redder. "Maybe not, but she don't want nothing to do with him now. She's been talking to me some. Says she needs somebody to talk to."

"Might be she should pick your ma for that."

"Ma don't condemn her or nothing, but I think Coralee can sense she don't exactly approve of what is either. Not that Coralee can do anything about that now. But if I married her, that would make an honest woman of her."

"Did she ask you to do that?"

"Oh no. I don't know that I could even get her to agree to it. But I'm wanting to ask her."

Caleb pulled in a breath. The boy was obviously sincere and maybe in love. Maybe not. Maybe just ready for any girl to catch his fancy, and now here Coralee was right across the dinner table from him. "Might be best to wait a while."

Josh frowned. "You sound like Tansy."

"Is that what she said?"

"Something like it."

"I don't say you have to wait forever, but Coralee has had a rough week. Let her settle in and get to know you better without having to say yes or no. That might put her in a bad spot living there in your house."

Josh hung his head and dragged his toe through a patch of snow. "I reckon you're right."

"She is going to have a baby."

"That's plain enough," Josh said. "I'm willing to be the babe's daddy."

"If she wants you to be his daddy." Caleb pushed that truth at him. Would he be willing to face that same kind of truth with Tansy if she told him she wasn't interested in him?

"Why wouldn't she want me to be the baby's daddy?"

"I don't know. That's something Coralee would have to answer. If she would or she wouldn't. But if I were you, I wouldn't make her answer it yet."

"You're thinking she might say no."

"She could. At least right now while she's still hurting from whatever landed her where she is."

"But what if somebody else comes along and asks her first?"

"I don't think that's likely, but if it were to happen, you'd be better off not baring your heart to her." Caleb picked up the axe. "Now we better help Arlie if we ever expect to get this tree trimmed up enough to drag back to your house."

Caleb chopped and sawed, and all the time he wondered if he'd given Josh the best advice. Maybe if he hadn't been so hesitant to show Tansy how he felt about her back when, she might be his right now. Instead, he might lose her once again, the way he'd thought he had before.

twenty-two

"LIVVY, DON'T CRIMP THE PAGE like that." Tansy swooped down on the little girl to scoot the book away from her before she could dog-ear it.

Livvy burst into tears and ran to Ma. "My book."

Tansy wasn't bothered by her tears as she smoothed down the page. "You have to treat a book right, Livvy."

Ma looked around at Tansy from the stove. "They aren't your books, Tansy Faith. The book woman left that here for Livvy." She emphasized the book woman as she hugged Livvy.

"I am the book woman." Tansy stared at her mother.

"Friday you were the book woman. Today you're Livvy's big sister, and there ain't no call in you acting so mean."

"All right. I'm sorry, Livvy." Tansy held the book out to Livvy. "But lots of other little girls will want to read this book too, and I didn't want it messed up."

Livvy sniffed. "Don't want it now."

Ma gave Tansy another hard look but didn't say anything as she patted Livvy's back. Being youngest, Livvy was some spoiled, but at the same time Ma was right. Tansy shouldn't have grabbed the book like that.

"Looks like somebody must have had grouchy beans for breakfast," Aunt Perdie spoke up from her chair. Prissy stood up in Aunt Perdie's lap, gave Tansy the same disapproving glance as her owner, then turned around and settled back down again.

It was true. Tansy hadn't gotten up in the best mood. Yesterday afternoon before she and Damien had parted ways, he had asked her to go down to Booneville today for that promised ride in his motorcar, but she had to say no.

She'd promised Caleb, but they could have gone to Aunt Perdie's cabin anytime. Who knew how soon Damien might move on to another town? She had ridden in a car before with Mrs. Weston to pick up some books last fall. That didn't mean she wouldn't like to ride in another car. With a fellow who looked like a character out of a book.

Damien hadn't acted disappointed when she told him she had other things to do. Tansy didn't know what bothered her most. Not getting to ride in Damien's car or that he didn't seem to care whether she could or not.

She pushed that out of her thoughts and attempted to make up with Livvy. "I didn't mean to make you tearful, Livvy. Here, I've got something for you."

Tansy fetched her saddlebags to find the bookmarks they'd made last week out of old Christmas cards. A few of the cards were so pretty, Mrs. Weston said they shouldn't cut those up but give them to children on their route if they turned in their books in good condition.

Tansy pulled out the very best card. She ran her finger over the picture of a cat sitting in front of a Christmas tree. The cat's black fur had a velvety feel, while the red ribbon around its neck felt like an ice-slicked rock. Tansy couldn't even imagine how much a card like that might cost. Probably as much as some books.

"Here, Livvy. I've got something special for you because of how I know you are going to take good care of any book you get from that book woman." Tansy winked at her and held out the card.

That made Livvy laugh. "You're the book woman."

"Not today. Ma says today I'm just your sister. But that doesn't mean I can't give you one of the book woman's treats. This is a card that people send to friends at Christmastime."

Ma came over to peer at the card. "Do they make them themselves?"

"Not this one. It has to be store bought," Tansy said.

"It's Prissy," Livvy squealed, her tears forgotten as she ran to show the card to Aunt Perdie.

"Imagine that. Buying a card like that merely because it's pretty." Her mother shook her head at the thought. "Is that only on loan to Livvy?"

"No, she can keep it. The judge's wife in town gave Mrs. Weston her old cards from years back. We made bookmarks from some of them but kept out the prettiest ones to give children on our book routes. Livvy's on mine. And you were right. I shouldn't have been cross with her."

"Well, she does need to treat those books right." Ma turned back to the stove.

Livvy climbed up beside the table, happy again with the book and the card. Tansy should make Livvy a book of her own the way she had for Junie. She'd been neglecting Livvy since she had taken the librarian job. Maybe she could take Livvy with her this afternoon when she went riding with Caleb. Give Ma a little rest.

"You want me to finish that soup, Ma?" Tansy asked.

"It just needs simmering a while. My hip isn't feeling so bad today. I think I'll make us some fried apple pies." She looked around at Tansy. "You should sit a while. You been going all week."

Tansy was ready enough to do that. She grabbed a magazine. One good thing about being a book woman, she always had something to read. She sat down by the fire with Aunt Perdie and opened up the *Saturday Evening Post.* She didn't often get to look at those since they were popular loaners, but this one had been a return at the last house she'd visited Friday.

She could feel Aunt Perdie eyeing her as she leafed through the magazine, but she didn't look up at the woman. No need giving her an opening to complain about something. But it appeared she didn't need an invitation.

"What put you in such a fine temper this morning, Miss Tansy?" Aunt Perdie rocked forward and back.

"Guess I got up on the wrong side of the bed." Tansy kept reading a recipe on how to make a coconut cake. If a person had coconut. Which she didn't. She'd rather have her mother's fried pies anyway.

"That rascal from the city do something to upset you yesterday?" Aunt Perdie persisted.

"No, we had a fine day. Talked to Miss Odella for a long time."

"She tell him about them feuds he's in such a stir to know about? She's probably old enough to remember them."

"Now, Aunt Perdie. Miss Odella can't be all that much older than you."

"I reckon she told you that and you believed her."

"I don't think she said anything about how old she was."

"That city writer came right out and asked me." Aunt Perdie's eyes narrowed as she stared over at Tansy. "Then he don't even ask Odella." She shook her head and blew out a breath. "I don't know why I bothered telling him anything."

"He did appreciate you talking to him. He wants to talk to you again."

"That's surprisin' after that sorrowful story I told him." Aunt Perdie stared back at the fire.

"That was sad."

"Had you crying, I reckon." Aunt Perdie didn't wait for Tansy to answer. Instead, she pointed toward the fire. "Put that log there on the hearth on the fire. No need saving it whilst we freeze. Your brothers is out there sawing up some more in the barn. I guess nobody told them it was Sunday."

"They're just getting me some cookstove wood," Ma said.

"I doubt they're working too hard." Tansy laid the magazine aside and fed the fire. "They'd get cabin fever if they had to stay in here on a sunshiny day like today."

Yesterday afternoon had been a fine day too. And today would have been fine for a ride too. In an automobile instead of on a horse. But a horse it would have to be.

"Where's Coralee?" Tansy asked. "She's not out in the barn too, is she?" If so, it could be Tansy should go make sure Josh wasn't doing anything foolish to show off for her.

"She's resting back on Eugenia's bed. It ain't easy carrying a baby. Said her back was paining her some." Prissy jumped down from Aunt Perdie's lap and sauntered away, her tail up in the air. Aunt Perdie fumbled around under the quilt she had over her legs to pull out a small chunk of wood Arlie had brought her. She studied it a minute before she went on. "Not that I would know the first thing about that except what I remember from when my ma was in the family way."

"I guess Ma can give her guidance. Or I might find a book that gives some help with what to expect."

"Books don't hold all the answers. Best you not be thinking they do."

"Unless you're talking about the Bible." Ma looked over from the table where she was rolling out pie dough. Tansy was relieved to see Livvy had her book well away from the drift of flour. "A body can find the answers she needs there. I sure do wish we had a churchhouse near enough to go to on a Sunday morning."

"Sunday. Saturday. The day don't make no difference. Hiram was here preaching to us yesterday."

"I've never heard anybody call Preacher Hiram," Tansy said.

"You don't think his ma and pa settled the name Preacher on him, do you?"

"I guess not."

"Other folks did that to him. But he appears to like it."

Ma spoke up again. "I never could figure why he didn't get us to build him that churchhouse. So's he could preach regular to us."

"He says he wasn't called. That he just likes telling Bible stories," Tansy said. "If people get comfort from that, he's happy."

"He's always been a storyteller," Aunt Perdie said.

"He does speak words at a funeral when he's needed," Ma said.

"He says he doesn't consider that preaching. That he can simply tell about the person and read some Bible verses and let the Scripture do the comforting," Tansy said.

"Like I said, he's always been a storyteller and he's told some fine ones at them funerals I've been to." Aunt Perdie made a face, as though not all the stories Preacher told could be believed.

"It's nice to say something good when somebody passes on," Ma said.

"I ain't arguing against that, Eugenia. If'n he outlives me, I'm hoping he'll make up something good to say about me."

"He won't have to. We'll just let Coralee talk." Tansy looked over at Aunt Perdie. "She doesn't have anything but good to say about you."

"She ain't known me overlong. And I'm not planning to kick the bucket just yet. She's got plenty of time to see my worser side."

Tansy had to laugh at that. Aunt Perdie did have a way of saying something smile worthy, whether she was poking at Tansy or herself.

"Ain't nothing to laugh about." She screwed up her mouth into a tight circle, but Tansy saw a sparkle in her eyes.

"I think you could tell Damien some good stories if you wanted to."

"Damien? That the city feller's name? Funny sound to it. But he probably thinks my name has a funny sound too. Maybe it does. So you think he's wanting made-up stories?"

"He'd like to hear some tales passed down through the years from one generation to the next. Like the Jack stories."

Livvy looked up from her book then. "I want to hear a Jack story."

"Of course you do." Ma smiled at Livvy. "Everybody likes a Jack story. Do you know any, Perdie?"

"I might give it consideration and come up with one," Aunt Perdie said. "My pa used to tell them now and again. Said he hear'd them from his grandpappy."

"Maybe that's where Joshua heard them too. The children could sometimes get him to tell one when he was in the right mood." Ma looked up from her pie making and stared toward the window. She looked sad.

"He might still come home," Aunt Perdie said.

Ma blew out a breath and went back to crimping the dough around the apple filling. "I reckon until I know for sure he can't, there's always that maybe can do."

Livvy deserted her book to come sit on the floor beside Aunt Perdie. "Tell me a Jack story, Aunt Perdie."

"It's been a long time since I heard them, child. Not sure I can remember them proper."

"Please?" Livvy begged.

"Now, Livvy. Don't be pestering Aunt Perdie," Ma said.

"Leave the child be, Eugenia. I ain't had people wanting to hear what I have to say for some time, and now looks to me like as how everybody does." This time Aunt Perdie didn't even try to hide her smile as she reached to rest her hand on Livvy's head. "But all right, I'll tell what I can remember of one. Long as you promise not to laugh if I get it mixed up some." She looked over at Tansy. "That goes for you too, Miss Tansy."

Her calling Tansy "Miss" made Livvy giggle.

Ma dusted off her hands and came over to sit in her rocker. "I'm not missing out on hearing this. I got plenty of time to get those pies done before eating time."

"Well, seems I has listeners so I better figure out what to say." Aunt Perdie rocked back and forth a couple of times. "There was this boy named Jack. He weren't a big boy. He weren't a little boy. He was somewhere in between. And he had a curious bone." She peered down at Livvy. "Something like that brother of yours. Arlie. Was always wanting to know about ever' livelong thing. He asked his ma so many questions one morning that she finally shoved him out the cabin door and barred it to keep him from coming back in."

Ma laughed. "You think I'd do that to Arlie, Livvy?"

The little girl giggled, and Tansy said, "There were times when he was your age, Livvy, that I was ready to if he asked one more question."

"That's true," Ma said. "As best I remember, you went to making up answers that didn't make a lick of sense, but he still halfway believed them."

"That's how our Jack was then," Aunt Perdie said. "Anyhow, once his ma put him out of the cabin, he figured he might as well head on down the hill to see what he could find. That was Jack. Always hunting for something, and what do you think he generally found?"

"I don't know," Livvy said. "What?"

"Trouble, that's what. And this time weren't no different. He come up on this big hole right under a rock on the side of the hill. Now you best be mindful if you ever come up on a big hole like that when you're out poking around. All sorts of things might be laying wait for you in there. Just what do you think could be in there?"

"Snakes." Livvy shivered.

"Bears." Tansy joined in the fun of imagining.

"More like a wildcat ready to spring," Ma added.

"Coulda been any of those or something even worse, but did that stop Jack from getting down and sticking his head in to ask whether anybody was there? No, indeed. Well, snakes or bears or wildcats ain't likely to answer in any words a feller can understand. But Jack was just nigh on sure he heard somebody whispering his name. 'Jack. Jack.'"

"Seems like that would be a good time to run on home," Ma said.

"For a sensible boy it would be for certain, but Jack sometimes lacked the sense he was born with. Instead of running on home, he crawled right into that dark hole to see who in there knowed his name."

Livvy pulled her knees up and wrapped her arms around them as she stared up at Aunt Perdie with big eyes. "Was it a ghost?"

"Weren't no haint." Aunt Perdie shook her head. "Not no bear neither, but maybe something worse. He scrunched on deep into that hole and came upon this little man sitting by a fire. He was a growed man, but he weren't no taller than Jack. Maybe not quite as tall. That stirred up Jack's curious bone even more. So he asked the man how he come to know Jack's name."

"Seems a reasonable question," Tansy said.

"Reasonable enough. But the little man didn't appear to want to answer. Instead, he took a gold coin out of his pocket and began flipping it up in the air. Jack liked the look of that gold coin and wondered if he could figure a way to trick the man into giving it to him. The way the man was playing with it and looking out of the side of his eyes at Jack made Jack think he wanted him to try to grab the coin. But Jack didn't think he oughta do that. Instead, he remembered his ma had been making some apple pies, just like your ma is doing here, Livvy. His ma made good pies. Good enough to be sold for a gold coin, in Jack's mind. So he told the man he couldn't stay to talk. That he had to get on home to eat an apple pie that might make your tongue beat out your brains it was so good."

"A tongue can't beat out brains." Livvy giggled.

"Well, maybe not, but hunger can make a man do some foolish things when he gets to thinking on how good an apple pie could be. He told Jack he'd give him that coin for one of his ma's fried pies. Now that seemed a good deal to Jack. So's he crawls out of that hole and runs home to see if his ma had unbarred the door so he can get inside and sneak one of her pies. Could be if he'd a-told her what he wanted and why, she might have give him a pie. But that Jack, he tended toward doing things the sneaky way, and while he knowed he'd give his ma that gold coin eventually once he got it, he didn't see the need in being in no hurry. He might want to keep it in his pocket for a spell so's he could pull it out to admire its sparkle."

Aunt Perdie rocked back in her chair. "All this talk has done got my mouth dry as a rocky creek in the summertime. Not sure I can finish up."

"You got to tell us," Livvy said.

"Fetch her a drink, Tansy." Ma sounded almost as eager as Livvy.

Tansy got the water and watched Aunt Perdie sip on it while she wondered if the old woman had wandered afield in the story and didn't know how to end it.

With a look at Tansy as if she knew what she was thinking, Aunt Perdie started talking again. "And so Jack took the pie and ran back down the way until he was where he was sure that hole was. There was that big rock, but the hole was gone. Disappeared. Jack was some disappointed and was about to eat that pie when he heard his name whispering down from a pine tree. And there was the same little man sitting up high in the tree. He must of smelled that pie. Anyhow, he shinnied on down out of that tree and reached in his pocket to pull out that coin. It shone even brighter out there in the sunlight. Jack handed him the pie and the little man flipped the coin over to him with a laugh. The man started eating that pie as he headed on up the hill. Jack wrapped his fingers around the coin, hardly able to believe his fine fortune. He watched till the little man disappeared in amongst the trees so's he could be sure the feller wouldn't try to come back and steal the coin away from him. Weren't he the happiest boy ever as he headed on back home?" She paused and took another drink.

"Is that the end?" Livvy asked.

"Not quite yet. Fact is, when you think on it, a gold coin is big pay for a pie, even a pie as good as the ones your ma makes. But the little man had made the deal. He appeared happy. Jack was extry happy. He started flipping that coin up in the air just to watch it sparkle. He kept flipping it higher and never once dropped it. But all of the sudden, he heard the swoosh of wings and an old blackbird, bigger than most, swooped down and grabbed that coin in its beak. Then it flew right up the hill the same way that little man had went. Jack was nigh on certain he heard that rascal laughing at him."

"Poor Jack," Livvy said.

"Naw, not poor Jack. He weren't out nothing but one of his ma's pies and he had a story to tell. Course nobody ever believed it. But that didn't bother Jack none. He knowed it was true. Never saw the little man after that though. But this big old blackbird would fly down and sit on Jack's shoulder now and again. Whenever that happened, you can be sure Jack kept any coin he had deep in his pockets, but sometimes if he had pie, he'd give that bird a bite. That bird did evermore like apple pie."

Livvy clapped her hands and Ma laughed as she got up to finish her pies.

Tansy said, "You think you could tell Damien that one, Aunt Perdie?"

"Ain't no need in me telling him. You know it now. You tell him, and that way I won't have to be pestered by his nosy questions."

"I couldn't tell it as good as you."

"Well then, I might consider it if he don't ask things he ain't got no business knowing. He coming around again today?"

"Not today. Caleb's coming over."

"Caleb is coming?" Ma said. "Then I better make an extra pie. He always did like my fried apple pies."

"He's not coming for dinner, Ma."

"Don't mean he won't want one of my pies. He never turned one down when he used to come see Hilda."

"He ain't coming to see Hilda now." Aunt Perdie gave Tansy a knowing look.

"It's nothing like that," Tansy said. "We're going over to your cabin to see if anything can be saved out of the ashes."

"Be a fool's trip, I'm thinking."

"Maybe." Tansy looked over at her mother. "You want me to take Livvy with me?"

Livvy looked hopeful, but Ma squashed the hope. "Don't think that would be good. Her ear's been paining her some. She best stay in by the fire." When Livvy looked disappointed, Ma went on. "Come on over here and help me crimp these crusts together with the fork, Livvy."

Livvy jumped up. She liked to help Ma cook.

"What's the matter, Tansy?" Aunt Perdie asked. "You 'fraid of being alone with that Barton boy?"

"No, of course not."

"That's good. Because it's not him that should give you worry. That's the other one. That city feller. That's the one to fret about."

twenty-three

As Caleb rode Phoebe toward Tansy's house, it felt like old times when he and Reuben would ride over to visit. Mrs. Calhoun always had sugar cookies or fried pies for them. Since Josh was still only a boy back then, Mr. Calhoun sometimes needed help with some task. Then he'd come show them how to do things they hadn't learned from their pa before he died.

They weren't the only ones to make a path to Robins Ridge. School friends gathered there most every Sunday afternoon along with the teacher, Bill Raymond. He pretended to like the extra teaching time, but what he really liked was Hilda. She didn't pay him much mind until after Reuben up and married Jenny Sue.

Funny how you could look back on things and understand more than you did then. Caleb was glad when Hilda decided to marry Mr. Raymond. He liked the teacher and never entertained the first desire to pop him in the mouth the way Reuben told him he should once Hilda took notice of the man. That feeling was some different with Damien Felding. He would take pleasure in wiping the smile off his face.

But Bill Raymond had become more friend than teacher the last year Caleb went to his school. There were no more than five years between them, and the man had found ways to keep challenging Caleb by letting him teach some of the younger ones. Told him he should consider teaching. Caleb had considered it. That was why he kept on with school longer than most. Reuben quit when he was sixteen, but Caleb went on to high school in Booneville after Mr. Raymond found a family willing to let him board with them during the week.

Thoughts of teaching were shoved aside when he left home to find a spot with the Corps. Shoved aside, but not forgotten. So when Ma said the teacher who started the term in July had packed up and left at Christmas, that seemed a sign. Caleb planned to go to Booneville to see if the county superintendent would let him take the teacher exam. The neighborhood kids needed a chance to get schooling.

Teaching wouldn't make a man rich, especially in these hard times. But Caleb had a place with Ma and, if he could convince Tansy to marry him, room on the farm for another house.

Not if. When. He needed to think positive. Not that he was going to propose today. Unless the opportunity opened up to him. He had to smile at that thought.

But it was better not to get ahead of himself. Today he'd simply tell her about wanting to be a teacher. That might be as good as some city writer.

If he passed the test and got to start right away, he'd have a couple of months before the school sesssion was over to see if he was the born teacher Bill Raymond said he was. Farming was fine. He liked planting and harvesting. Not that their hill farm would do more than keep food on the table. Especially now that the chestnuts were gone. They'd been that added boost to a mountain family's larder while putting money in their pockets as well. Felding should have asked about the chestnuts if he wanted to know about things up in the hills.

That would be of more use than writing about feuds that did nothing but cast a shadow over the area.

Caleb pushed thoughts of the writer away, but in spite of his best efforts, he kept seeing Tansy riding off with the man as though she'd been given a gift. Felding wasn't a gift. The only good thing about him was he wouldn't be around long, but would it be too long? Tansy surely had more sense than to fall for the man. The question was, could Caleb convince her to fall for him?

He'd never had eyes for any girl but Tansy, even when everybody thought he was struck on Hilda. Especially Hilda. She'd thought she was shoving him off on her little sister, when Tansy was the one he came to see. He'd planned to declare his intentions when she turned sixteen, but then she'd started going around with Jeremy Simpson.

The timing hadn't ever been right. Could be it wasn't right now, but it was the time he had. He didn't aim to keep his hand hidden now. She'd have to give him an answer, and until he knew different, he'd think on yes.

When he rode up to the Calhoun cabin, he was glad not to see any horses tied to the porch. Of course, Felding could have left his horse in the barn, but Caleb had sent some sincere prayers up to the Lord to lead Felding along other trails today.

Inside, just like old times, Mrs. Calhoun had a fried apple pie waiting for him. Aunt Perdie surprised him with a smile as she asked about his mother instead of poking him about something.

Arlie was whittling on the piece of wood he'd decided might have a bear inside it. If so, he hadn't found it yet. Something he cheerfully admitted when Josh pointed that out.

"Where you'uns thinking on riding to?" Arlie asked Caleb. "I could maybe go along."

Caleb was relieved when Josh frowned at Arlie. "What do you think you'd ride? Our old mule ain't going out of the lot today. Dragging that tree in yesterday has Sully hanging his head some."

"Poor Sully is piling up the years," Mrs. Calhoun said. "Your pa talked about getting another one and putting him out to pasture, but that was before hard times hit."

"Could be he'll bring one in with him when he finds his way back home," Tansy said.

"Could be." Mrs. Calhoun looked down at her hands clutched in her lap. She didn't appear to believe that had much chance of happening.

Tansy looked as sad as her mother, and Caleb feared she might back out of going with him. She hadn't exactly appeared overcome with eagerness to see him. No doubt she was tired after riding her book routes last week, but she hadn't been too tired to take Felding around yesterday. He pulled in a slow breath and hoped she wasn't about to offer Shadrach to Arlie. That definitely wasn't what Caleb had in mind.

Aunt Perdie peered over at Arlie and came to Caleb's rescue. "I'm guessing you weren't the one to get the invite, Arlie. So's you better just sit here and whittle with me till you find that bear you're looking for in that wood."

"He don't seem to be showing up too fast," Arlie said.

"Some things take a little extry time." When she looked over at Caleb, he was almost certain she winked. "Speaking of time, if'n these two aim to go on a fool's journey over to sift through the ashes to see whether they can find anything worth saving, they best be on their way."

"That's true. Thank you for the pie, Mrs. Calhoun. It was every bit as good as I remembered." Caleb stood up. "Are you ready, Tansy?" He wanted to reach out a hand to her. Not because she needed help getting up, but because he wanted to touch her hand. But he didn't.

At the barn while he helped her saddle Shadrach, she did smile and talk easy about the weather and thanking him for helping the boys the day before. His talk back was as easy, even while roiling around under his words was what he really wanted to say but wasn't sure she was ready to hear.

On an easy stretch of the trail where they could ride side by side, she said, "This feels like old times."

"Except when we went woods exploring back then, we rode double." He wouldn't mind riding double now.

"Or walked. Remember that time we went all over this hillside looking for wildflowers?" She laughed. "We found more beggar lice than flowers."

"Oh, I don't know. Seems like we found a few beauties." He was looking at one right now. With the air warm for January, she hadn't worn her wool cap, and strands of her golden-brown hair had escaped from the ponytail at the nape of her neck to curl around her face. What he wouldn't give to loose her hair and run his fingers through it.

"Ma was always disappointed that Hilda didn't encourage your courting," Tansy said.

"I wasn't."

"What?" Tansy looked as though she wasn't sure she'd heard him right.

"I said I wasn't disappointed Hilda never wanted me to court her. Actually, I never did."

"But you came over all the time."

"So I did. Hilda and I were friends, but she wasn't the only girl at your house."

"Which of Hilda's friends then caught your eye?"

This girl just wasn't going to believe she could have been the one he wanted to see. "I liked talking with you, Tansy. Don't you remember that?"

"I do remember. We talked about everything." She looked back toward the trail. "But I figured you just talked to me because Hilda wasn't giving you the time of day."

"Hilda was struck on Reuben."

"I know. I don't think Bill would have ever talked her into marrying him if Reuben had stayed single." Tansy shook her head. "I miss Reuben. He had a way of lighting up a place, didn't he?"

"He was one of those people everybody liked. The two of us weren't much alike. He always ran toward whatever he wanted in life, and me, I tend to stand back and study things a while."

"Nothing wrong with that. And everybody liked you too, Caleb."

"Did you, Tansy?" He hesitated for a bare second, then added, "Do you?"

She appeared to know what he was trying to ask, but she sidestepped it. "Of course I did. I do. We've been friends forever."

The trail narrowed a bit, but not so much they couldn't ride side by side. Still, she kicked Shadrach to move out in front of him. His question had made her look at him in a new light. Maybe in a way she wasn't ready to consider. And what was he doing? Standing back and studying out what to do next.

Tansy kept moving down the trail. That last question had surprised her. Had left her not sure how to answer. She did like him. But exactly how did he want her to like him? Aunt Perdie said Caleb was looking at her with a courting eye, but she had waved away her words. They were friends. Very old friends. He had loved and been disappointed by Hilda.

Or so she had thought. There was a time when she had followed Caleb around like a puppy dog, hoping to be noticed, and then had felt as though she'd won a prize when he left Hilda and the others his age to talk to her. But she thought he was just being nice to the little sister of the girl he hoped to court.

Everything shifted around in her head. All those wrong

assumptions. Why hadn't he told her? Why had he let everybody think he was after Hilda? Tansy should ask him. Right now. Slow Shadrach down to let him catch up with her and ask him straight out. *If you were coming to see me and not Hilda, why didn't you say so?*

But she didn't have to ask him. He'd just told her. He was the kind of person who thought things out. He had probably thought out what he'd said to her today even though he couldn't know what her response might be. She wasn't sure what her response was. She liked Caleb. Had been sure she was in love with him when she was fourteen. Then she had given up on him and started seeing Jeremy. Caleb seemed out of reach, and now here he was right behind her, reaching out toward her.

Then again, maybe she was making his question of whether she liked him more than it was. He hadn't asked her if she loved him. The word *like* was not the same as the word *love.*

She blew out a slow breath. Time to slow down her runaway thoughts. Be like Caleb and step back to look at what would be best to do next. She didn't need to go off half-cocked and ruin a good friendship.

She glanced back at him. He sat his horse with easy confidence as he followed along behind her. He wasn't trying to catch up with her. He was giving her time. He looked the same as he had when he used to come visit, but different too. Then he'd been like Josh. Almost filled out into his man's body, but not quite. Now Caleb looked strong and capable, as if he could tackle any challenge that came along.

He was nice looking too. Nothing at all like Damien Felding. But there were all different ways to look good, and Caleb had those great eyes always ready to smile. He was a man who found ways to make the world around him better.

He might not look like a character in a Jane Austen book, but she could fit him into other stories she'd read. A Mark Twain story or one of those rugged heroes in a western novel.

She pushed aside her silly thoughts. Her father might have been right. Maybe she did read too many romantic stories that had her comparing everybody to somebody in a book. As Aunt Perdie told her just that morning, all the answers weren't in books. Except the Bible, as her mother made sure to point out. She needed some of those answers now.

She shuffled through her memories of Bible verses and thought of the Beatitudes. Blessed are the meek, those hungering after righteousness, the peacemakers, the merciful, the pure in heart. Nothing there about the confused or embarrassed. Her mother probably would be able to come up with a Bible verse with an answer. Seek and you shall find.

When Caleb rode up beside her, she didn't try to move ahead again. No need in her acting like a skitterish horse.

"I'm sorry, Tansy," Caleb said. "Wasn't very nice of me asking you that. Sounded like something a boy Junie's or Arlie's age might ask. No wonder you weren't sure how to answer."

"I answered you. We're friends. Longtime friends."

"That we are. Friends." His smile didn't completely hide that her words weren't what he'd hoped to hear.

They rode a ways without saying anything. The horses' hooves pounded into the ground and their saddles creaked. Air whistled through the pines. It should have been an easy silence between friends, but it wasn't.

Tansy searched for something to say that would get them back on better footing. "Do you think we'll find anything to salvage out of Aunt Perdie's house?"

"Hard to say. Did she mention anything she hoped we'd find?"

"I don't think she expects us to find anything. You heard her call it a fool's errand."

"A fool's errand." He looked off at the trees. "I guess we all set off on one of those sometimes."

Silence fell over them again for a minute before he turned back to her with a smile. "Then again, a treasure might be in the ashes. Aunt Perdie looked happy enough at your house. Seems right taken with Arlie."

"Oh, Arlie's one of those people like Reuben was. Everybody takes to him, except sometimes Josh, who wants to choke him now and again." She smiled over at Caleb, glad the tense feeling between them was gone. "Did you feel that way about Reuben?"

Caleb laughed. "You have to remember I was the little brother. So it was probably Reuben wanting to choke me. He was always after me to hurry up and get things done. I like to drove him crazy when we were building that house for Ma."

"How's that?"

"I kept heading off to Preacher Rowlett's to ask the best way to do something."

"Really?" Tansy glanced over at Caleb with a puzzled frown. "Pa liked Preacher. Don't get me wrong. But he used to say Preacher was more talk than do except when it came to preaching or horses. That he mostly stood around giving advice instead of pitching in and getting his hands dirty."

"Maybe so, but they should have listened, because Preacher was probably telling them a better way. Preacher has read up on just about everything. You know how he made those trips over to Jackson to that college library. Preacher can remember most everything he ever read. So he knows some good ways to try even if he's never done it."

"So did it help when you asked him for advice?"

"Ma still has a roof over her head." Caleb laughed. "And the chimney draws smoke out of the house the way it's supposed to."

She laughed with him. This was how she remembered times with Caleb. Easy. Fun. No hard questions to put a shadow between them.

Being swallowed up by his smile made her remember how much she did like him. In fact, she no longer felt as though she'd missed out on a thing by riding through the trees with him instead of taking that motorcar ride with Damien. With Caleb she could be herself without worrying whether something she said would have him smiling at her as though she were one of those backward hillbillies they made fun of in the city papers. But then again, she was a mountain girl. If that made her a hillbilly, so be it.

She gave Caleb another look. He was a mountain boy. Even if Aunt Perdie was right and he had come courting, he'd give her time to think it through and figure out what she thought about it. What she thought about him.

twenty-four

THE ROCK CHIMNEY SHOT UP out of the ruins of Aunt Perdie's cabin. A charred tin cabinet lay crumbled on its side. Shards of bowls or plates littered the ashes around it. In the corner where the bed had been was a thick layer of burnt rubbish. With a stick, Tansy stirred around in the pile and was surprised when warmth radiated up from the ashes. She hit something hard and fished out a chamber pot. No treasure there.

A plume of smoke drifted up from the ashes. Caleb looked over at her. "Watch out. Don't scorch your boots."

"I didn't think it would still be hot."

"Logs take a while to burn. And if Aunt Perdie had a feather bed, that might have smothered down the flames and kept the fire smoldering." Caleb kicked aside a tin can. "We should have brought a shovel or a rake."

"Might be something like that in the barn," Tansy suggested.

"She did have a hoe and a shovel needing a handle out there, but I took those on over to your house with her other things." Caleb shook his head. "Wasn't much. A few pans, a rocking chair, some quilts. Nothing much left here now." He stared around at the cabin's ruins.

"So I guess we are on a fool's errand."

"Aunt Perdie said that, but I'm guessing she wanted us to look. Her trouble is she's been contrary so long she doesn't know how to be any other way."

"Then it's time she learned."

"Maybe she is. The old lady was practically nice when I was there a while ago."

"She's like her cat. Nice if she wants to be, which isn't often when it comes to me." Tansy poked in the ashes again. "But she's some nicer to everybody else. Especially Coralee."

"What is the story with her?"

"You know about as much as I do. She doesn't seem to want to talk about her 'situation,' as Ma calls it. Ma doesn't pester her about it, but she did ask. Seems Coralee's fellow, if you can call him that, took off when she told him about the baby. That was some months ago and she hasn't seen anything of him since. Ma said she wouldn't name him. I guess she loved him."

"That's reasonable to assume." Caleb picked up an iron pot and dumped ashes out of it. "This might still hold water."

"Look and see if you can find her teakettle. She might like having that."

"Found it." Caleb dug the black teakettle out of the rubble. "You said loved. Not loves."

"Hard to keep loving a man who misuses you like that," Tansy said.

"You'd think so, but it happens all the time. What about those bruises?"

"She told Aunt Perdie she ran into a tree, but Aunt Perdie figures it was Coralee's father and that's why she came hunting a place with her."

"You'd think Perdie Sweet would be the last person anybody would expect to take them in."

"I was of the same mind," Tansy said. "Coralee claimed her sister prayed about it and Aunt Perdie was the answer she got."

"Sometimes the Lord's answers can surprise us."

"I guess so, but Coralee and Aunt Perdie are doing fine with his answer." Tansy straightened up and stretched her back. "Too bad about the cabin though."

"They found a place with you."

"That might turn out not so good for Josh. He's ripe for romance, but Coralee's not showing much interest."

"Yeah." Caleb smiled. "Josh asked me about being in love yesterday."

"I hope you didn't encourage him."

"A boy doesn't need encouragement to be in love." He stared across the ruins of the cabin at her.

Tansy looked down and began poking in the ashes again. "Or to get his heart broken."

"That can happen." Caleb picked up a broken bowl and then tossed it aside. "Did your mother say whether Coralee had hope her baby's father might come back?"

"She didn't say, but I don't think Coralee's expecting that to happen."

"Then who knows? Josh might have a chance."

"Maybe eventually."

"Eventually is better than never." He looked as if he might be talking about more than Josh.

"I suppose." Time to step away from talk about romance. "Coralee's family lives down the way. A fair walk, but not too far by horse. I hoped to go by there yesterday after I left Miss Odella's, but that woman likes to talk as much as Preacher. I finally said I had to leave to get in before dark."

"You ride Felding back to Booneville first?" Caleb didn't look at her.

Tansy pretended not to notice the tightness in his voice. "I pointed him in the right direction."

"A city man like him might run into trouble up here in the hills," Caleb said.

"From you?"

"No, not from me." Caleb threw a glance her way before he went back to scratching through the cabin's rubble.

"Anyway." Tansy scooted the conversation away from Damien. "I thought maybe we could ride on over to Hardin's Creek. Coralee didn't ask me to, but I'd like to let her sister know she's okay."

"Sure. We can go on down there now," Caleb said. "Not much worth saving here."

"Doesn't seem like it." Tansy gave a last push with her stick and hit something solid. "But wait. I think I found something." She probed with her stick. "I think it's a book."

"One of your library books?" Caleb came over to look.

"No, Aunt Perdie made sure I knew she saved that book from the fire." Tansy scooted some debris aside with her boot. "It looks like a Bible."

Caleb squatted down to brush away the ashes.

"Are they still hot?" Tansy asked.

"Not very. I can dig it out." He dug down into the ashes to pull out the Bible. When he opened it up, the words were as easy to read as ever.

Tansy reached to touch the tissue-thin pages. "I've always heard the Word of God can't be destroyed, but I never thought a Bible wouldn't burn up in a fire."

"I don't know about every Bible, but this one didn't." He flipped through some of the pages. "Here's her family page. The names of her parents and sister and brothers and the dates of their births and deaths." Shutting his eyes, he smoothed the page with his hand.

"Are you praying over them?"

He opened his eyes and looked from the Bible page to her. "I guess in a way. After hearing her talk about losing her brothers and little sister yesterday, these feel like more than names."

Tansy put her hand over his on the Bible. "They do." Caleb's hand felt strong and solid under hers.

Silence wrapped around them as they shared the wonder of the Bible not being destroyed by the fire.

Finally, Tansy pulled her hand away. "Do you think she'll call this a miracle?"

"I don't know. Do you?" Caleb closed the Bible and dusted off the cover.

"Seems something like one."

"You found it." He handed her the Bible. "It can be your miracle too."

She stuffed the Bible into the waist of her riding britches inside her jacket where it felt warm against her shirt. "Should we look for more?"

"I think we've found what we were meant to find." He looked at her and laughed. "You look like you've been playing in a coal pile." He pulled a handkerchief out of his pocket. "Here. You can wipe the ashes away from your eyes."

She took it and rubbed her face. "You have a few black smudges too."

"Won't be the first time."

"If I have to clean up, so do you." Tansy turned the handkerchief to a clean spot and reached to wipe a smudge off his cheek.

He stood still, his eyes burning into hers. When she started to step back, he caught her hand in his gently, as though he had snatched a bird out of the air. Her heart pounded as she looked up at him. His breath ruffled the stray curls on her forehead. He was that close, but she didn't pull away. Her whole being felt as captured by his eyes as her hand in his.

She waited for what would happen next with such a jumble of feelings she had no idea what she hoped that might be. For one insane moment she was almost sure he was going to kiss her.

But then he pulled in a breath and let her go. She kept her hand in the air for a second before she folded her fingers tight around the handkerchief and let her arm drop to her side. She stuffed her hand in her pocket and felt the Bible inside her coat.

"I guess we should go." Her voice sounded as though it had been hours since she'd said anything.

Caleb had done some hard things in his life, but he didn't know if he had ever done anything that took more strength of will than letting go of Tansy's hand and stepping away from her. Standing there looking down into her beautiful eyes, he could almost feel her lips against his. Feel her surrendering to his embrace.

He could have kissed her. Maybe even should have kissed her. She would have let him, but at the same time he caught a flicker of uncertainty in her eyes. A kiss might have convinced her. Made her forget Damien Felding.

He could almost hear Reuben telling him that. *You think too much,* he'd once told Caleb. Perhaps he was right and Caleb should have taken advantage of the moment without hesitation. But he wanted her to come to him willingly. He wouldn't force himself on her. Ever. He'd give her time and hope fervently that she would eventually take that step toward him.

"Right," he said now. "We need to get started if we're going to Hardin's Creek. You have the Bible?"

She patted her coat. "I've got it. Can you bring the teakettle?"

"Sure. I brought a flour sack to carry back anything we found." He picked up the kettle. "You want me to take the Bible too?"

"No, I'm used to carrying books." She smiled.

"Do you have Bibles in your library books?"

"Some. And plenty of Sunday school papers. Those are popular, but magazines are most in demand. Especially *Woman's Home Companion.*"

"I saw Ma looking at that the other day. When I asked what she was reading, she said recipes. Why, I don't know." Caleb shook his head. "I've never known her to use one."

"I guess she has her recipes memorized. Wonder if she'd write one down for me. We're making cookbooks to loan out."

"She probably would." Caleb pushed the teakettle down into the sack tied to his saddle and stroked Phoebe's neck.

He waited until Tansy mounted Shadrach and then lifted himself into the saddle. The day was passing, and he wasn't sure he'd done the first thing to win Tansy's favor.

As they headed away from Aunt Perdie's place and before the trail narrowed, he kept Phoebe even with Shadrach. "I'm going to take the teacher exam next week," he blurted out. At least he could make sure she knew that.

"Really?" She looked impressed. "That's wonderful, Caleb. You'll make a great teacher."

"How do you know?"

"Because you love reading about history and more, plus figuring out how to fix things." She smiled over at him.

"You can't fix everything. We never found a way to fix the chestnuts." He pointed toward a dead tree near the trail.

Her voice softened. "Nobody else did either. We were just kids back then when we hoped for a cure for the trees."

He grinned at her. "You're still a kid."

"Oh, come on." She laughed. "I'm twenty. Most girls my age already have one or two kids. At least."

"Are you sorry you don't?"

"Not a bit. I'm glad I don't have to worry about kids while I do my book routes. But Ma's afraid I might end up alone like Aunt Perdie."

"Mothers do find things to worry about." He wanted to say he could ease her mother's mind whenever Tansy said the word, but he'd already pushed her too much today. "But what about you and Jeremy? I thought you two were headed to the altar when I joined up with the Corps."

"I thought we might be on that path myself." Tansy shrugged a little. "But I didn't care much when it ended. It just never seemed right, you know?"

He was glad to hear that. "So no broken heart?"

She actually laughed. "Far from it. Everybody thought so. Everybody but me. Pa wasn't happy about it. Said I read too many books. That I wanted a knight on a white horse like in the stories. At least he thought that was what was in the books I was reading, but I'd outgrown fairy tales a long time before."

"Oh, I don't know," he said. "I've always liked the line at the end of those stories. That 'happily ever after' line."

"But do you think that's ever true? Except in books?" She glanced over toward him.

"I do. Maybe not in a storybook way, but in a more everyday ordinary way. Where two people have love between them and they make a good life together. Like my ma and pa. Like yours."

"I guess, but then Pa took off." Tansy sighed. "To find work, I know, but Ma didn't want him to leave. She's worried about him. She thinks he might have come to a bad end."

"You'd know if that happened."

"I tell her that, but we haven't heard a word from him, and he's been gone since September. He never learned to write more than his name, but he could have gotten somebody to write for him."

"He could have." They were almost in sight of the Embrys' house, so Caleb slowed Phoebe to give them more time to talk. "Do you watch the papers that come into your library place for news of anything like that?"

"I do, but the papers we get are usually weeks old. Besides, I'm not sure the news would bother mentioning the loss of one vagrant when so many are out there looking for work." She sighed. "And not finding any."

"That might be why you haven't heard from him. He hasn't got money to send a letter."

"That could be." She smiled a little. "Thanks, Caleb, for letting me talk about it."

He had the feeling then if they hadn't been on horses, she might have reached over to touch his hand. Instead, she flicked her reins to move Shadrach out in front.

A child's wails pushed through the trees to them. Then a girl's pleading voice. "No, Pa. Please! Wait."

"Stop yer caterwauling." A man let loose a volley of angry words when a dog started a high-pitched yelping.

Caleb urged Phoebe into a trot to move past Tansy. He had no idea what they were riding into, but he wasn't about to let Tansy be caught in it first.

twenty-five

TANSY HESITATED BEFORE she flicked Shadrach's reins to follow Caleb. It might be better for them to go the other way. The man sounded upset enough without them riding up uninvited. Mrs. Weston had warned the packhorse librarians not to get involved in family disputes. That if someone didn't want books or had a complaint, they were to listen and not argue.

But this wasn't anything to do with books. No stopping Caleb anyway. She saw the determined set of his mouth as he pushed past her. If something needed doing, Caleb wouldn't back away from the task. No choice but to follow him, even though she had no idea what Coralee's father might do. She didn't know him, but she had been around Mrs. Embry a few times. She'd never had a harsh word for anyone.

Obviously that wasn't true for Mr. Embry. He was saying plenty of harsh words. Words that got a little harsher when they rode up. Two little boys were sitting on the ground behind him. The bigger boy had his arms around the younger boy, the one wailing.

Coralee's sister had a hand reached out toward her father as though to stop him. He wasn't paying her any mind. Even with a gun looped through his elbow, he managed to keep hold of a long-legged pup making pitiful yelps.

The girl glanced at Caleb riding up, but whether with relief or new despair, Tansy couldn't tell. She didn't have the same uncertainty about Mr. Embry's look as his scowl grew fiercer.

Caleb acted as if he didn't notice. "Afternoon." He nodded at the man. "We were riding by and wondered if you might need some help."

"This ain't a bit of your concern." The man's eyes narrowed at Caleb before he glared over at Tansy. "Either of you. The best thing you can do is ride on by to wherever you had intentions of going."

Tansy tightened her hold on Shadrach's reins. She wanted to ride away. Anything they did might simply make things worse for the girl and boys, but Caleb kept an easy smile on his face as he swung down off his mare.

"It's been a while since I've been down this way." Caleb stuck out his hand as if expecting the man to shake it. "You might not remember me. Caleb Barton."

"I knowed you. You're the brother of that Barton man what got kilt dead by a tree last fall whilst you were off somewhere."

"Hard things happen." Caleb dropped his hand when Embry made no move to shake it, but he didn't drop his smile. "Like your wife passing. I was real sorry to hear about that. Your missus was a fine woman."

Sorrow deepened the lines of the man's face. "I reckon it was her time."

New tears glistened on the girl's cheeks, and behind their father, the bigger boy was hugging the little one tighter. The boy had stopped wailing, but his face was no less tragic. Even the pup stopped wriggling as it eyed Caleb.

"That don't make it feel right," Caleb said. "We were just up at Perdie Sweet's place. I guess you knew her house burned down."

"'Twas a shame." The man shook his head, anger gone from his face now. He turned to his daughter. "Here, girl, take this here pup. But don't you be letting it loose. I ain't changed my mind none about it."

"All right, Pa." When the girl took the floppy-eared brown pup, it cuddled against her as if trying to hide as she stroked its head.

The man turned back to Caleb and held out his hand.

Caleb gripped it as he nodded toward Tansy. "This is Tansy Calhoun from over on Robins Ridge. She's one of the book women."

Mr. Embry stared up at Tansy. "I've heared about those book women." He didn't look as if what he heard was to his liking.

Tansy slid off Shadrach. "Does one of them come by here to bring books for you?"

"I ain't got time to be reading some fool book."

"Most youngsters like books." Tansy kept her tone easy. "I can come by next time I'm over this way and let your boys or Saralynn pick out one. They'd have it for a couple of weeks until I came back by. Then they could trade that one back in for something different."

"Sounds like a waste of time." He looked over at the girl holding the pup. "You don't want to read no books, do you?"

"I wouldn't mind, Pa." A little smile turned up her lips. "I might could read to the young ones."

The man appeared to consider what she said before looking back to Tansy. "I'll think on it."

"They loan out more than storybooks," Caleb spoke up. "My ma had a magazine with recipes and other hints on sewing and such."

Tansy picked up on that. "We've been collecting recipes from women around here to make a cookbook with our own mountain recipes."

"I sure would like to see that, Pa. Ma knew how to make

everything, but I reckon I didn't pay the proper mind to how she stirred things together." The girl sniffed and rubbed her cheek on the pup's head. It whined and licked her cheek.

"You want to bring books over this way, I'll let Sarielynn look at them. Long as she don't get anything that messes with her head."

"We're careful not to have books and magazines we think folks will take exception to," Tansy said.

He gave her a look through his eyebrows. "We'll see."

"That's a nice-looking pup." Caleb looked ready to settle in for a neighborly talk. She thought he had to be pretending to feel comfortable talking to this man she was sure had mistreated Coralee and perhaps his other children too, from the way the boy had been crying. But then he seemed to care about what Saralynn wanted.

The man swore and spit on the ground. "You ain't got much of an eye for critters if'n you like the looks of that mongrel."

As if the man hadn't said the first word, Caleb asked, "Does it belong to your boys back there?" He smiled at the boys standing up now and watching with big eyes.

"Don't belong to nobody, I don't reckon. Followed my least boy home and ain't been nothing but trouble. Pestered my old mule and chewed on the porch step. It ain't worth shooting, but I tried taking it off yesterday to leave it to find some other place, and the fool pup beat me back here. Nothing to do but waste lead on it."

The youngest boy looked ready to wail again, but his brother put his hand over the boy's mouth. Saralynn shot a look over at them.

"Pups can be a bother for sure." Caleb sounded agreeable. "But this one must have a good nose to come on back here when you took him off."

"Good nose or not, it's got to go."

Caleb turned to look at the boys again. "Which one of you claims ownership of this pup?"

When the littlest boy looked afraid to say anything, his brother answered. "The pup followed Sammy here. Didn't hardly want to get away from him."

As if to prove it, the pup looked over at the boys and whimpered.

Saralynn nodded. "That's right. He aimed to be Sammy's dog."

Tansy didn't know where Caleb was going with this, but she hoped whatever he had in mind, it would work out good for the boys.

"You know, I've been looking for a pup. I've got this nephew that's been punying around, and I'm thinking a pup might be the very thing to cheer him up. Tell you what, Sammy." Caleb reached in his pocket and pulled out a couple of coins. "I'll give you a quarter for him. Would that be enough?"

The little boy started to shake his head, but again his brother spoke up. "That's mighty generous, sir."

"But I want to keep him," Sammy said.

"You can't." Saralynn's voice sounded stern.

The little boy teared up again. "I done give him a name."

"What's that?" Caleb asked. "Puddleby? Chee-chee? Dolittle?"

Tansy had to bite her lip to keep from laughing when he said those names from the Doctor Dolittle book. The silly names brought smiles to the boys too.

Even their father looked close to a smile. "Do Little sounds right."

"No," Sammy said. "He's Rusty."

"He doesn't look all that rusty to me," Caleb teased.

Now Sammy actually laughed. "He's not rusty. His name is Rusty."

The pup's tail whipped back and forth.

"Rusty's a fine name. So can I buy him from you?" Caleb held out his hand with the coins in his palm. "I'd count it a fine favor because I need a dog."

"You don't have to pay the boy. It ain't nothing but a stray." Mr. Embry frowned as he hitched the gun up in his elbow. "Just take it. Be doing me a favor."

"I'd feel better about it if Sammy would let me pay him."

The boys had edged closer and Caleb squatted down in front of them. "I promise Rusty will have a good place and maybe I can bring him by to visit sometime." He reached over and placed one quarter in Sammy's hand and the other in his brother's. "Now you tell Rusty that he needs to be a good dog for me, won't you? I know he'll listen to you."

Saralynn leaned over to let the little boy rub the pup's head. The pup's whole body wiggled as it licked the boy's face. The girl had to grasp the loose skin on the pup's neck to hold on to him.

Tansy blinked back tears.

Even Mr. Embry's face softened a little, but in a flash, the hard lines came back. "That's more than enough. Give the man the no-account dog and get on in the house."

Caleb looked at Tansy. "Think you could hold the pup while I get a rope to make a lead?" He grinned over at the boys. "I wouldn't want to lose him after I bought him."

Tansy moved in front of Shadrach to take the pup.

The girl stepped past her father to hand over the dog. "You were talking about Aunt Perdie's house." She kept her voice low. "Do you know what become of Aunt Perdie? I've been some worried about her."

"I do know." Tansy met her eyes. "Aunt Perdie is my pa's cousin, so we took her in along with her cat. And more. They're settling down nice enough at our place."

Relief washed across Saralynn's face. She was a pretty girl but young to have to take on a mother's role in the family.

"That's good to hear. I hated to think about her not having a place."

Tansy knew which her she meant. "She's fine. Feeling pretty good in spite of everything. But she's missing some things."

"I guess that's to be expected after a fire." Saralynn handed her the pup. She looked over at Caleb. "It's kindly of you, sir, to take Rusty. The boys are some attached to him. Especially Sammy. But Pa's right. He did take an eye on him all the time to keep him out of trouble. And I reckon I didn't have eyes enough."

"Come on along, daughter," Mr. Embry spoke up. "Don't be jabbering their ears off. I expect these two need to get on their way."

"Yes, Pa." She nodded toward Caleb and then Tansy. "Much obliged to you both." She went back past her father to gather the two boys and head toward their house.

Tansy called after her. "I'll be by with some books. Not tomorrow, but the next Monday. Got any particular thing you might like?"

"Any book would be a pleasure to read. But one with a little news in it might be extry good." She looked around at Tansy.

"I'll see if I can find one like that." A note from Coralee would be better than any book Tansy could bring.

The pup seemed to realize his fate and stopped trying to get away. Perhaps because Caleb was talking to him as he worked on the rope. His eyes were on the dog, but he was so close she could feel his breath on her arm. The same as when he'd grasped her hand while they stood in the ashes of Aunt Perdie's house, her heart began beating a little too fast.

This was crazy. Caleb was a friend. She wasn't even sure she wanted him to be anything other than that.

"Easy, Rusty." Caleb kept his voice calm as he slipped the rope around the pup's neck. "This won't hurt. Good boy." He stroked the dog, which wriggled around to lick his hands.

Caleb's fingers brushed against Tansy's arm and shot her heart into an even faster beat.

"You're liable to choke him down with that and finish off what I was intending," Mr. Embry said. "But I'm just as glad my boys won't have reason to hold it agin me. Not that they don't need to learn life ain't always easy."

Caleb looked at him. "I'm thinking they already know that from losing their mother."

Again the sorrow flickered across the man's face before it hardened into a stone mask again. "We take whatever the Lord sends us."

"We do. But I'm not aiming to choke the pup. I'll pack him home. The rope's just in case he gets loose from me and tries to run off."

"That's good. The ornery thing would be liable to show up on my doorstep again and start this whole mess all over." The man almost smiled. "And you'd be out your quarters."

"You send word if that happens and I'll come get him." Caleb turned to face the man. "And if you need help doing something around here, let me know. I'd be glad to lend a hand."

"Neighborly of you, but I make out fine on my own."

"Everybody needs help now and then," Caleb said.

"You ain't a preacher, are you?" The man frowned at Caleb. "You're startin' to sound like one."

Caleb laughed. "I've never been accused of that. But my ma will be pleased to hear that you wondered if maybe I was."

"Womenfolk is always pushing a man toward church. My ma was the same and so was my woman before the Lord took her." Another flicker of sadness that this time lingered a few seconds longer on his face. "Never could figure why he'd take her and leave an old sinner like me here."

"I don't have an answer for that," Caleb said. "I guess nobody does."

The man looked over at Tansy. "You think one of them books you want to bring out here has answers?" There was challenge in his words.

"We do have some Bibles. You can find answers there," Tansy said.

The man made a face between a smile and a frown. "I done feel like I've had a visit from some missionaries."

Tansy had to smile at that. Caleb winked at her before he swung up on Phoebe and reached down for the pup.

Tansy was glad to hand him over and get back on Shadrach. Saralynn and the boys had gone on into the house, but Mr. Embry was still eyeing them with a mixture of neighborliness and suspicion.

"Good day, sir." Caleb nodded at the man and turned his mare away from the house. He gripped the pup with one hand and handled the reins with the other.

Tansy followed and caught up beside him. She was surprised the pup didn't look too concerned about being on a horse. "Will you be able to hold on to him? You could throw out the teakettle and put him in the bag."

"That's a thought, but right now he's being still enough."

"He's lucky to be alive." Tansy looked over her shoulder. They were out of sight of the house. "That was such a nice thing to do." Just thinking about it made Tansy's heart warm.

"The boy still lost his dog."

"But not in a way that would have been harder. If we hadn't shown up, his father would have shot the poor puppy."

"Ma may be ready to shoot me when I get home. She won't like a pup underfoot, but I think Junie will. Maybe you can find a book about dogs to bring them next time."

They rode along in silence for a minute before Caleb grinned over at her. "How'd you like being taken for a missionary?"

She bounced the question back at him. "How did you like being accused of being a preacher?"

"I don't guess that's such a bad accusation, but I'm like Preacher. I haven't felt the call. I did once hear a preacher claim how the Lord can hit a mighty straight lick with a mighty crooked stick." Caleb laughed. "Maybe we were those crooked sticks the Lord used today. In a missionary role to help those kids."

"I do sometimes feel like a missionary. A book missionary." Tansy looked over at him. "Do you think that's wrong?"

"Nope. A book woman ought to want to spread the love of books around." His smile over at her was good. Better than good.

twenty-six

Caleb wanted to see Tansy all the way to her house, but by the time they came to the place where the trail to his house veered away from the trail to hers, his hand was already cramping from the hold he had to keep on the pup. Maybe he should have stuffed him down in the sack, but the poor pup had already been traumatized enough for one day.

At least they had been able to take the shorter trail home along Manders Creek since they had no need to go back by Aunt Perdie's. When they rode by Manders School, empty now, Caleb pictured it teeming with youngsters looking at him with expectant faces. If he passed the teacher's exam.

He was doing a lot of imagining about things he wanted to happen. He even imagined Aunt Perdie's face when Tansy showed her the Bible they'd found. He was so full of imagined things, maybe he was the one who should be writing stories to impress Tansy.

After they reined in their horses before they parted, things to say skittered through his mind. *Thanks for being with me. Hope you had a good time. When can I see you again?* But he didn't say any of them.

She didn't seem to know what to say either as the silence

between them started to feel awkward. He couldn't let that happen.

"Aunt Perdie is going to be surprised when she sees her Bible. I'll be interested to hear what she says." He managed to undo the sack with the teakettle in it with one hand while still holding the dog with the other.

Tansy took it from him. "Could be she might finally have something good to say about me." She smiled. "And it will be fun to see what Junie thinks about that pup next time I bring books your way."

"I hope to see you before then," Caleb said.

"Well, sure. Come over anytime. We'd all be glad to see you." She picked up the reins to turn her horse away.

He threw some words after her. "Don't forget what I said." When she glanced back at him, he went on. "It was never Hilda I came to see."

When she just lifted a hand in farewell as though she either hadn't heard him or wished she hadn't, he wanted to chase after her. Make her say something to give him hope. Instead, he sat still on Phoebe until she went out of sight.

"Well, pup, guess it's time to go let Ma yell at us some." The pup licked his hands. "You are a friendly fellow."

On up the trail, he dismounted to walk the rest of the way. The pup needed to stretch his legs and take care of business before they got to the house. Peeing on the floor wouldn't be the best first impression. He kept the rope on Rusty to make sure the pup couldn't take off for the Embry place or who knows where. His long ears flapped against the ground as the pup sniffed here and there, but he didn't try to pull free.

"Rusty's a good name for you." The pup looked up as though recognizing his name. So maybe he wouldn't be as useless as Embry thought. Then again he might live up to his name.

He'd accused some of his fellow workers down in the Smokies of pulling a rusty after they tried to get a joke on somebody.

Calling a joke a "rusty" had proved to be funnier to them than whatever their original prank had been. They had laughed about it every time they were around Caleb while getting him to share more mountain sayings.

That didn't bother Caleb. Mountain talk had some poetry in it. Like now, the edge of night with the sun down and shadows deep in among the trees. If they wanted to laugh at what he said instead of hearing the music in the words, that was their loss.

That Felding man probably laughed at their mountain talk whenever nobody could hear him. Then again, he might laugh right in their faces. Caleb had to admit he thought Felding said some odd-sounding words. People took comfort in the familiar.

Like this ride home. All new to Rusty, but as familiar as the back of his hand to Caleb and Phoebe too as she got more energy in her step nearing home and her feed.

By the time he took care of Phoebe and did the night chores, it was full dark. Supper would be long over. Ma had been fixing eats at Jenny Sue's since she said the poor girl couldn't find enough focus to cook a meal. Then again, the way Ma took over might have stolen Jenny Sue's want-to. But no way could Ma not make sure Reuben's children were fed proper.

Even with the pup prancing along beside him toward the house, a pall fell over Caleb. He tried to shake it off, but Jenny Sue's house reeked of sadness. He wouldn't skip going in to see the kids and read to them. The silly book Tansy had left them did cheer the air a little. Talking animals. He'd noted Tansy's smile when he spoke names from the book at the Embrys. Puddleby might be a good name for the pup. He looked the color of a mud puddle.

Clouds had moved in to hide the stars, and the smell of snow was in the air. He might have to wait until Tuesday to go into town to see about the teacher test. He smiled. That might work out well, since Tuesday was Tansy's day at the library room in Booneville.

Besides, tomorrow he'd have to build the pup a pen. Until he got that done, he'd keep Rusty in the house even if Ma was against it. If his plan worked, Junie could see to the pup during the day. A good way to get the boy's mind off his sorrow.

A lamp was still lit in Jenny Sue's house. Maybe Junie and Cindy Sue wouldn't already be abed. Plus, his own supper might be in the warming oven and he could rustle up some scraps for Rusty.

"Whatever are you bringing in here?" Jenny Sue asked when he went in the kitchen door. She was at the table with a cup of coffee. She looked better than she had for a while, with her hair combed and wearing a dress and a sweater instead of one of Reuben's shirts.

"It's for Junie and Cindy Sue," Caleb said. "A pup. Thought it might bring Junie out of his doldrums."

"Doldrums." Jenny Sue echoed his word. "They ain't so easy to push out of. What are doldrums anyhow? I mean really. Not the sad-down feeling we mean when we say it."

"From what I've read, the doldrums are when a sailing ship is out on the sea and the winds stop. The ship floats there without getting anywhere until the winds pick up again."

"That's a feeling I know for sure." Jenny Sue patted the table. "Sit down. Ma Vesta went on home a while ago, but she left your supper here. She made me promise I'd see that you ate."

"Sounds like Ma." While he tried not to show it, he didn't like being alone with Jenny Sue. But the children would be in bed in the front room. "Are the children asleep? I want to show them Rusty."

"I wouldn't have no way of knowing." Jenny Sue stood up to set a plate piled up with potatoes and ham and biscuits in front of him. "Ma Vesta took them on home with her for the night. Said looked like I hadn't been sleeping good enough. Your ma, she has an eye on everything."

"Had you?" He gave the pup one of the biscuits.

"Had I what?" She sat back down across from him.

"Been sleeping good enough."

"Who can sleep in a lonesome bed? You ever slept with anybody, Caleb?"

Caleb thought about picking up the plate and carrying it on to Ma's house. "Not since me and Reuben shared a blanket." He dug into the potatoes. He didn't know who was watching him more mournfully. The pup hoping for more to eat or Jenny Sue. He wasn't sure what she might be hoping for. He was afraid to think too strongly about that.

"Well, I reckon we're some the same there." She smiled, but it wasn't an easy smile. "I shared some blankets with him myself."

Caleb concentrated on finishing off his supper. He dropped a few of the potatoes on the floor for the pup.

"Your ma wouldn't like you feeding the supper she fixed to a dog."

"I expect you're right." He gave the pup a morsel of the ham. "But the poor fellow's had a hard day. You think Junie will like him?"

She shrugged. "Why wouldn't he? He's a boy. Boys like dogs. Reuben did. He about took to his bed when he found Clem gone one morning last summer. I reckon he'd had him since he was a boy."

"Yeah. Clem was a fine dog. I missed him being here when I came home."

"Missed your brother more, I'd think."

"We all miss Reuben."

"Sometimes I think my heart is tore in two." She looked away at the wall. "You ever think about getting married, Caleb?"

"I've thought about it." Caleb pushed the plate away. The

food he'd bolted down sat heavy in his stomach. The pup, giving up on more food, settled on top of Caleb's foot.

"But you haven't said the vows with anybody."

"No." Caleb let the silence sit between them a moment before he went on. "Ma should have taken my plate on home with her so you could have gone on and gotten that extra rest without the kids here."

A ghost of a smile lifted the corners of Jenny Sue's mouth but didn't seem to make it to her eyes. The light from the oil lamp on the cabinet behind her put her face in shadows. "Oh, Ma Vesta had this all planned out. You know your ma. She intends things to happen the way she wants."

Caleb tried to tiptoe away from the path their talk was taking. "She aims to be a help to you."

"A help." She leaned her elbows on the table and stared straight at Caleb. "You know how she's intending to help, don't you? Or has she not told you how she figures to keep the family together?"

"Junie and Cindy Sue and you will always be family."

"I could marry again."

"You could. You should." Caleb didn't slide his eyes away from her. She was a pretty woman in a sweet way. A soft way. Nothing at all like Tansy with strength sitting easy on her face. Jenny Sue needed somebody to do for her. "In time you might be ready to marry again."

"That's what your ma is feared of." She leaned back in the chair and played her fingers along the table like it was a piano.

"She wouldn't deny you a fresh start."

"Oh, she ain't wishing permanent widowhood on me. She's just intending me to always stay a Barton. The next brother up. She told me such is spoken of in the Bible. But I read where she said and that was if the poor widow woman didn't have a son. I'm thinking Junie, poor little feller, is supposed to be saddled with taking care of his ma. Like as how you and Reuben did for your ma."

"Look, Jenny Sue." Time to quit sashaying around what the woman was saying and be straight out about it. "I know what Ma wants and it's not happening. I'll be here to help you and Reuben's children any way I can, but I've already told Ma she needs to quit thinking on this. I'm sorry if she's led you to believe something different."

"You don't think I'm pretty enough?"

Even in the dim light, he could see Jenny Sue batting her eyes and pretending to flirt with him. This wasn't Jenny Sue's fault. Ma could be insistent with what she wanted to happen, and Jenny Sue lacked enough pushback.

"You know that's not a problem, Jenny Sue. But I can't take my brother's place."

She let out a long sigh. "Nobody can. Ever. Might as well be you giving it a try as anybody. Leastways that's what Ma Vesta thinks."

He felt sorry for her then and reached across the table to put his hand over hers. "Ma is wrong. You are wrong too. Someday you'll fall in love again. It won't be the same as how you loved Reuben, but it will still be good. Ma's just worried you'll take the kids off where she won't see them every day. But if that happens, it happens. You deserve your happiness."

"I don't expect to ever be happy again." Tears were in her voice now. "Sometimes I think about just walking until I get to the river and then go on walking until the water is over my head. I never could swim, you know. The other girls thought I didn't want to strip down because I was embarrassed, but it was because I was afraid of the water."

Caleb shot a little prayer up to the Lord that he would say the right words. "Jenny Sue," he started.

She pulled her hand out from under his and held her palm toward him to stop his words. "And don't be telling me to pull myself together. Or how the kids need me. They would be fine with Ma Vesta. I'm the one who ain't fine."

He had been going to say that about Junie and Cindy Sue. How could he not say it? "That's not true. They won't be fine without you. Ma's getting old. She might not have enough good years left to take care of them and then what? Orphaned by father and mother and granny too. That's a lot to put a child through, don't you think?"

"I don't know what I think." Her face looked tragic in the shadow of the lamp as she let her hands sink down on the table.

"It hasn't been that long since you lost Reuben. It's only natural to feel sad."

"Sad." She looked away toward the wall again. "Such a short little word for this dark place I'm in."

Caleb had nothing to say to that and the silence built. Not a good silence but one full of that black, creeping sorrow. He wished Ma was there with them. She might know what to say, but then she'd been the one to make sure they were alone as if she could force her plans on them.

Jenny Sue turned her head to settle her gaze on him again. This time she reached across the table to touch his hand. "Would you marry me, Caleb?"

He had to force himself not to jerk away and run out of the house, but this was his sister or near to it. She deserved an answer. "I told you already I can't be my brother for you."

"I wouldn't expect you to be, but you could be a good father for his children. I'd be a wife to you and we could have more children."

He sat very still and prayed for the right words. "Jenny Sue, I don't love you in that way. You need somebody who will love you proper."

"I've had that. With Reuben. Now I just need an anchor to keep me from floating off into this blackness that keeps grabbing for me."

"I can anchor you best I can, but I can't marry you. I'm in love with someone else."

An odd little smile slipped across her face. "I told Ma Vesta that. I saw your face when you found out that Tansy Calhoun wasn't married like you thought. Sometimes Ma Vesta doesn't know as much as she thinks." That seemed to please Jenny Sue so much she actually laughed. Then she went on. "But does Tansy look with favor on your attentions?"

"That I can't say for sure." Caleb had to be honest.

Jenny Sue patted his hand then, before she pulled away. "She should. You're a good man, Caleb." She blew out a long breath. "Could be she'll turn you down and then we can consider a convenience arrangement again. But whether that happens or not, I can tell Ma Vesta I offered and you didn't want to hear it."

"I'll make sure she doesn't bother you about this again," Caleb said. "Ma meddles more than she should."

"Reuben was always ready to follow whatever trail she pointed him toward."

"She means well." Caleb needed to remember that before he headed over to confront her. At least now he'd have a trade-off. He'd forgive her setting him up like this with Jenny Sue in exchange for the pup getting house privileges. She'd hate that, but it would serve her right.

"I never said she doesn't. Don't mean she's always right."

"True enough. But I guess I better go on over and spring this pup on her." Caleb stood up.

"She won't be happy about that."

"Nope. But I got the pup for Junie and Cindy Sue. So she'll give in."

Jenny Sue stood up too and walked to the door with Caleb and the pup. At the door she put her hand on Caleb's arm. "I know you're feeling uneasy with all this, but do you think you could step into your brother's shoes long enough to give me a hug? I surely do need a hug."

What could he do but open his arms to her and hug her as a brother would? She started sobbing and he stroked her head the way he might have Cindy Sue's and let her cry.

She finally backed away and grabbed a tea towel off the back of one of the chairs to rub off her face. "Sometimes I think I could cry a river to drown in and save me the walk."

"None of that. Folks need you."

Rusty whined and jumped up on her.

"Even this pup needs you. You have to show Junie how to take care of it."

She sniffed and leaned down to pet Rusty. When he licked her face, she laughed and looked up at Caleb. "I know I done asked too much of you tonight, but do you think you could leave the pup here with me? Let me be the one to give him to Junie and Cindy Sue in the morning when they come back from your ma's house?"

He was disappointed to not be the one to bring the pup to the children, but he couldn't refuse her. "That would be a fine idea. Looks like the pup's right taken with you anyhow. His name's Rusty." He held out the rope still around the pup's neck. "He could make a mess. I doubt he's used to being in a house and he might chew up your shoes if you leave them where he can get them. Or who knows what else."

She took the rope. "I can clean up messes and I'll put things where he can't get to them." She reached down to ruffle the pup's ears before she looked back at Caleb. "Thank you, Caleb. I do think we might make things work if you were to change your mind."

twenty-seven

PERDITA SLIPPED OUT OF BED before any sign of daylight was edging into Eugenia's bedroom window. It was a fine thing to have a separate room for sleeping with a door to close when a body needed some alone time. While she didn't know from experience, she expected that was especially good for a man and wife, but from what she saw of her neighbors, most of them found plenty enough private time to make a pile of young'uns.

After pulling on one of Joshua's old flannel shirts over the nightgown Eugenia had give her, she crept into the front room to poke up the fire and lay a fresh log on the coals as quietly as she could. Cold had crept in overnight. Whilst yesterday had been a gift of January mildness, her rheumatism told her bad weather was coming.

She peeked over at Coralee, who gave no sign of stirring. That didn't mean she wasn't awake. The girl sometimes tried to disappear and be no trouble by playing possum. But if she was asleep, that was good too. From the looks of her, she wouldn't have many more weeks before a baby's needs stole her nighttime peace.

The thought of that brought a smile to Perdita's lips. Heaven only knew when she'd ever smiled as much as she had in the last week. Being in the middle of a family was a fair thing. A reason for smiles. And she'd soon be a granny of sorts. Granny Perdie sounded some better than Aunt Perdie. She hadn't never been a born-to aunt to them that called her that. Wouldn't be no different being a decided-on granny except she would be a heap happier to be called that by Coralee's little sprout.

The mantel clock chimed out four times. She figured Tansy wouldn't be climbing down from the loft until it was nigh on ready to chime five. She was always up first to be off on her book deliveries. That girl did favor those books.

Sometimes Josh came stumbling down after her, working so hard not to peer over at Coralee that he often as not tripped over his own feet. Easy to see he wanted to do some sparking, but the girl wasn't up to it yet. Might never be up to it with him. Hard to say about sparking.

Just like Tansy not knowing the Barton boy had calf eyes for her. He had it bad for sure, while the silly girl appeared to be hankering after that city fellow. She didn't need no city fellow.

Perdita shook her head and shoved all that thinking aside. Wasn't likely anybody would listen to romantic advice from her. She'd got up early to have private time to look at her Bible. She picked it up and settled down in the rocking chair. A fine wonder it was. Coming practically unscorched through the fire Tansy said left nothing but rubble and her iron teakettle. Perdita glanced at it on the hearth. It would need a good cleaning before it was useful for anything. But the Bible was useful right now.

She ran her hands over the cover, imagining the heat that must have bounced off it. The Lord had protected his Word the way he'd protected Shadrach, Meshach, and Abednego in the fiery furnace. Of course, the good Lord had marched around in that fire with those believing men. She felt another smile coming on when she thought about the Lord putting his mighty hand over her Bible while the cabin burned.

The enduring Word of the Lord. Not that his Word wouldn't endure just the same even if this Bible was naught but ash, but that it lay warm in her lap was a gift.

She shut her eyes and threw a silent prayer up. *Dear Lord above, I thank ye for the girl you sent me to be a daughter for however long and for burning down my house so's we could find a place with food to keep her healthy. Her and the babe she's carrying. I know she asked your forgiveness for whatever sinning she did to get in this family way, and so I know you has wiped her slate clean. I thank ye for that. And pray ye will wipe my slate clean when I ask forgiveness for sometimes forgetting the way I should be. I do appreciate how you has surrounded me with all this good that's helping me keep my promise to you about being less contrary. And thank you for this Good Book you let Tansy find in the ashes. Amen.*

Touching the cover of the Bible made Perdita remember her mother's hands that had held this book so many times. She let the Bible fall open in her own hands, and there were the names of Perdita's brothers and sister. The fire flickered enough light to let her see the dates they were born and the dates they died recorded in her father's same steady hand. The older brothers that went away and never were heard of again had dates for their births, but nothing was known about them dying or not.

She sort of hoped at least one of them was sitting somewhere by a fire with a grandchild crawling up in his lap, but she reckoned that was something she'd never know until she went on to glory, where a person would know most everything. Leastways, she'd once heard a revival preacher claim that for a truth. She couldn't remember now if he'd proved it with Scripture.

She'd asked Hiram about it once. Back when he used to come visit Ma. He'd said that whatever we knew, it would all be happiness. Joy and wonder. Perdita figured that meant if any of those brothers had come to a bad end, she'd never know it. When she had said something like that to Hiram, he'd given her a funny look and told her she shouldn't be trying to dig under the Scripture words.

But she didn't have the first doubt the good Lord didn't mind her thinking under the Scripture words. That he could handle any question or doubt she had and give her an answer. On top of that, she believed he had a goodly amount of patience with her contrariness, even while prodding her to change. Like now with giving her family. And then saving this Holy Book that held her mother's fingerprints all the way through it in verses she'd marked.

She leafed through the Bible to where Psalm 46:1 had lines under the words. *God is our refuge and strength, a very present help in trouble.* She traced the verse with her finger, thinking how her ma had surely done the same when she was confronting the sorrows that had come her way.

Coralee got out of bed and came over to the fire. "You're up awful early, Aunt Perdie. You ain't feeling poorly, are you?"

"No, child. Folks add on a few years, they don't always sleep so easy. I'm sorry to be waking you."

She smiled. "I reckon carrying a babe inside you makes sleeping less than easy too. Can I sit a while with you? I'll be quiet as Prissy when she's good so's not to disturb your Scripture reading."

"Where is that Prissy?"

"She give me a look when I got up but burrowed back down under the covers."

"Pull that chair over here close and we can look at this here Bible together. It was my ma's Bible."

"She must have asked angels to hover over it to keep it from burning." Coralee put a straight chair next to Perdita's rocker. "Thinking about that makes me happy."

"Do you have a Bible?"

"Ma did. Saralynn—that's my sister—wanted me to take it, but I told her she'd need it more than me. Staying there to mother my little brothers and tend house for Pa."

"Your ma was a believing woman and so was mine."

"But you are too, aren't you?" Coralee looked over at her.

"I am. But not as sweet in the spirit as my ma was or as you."

Coralee looked sad as she shook her head. "I don't reckon I have any right to claim much spirit after what I've done."

"None of that, child. There ain't a one of us that ain't sinned and come short. A verse in this here Bible says so. And you asked for forgiveness. Whilst you might be remembering your wrongs, the good Lord has done sunk them down deep in the sea where they can't never float back to the top. Now it's time for you to forgive yourself the same as he's done."

"I'll ponder on it and do my best." Coralee peered over at the Bible page. "What's underlined there?"

"Ma said everybody ought to have a favorite verse that could help them through thick and thin. This one here was hers." She held the Bible over to let Coralee read it.

"'God is our refuge and strength, a very present help in trouble.'" Coralee looked at it a minute longer. "Your ma did have her load of trouble and you too, from what you told that city feller."

Perdita sighed. "Most everybody has troubling times now and again."

"Do you think your ma would mind me taking on that verse for my own comfort for a spell?"

"I think she'd be happy as a honeybee in a patch of clover."

Coralee whispered the words of the verse again as if committing them to memory. "Do you have a favorite, Aunt Perdie?"

"Ma wanted me to, and there was a spell where I searched through the Bible looking for just the perfect one. I'd find one and then another. I regret to say I've always been something of a jumpy person. Especially back then. I think I 'bout drove Ma to distraction. She kept saying I needed to settle on something and that might rest my spirit some. So one day I was reading hers here at the beginning of this psalm, and my eyes drifted down a ways to this other verse. 'Be still, and know that I am God.' Those words spoke to me and somehow let me know that my way of being still and knowing the Lord weren't the same as Ma's way. She could sit and ponder for the longest time without moving, but I had to be doing something with my hands whilst I pondered what the Lord wanted to say to me, else I'd feel like ants was biting me through my drawers."

"Is that when you started your whittling?"

Perdita smiled over at her. "I'd whittled on things before. It was something I used to do with Pa, but we just made shavings good for naught but starting fires. Once I started pondering the Lord whilst I cut into the wood, things changed somehow. I still couldn't make nothing that looked exactly like any of God's creations. No bear or bird. But the wood seemed to tell me something whether other folk could see it or not." She smoothed her hand over the page. "Ma was always after me to make something. If not something she could recognize, then something useful like a spoon or a bowl."

"Did you?"

"Now and again I could get a spoon chiseled out or a wooden edge good for scraping something. But that wasn't in my pondering times. Then it was always those crazy shapes that folks have to look at and figure out, I reckon."

"But you always knew, didn't you?"

"Not always. Leastways not till I was nigh done. Then the feeling would rise in me and I'd know."

"Like that sorrow one up there on the mantel." Coralee pointed at it. "Caleb knowed what it was."

"That he did. I reckon from the sorrow of losing his brother. Did you?"

"I saw it once he spoke it, but before that I was trying to push something happier on it. Hope, maybe. Or promise."

"Promise sounds good to think on. Maybe I can whittle out one that'll look like promise to you, seeing as how you is full of promise right now with that baby to come."

"Well, I'd be happy with anything you might carve for me, but if I could be pickin' I might think on strength." Coralee pointed toward the Bible. "If'n you felt you could do that. Strength is something I'm gonna need in plentitude."

"I'll give it a try, but I can't make any guarantees. But I still have that wood Arlie brung me, and could be if I think about this verse of my ma's and now yours whilst I'm whittling on it, strength will be what shows up. Besides, what I see and what somebody else sees don't have to be the same."

"You are surely a tower of strength for me, Aunt Perdie."

Coralee leaned over so close to hug her that Perdita felt her baby moving under the skirt she must have pulled on when she climbed out of bed. It was on Perdita's lips to tell the girl what a promise she was to her. An answered prayer, but she wasn't sure what the girl would think about that.

So instead she patted her arm. "You'd best go lie down and get you a bit more rest afore the house starts buzzing."

"Yes'm. A few more minutes under the covers will feel right pleasant. Do you need another quilt tucked around your shoulders?"

"I'm fine with this'n I've got on my knees and the fire burning."

When the girl got back in bed, Prissy got stirred up again. This time the cat jumped down to come sniff the Bible in Perdita's lap, but she didn't try to climb on top of it. Instead she curled up at Perdita's feet. Seemed even Prissy had respect for the Scripture.

Perdita turned a few more pages and found the letter she'd stuck there years ago. She gingerly unfolded it. The paper was old. She'd written it after her ma died. Hiram was a widower by then, losing his pretty Mary Ellen some while before.

Even now it turned Perdita's face red to read her words.

Dear Hiram,

I do know how much you surely must be grieving Mary Ellen. I know how you carry her in your heart. I've been praying that the Lord will send you comfort. And it seems he's spoke to me to offer to be that comfort.

I know you haven't never had loving feelings for me, but we've had some friendly years of getting along. I think we could move those friendly feelings a little deeper if you were to be willing. I ain't a bad cook and I keep a fair house. I'm even right fond of horses since I know that's your business aim. I read the Bible regular and I know we could have some good talks about what the Lord intends for us to see in those verses.

I figure it ain't no use both of us sitting by our lonesome in two different houses when we could be proper company for one another.

Respectfully yours, Perdita

She looked up from her words at the fire. Could be time to pitch the letter in the flames. She'd considered doing that so many times. More times than she ever considered actually sending it to Hiram. Some things a woman just shouldn't do, and one of those was to beg a man to marry her.

She folded the letter and gently put it back in the Bible, right at the beginning of Ecclesiastes. *Vanity of vanities, saith the Preacher, vanity of vanities; all is vanity.*

twenty-eight

FAT SNOWFLAKES STREAKED the early morning air when Tansy headed out on her book route Monday. Most of the last snow had melted away in yesterday's sun, but with flakes like these falling, snow would be knee deep by noon.

She was tired after her busy weekend. Shadrach lacked a little energy in his step too, but he'd get a rest in the stable tomorrow at Booneville. She'd have a break too. At least from the saddle. But tired or not, she couldn't regret her ride on Sunday. The look on Aunt Perdie's face when Tansy pulled that Bible out and handed it to her made a few tired muscles of no consequence at all. The woman had hardly said a cross word since. It was good, but kind of strange, to see Aunt Perdie smiling more than frowning.

Smiles kept sneaking on Tansy's face too every time she thought about Caleb and that pup. Maybe the pup would be just what Junie needed. Tomorrow she'd try to find a hound dog book or else make up a story.

Mrs. Weston would be fine with that. An added book, she'd said. Tansy was eager to show her Aunt Perdie's Jack story. Tansy wrote it out from memory and then asked Aunt Perdie to make sure she got it right.

Aunt Perdie waved her away. "There ain't no right or wrong to a Jack story. The teller or, like now, the one what writes it down can add or subtract whatever fits or don't fit."

"But I want it to be yours. See, up at the top, I put 'As Told by Perdita Sweet.'" Tansy pointed at the words.

"Well, look at that. I didn't have no idee that you even knew what my given name was."

"I didn't until I heard Preacher say it. It's a pretty name. Were you named for somebody in your family? Your grandma, maybe?"

"Nope. My name was Pa's doing. Ma used to say she didn't know what stream he fished it out of. Pa would laugh at that and say maybe a little fishie did tell it to him. Ma suspicioned it was an old girlfriend. I figure that's why she always called me Perdie, but Pa called me Perdita most all the time. He appeared to take pleasure in her thinking it was an old flame. He was a good one to pester Ma some."

Tansy smiled now as she thought about Aunt Perdie. In just the few days she'd been with them, she'd softened up until her cat was crankier than she was. And seeing her that morning sitting by the fire with her Bible in her lap had been worth all the ash soot Tansy had to scrub off her hands and boots the night before. She wondered if Caleb had the same trouble.

Caleb. She hadn't decided exactly what to think of him being sweet on her. No way she could deny that any longer. For a minute there in the ruins of Aunt Perdie's house, she'd thought he might kiss her. Now when she thought back, she wasn't sure if she was worried he would or worried he wouldn't. Shouldn't that be something a girl ought to know? At the time or even a day later?

She did like Caleb. Who wouldn't like such a good man? When he'd given that little boy a quarter for his pup, tears had popped into Tansy's eyes. And she had to admire his patience with Mr. Embry, who didn't deserve any patience at all. At least the man had said she could bring some books to their house next week. Coralee cried when Tansy told her she'd seen Saralynn and would be going back next week if she wanted to write her. That made her smile through her tears.

But what about Damien Felding? He made her curious about so many things. Made her wonder if she was missing something by not leaving the mountains. Was it like Pa said and the stories she read had her thinking different than others here in the hills? Not that girls from around about here didn't run off to the flatland with a fellow sometimes. Her own sister had done so and seemed to have forgotten the way home.

Tansy couldn't imagine that. Her feet belonged here on this hill or in stirrups riding these trails through snow-coated pines. What would snowflakes look like falling on tall buildings and a street full of cars?

No cars here. Only trails and creeks. Shadrach slid a little when they started down an incline. Time to stop woolgathering and pay attention to the trail. She needed to be timely since folks would be on the lookout for her.

For sure, Alma Riddle would be watching to see Tansy coming. She always came out on her porch to trade the book she had for whatever new one Tansy put in her hands.

When Tansy had first gone up to her cabin, Alma shooed her on her way and wouldn't take a book. Alma didn't know Tansy, although when Tansy named over some of her family, she did seem to recall Tansy's Grandmother Marlowe. That hadn't been enough to allay Alma's suspicions about somebody wanting to give her something, even if it was only on lend. She'd stared hard at Tansy and claimed nobody ought to trust a book other than the Bible to be decent. But on Tansy's second try, Alma agreed to try a book after her grandson came to visit.

"Little Jimmy says if I take one of your books, he can practice his reading when he comes by. He's a right smart boy with some book learning. So if you've got something fit for a boy to read, I reckon you can leave it off here."

Tansy had left her an illustrated Bible story. Since then, Alma had never missed taking a book.

"Gives a body a break from chores." But she never took a book without giving Tansy something in return. Sometimes a walnut. Sometimes a sorghum cookie.

This week Tansy brought her a *Saturday Evening Post* magazine. Tansy had looked for something with pictures, because she wondered if Alma had to wait until Jimmy was there to read the stories. Some on Tansy's route lacked reading skills, but even those seemed eager to have books or magazines. Especially the ones with pictures.

When they first started the packhorse library, people in town doubted they would find takers up in the hills for their books, but instead people were eager for more. Children would run out to meet her to beg for a book. Any book. Tansy knew how they felt. She was a child like that, ready to treasure every book that landed in her hands.

Even with the snow falling, Alma was on the porch when Tansy rode up to her cabin.

"Ain't these snowflakes a purty sight?" The woman caught one in her hand. "But you is liable to end up soaked afore your day is through."

"I've got a good coat." Tansy exchanged the magazine for the book Alma handed back in. "Would Mr. Riddle like a book today? I've got one here about horses. Preacher Rowlett liked it."

"That Preacher Rowlett. I'm guessing he's done read ev'ry one of your books."

Tansy laughed. "He's working on it, but we get new ones in from time to time. A woman all the way from New Hampshire stopped by to donate books and magazines last week."

"New Hampshire? Ain't that something? Folks in the East

bringing books down here to us." Alma's eyes went to the front of the magazine as if she couldn't wait to open it. She smoothed her hand across it and looked back at Tansy. "I reckon if you have that book handy, he might to give it a look. Little Jimmy will be by afore you come around this way again and he's some partial to horses. Wishing he had one of his own, like most boys."

"My little brother is the same way."

Tansy pulled the horse book out of her pack. She felt a surge of victory each time she talked a person into giving a book a try.

"I guess you're needin' to ride on. Regretful as I am, I ain't got much for you today."

"That's no problem, Miss Alma. These are just loaned books and you don't have to pay a thing for them."

"I know, Tansy. You tell me that ever' time, but nobody don't give Alma Jean Riddle nary a thing that she don't turn the givin' back to them. Only right way." She reached into her coat pocket and pulled out a fold of paper. "So the last time little Jimmy come over this way, I had him writ down this here little prayer I've been saying for you."

She spoke the words from memory as she handed Tansy the paper. "Dear God up in heaven. You keep watch on this book woman so's she won't come to no harm carrying these books all around. And thankee for the books she brings. Amen."

"Thank you, Miss Alma. This is the sweetest gift ever. Nothing better than a prayer. Maybe next time I come, you could give me the recipe for your sorghum cookies. We're putting together a recipe book to loan around to folks."

"I don't reckon I ever had a writ-down recipe. Not sure that I'd know how to make one such as that. Besides, everybody surely already knows how to make sorghum cookies." She frowned a little.

"No, ma'am. Not everybody. And I've never eaten any better than yours."

That made the woman smile. "Well then, I'll give it some consideration."

"I'd appreciate that. Now you'd better get inside and warm up and I'd better get on to my next house."

"If that's Beth Harley's place, you tell her howdy for me. She makes a fine hickory nut pie. If she gives you the how-to for that, I wouldn't mind giving that book you make a look-see."

"I'll put you at the top of the list." Tansy waved and headed out to the next stop.

The snow kept falling. Shadrach didn't pay it much attention except to shake his head and whisk his tail to get rid of the snowflakes hitching a ride. Tansy exchanged books at three more houses, all the time worrying about Gin Mayfield, who came to meet her at the farthest point of this day's route before her turn back toward home.

Even with all the routes Tansy and the other book women rode, they couldn't reach every area. Gin had taken that as a challenge and walked six miles out to meet Tansy to get books for her and her neighbors. Tansy had extra books loaded up to trade out with Gin, who kept an account of which of her neighbors took which book and gave Tansy a listing each time.

Gin waited in a natural arbor of grapevines that gave the woman some shelter. She appeared to be some older than Tansy's ma but a good deal spryer when she threw a load of books over her shoulder to make a fast track back up the trail to her house, stopping along the way to share with others.

Tansy was relieved to see Gin when she rode up. "Miss Gin, I was worried you wouldn't be able to make it today or maybe worried you would. Had to be a rough trek for you through this snow." After she dismounted, Tansy lifted off the oilcloth protecting the extra sacks of books for Gin and untied them from her saddlebags. She kept them in two sacks to balance the weight on Shadrach.

"This little pile of snow's not near enough to keep me away. It would be a lonesome two weeks coming up without more books to circulate up home."

"But it is getting deep."

"I've got my tall boots on," Gin said. "And I wore my slicker so I can pack the books under it and save them from the weather. I hope you brought us some good ones."

"There's one by Pearl S. Buck you might like. *The Good Earth*."

"Have you read it?"

"Not yet, but I gave it a look. Appears to be set in China. You can tell me if it's good when next I come this way."

"I like a book that stretches my thinking." A teacher in her younger years, Gin still gathered some of the children who lived near her for reading lessons when school wasn't in session.

Gin helped Tansy divide out the books she'd brought back into the two sacks to tie to Tansy's saddlebags before Tansy draped the oilcloth back over them.

"See you in a couple of weeks," Tansy said.

"I'll be here, God willing." Gin shrugged off one sleeve of her slicker and slung the sack of books over her shoulder, then pulled the slicker back on. She waved and headed back up the trail, looking like a book Santa Claus.

Snow kept falling as Tansy turned Shadrach to the east and started the homeward loop. By the time she'd stopped at her last house, the snow was a foot deep. She rode through a muffled world up Robins Ridge toward home. Tansy was ready to take a turn by the fireplace. The smoke rising up out of the chimney and the glow of lamplight through the window looked more than inviting.

She slid off Shadrach and led him into the barn. "Looks

like another ride through the snow to Booneville tomorrow, Shadrach."

After she took care of Shadrach and put him in his stall for the night, she slung the sacks of books over her shoulders and picked up her saddlebags to carry to the house. She thought of Miss Gin making that long walk home. She might just now be seeing her cabin too if she'd stopped to give out books on the way. Mr. Mayfield had passed on and her children were grown, so Miss Gin would go into a lonesome house. But with a book to keep her company.

No lonesome house here. Instead, it was bursting at the seams even with Pa gone. Tansy stood still a moment to picture those inside. Livvy would likely be in bed already. Aunt Perdie would be in her rocker by the fire and Ma in hers. Both would be busy with their hands. Aunt Perdie might be whittling while Ma would definitely be sewing. Coralee would be sitting between them, hemming soft cloth for diapers. Tansy could almost hear Arlie cracking walnuts on the hearth to dig out the meat. That boy was ever hungry. Josh would be in his favorite place on the steps, sneaking looks at Coralee. No telling what Prissy would be doing.

As Tansy headed up the porch steps to see if what she imagined matched reality, she wondered how it would be to come home to her own hearth instead of her mother's. First, she might need to settle on a man. She shook away the thought.

Instead, she thought about being in Booneville tomorrow and whether the snow would be too deep for that motorcar ride Damien Felding had promised if he was still in town.

twenty-nine

CALEB DIDN'T GET THE SOON start to Booneville he'd hoped for on Tuesday morning since he had to finish up the pen for Rusty to make peace with his mother. Ma had been every bit as cross about the pup as he'd expected, but no crosser than he was with her for orchestrating that time for him to be alone with Jenny Sue. He'd told her in no uncertain terms to never do anything like that again.

Neither of them tried to smooth over the rift between them with words. But Monday afternoon she made him a pie while he started building a dog pen in the middle of a snowstorm.

Not that the pup would need a pen of any kind, with the way Jenny Sue had taken to him. Leaving Rusty with her overnight seemed to have worked a near miracle, since Monday morning when Caleb took the children over to meet the pup, Jenny Sue had on an apron and was at the stove turning flapjacks. Her hair was tied back neat while a smile sat easier on her face. But who wouldn't smile watching Junie and Cindy Sue giggle when the pup jumped around them?

Even better was Jenny Sue insisting Junie sit at the table to eat breakfast. When he claimed to be dizzy, she gave him a mother look and told him to hold on to the table and things would be better in a minute. "You have to get your legs under you to take this pup for a walk."

Ma, who followed them in, frowned. "It's snowing."

"Junie has boots." Jenny Sue didn't shrink from the sound of Ma's disapproval. "That need wearing."

"I might fall down." Junie had a whine in his voice.

"You might," Jenny Sue agreed.

The boy looked over at Ma, obviously expecting her to overrule his mother. But Ma's face changed as she looked from Jenny Sue to Junie. Her voice was soft but without bend. "Your mother's right. If you have a dog, you have to take care of it. And if you fall down, you can get right up."

Jenny Sue gave Ma a little smile before she turned back to Junie. "That's right. Your uncle Caleb here could probably tell you how he fell down in the snow more times than he could count when he and your pa were boys." Jenny's voice stayed steady, but her shoulders stiffened a little. Just mentioning Reuben appeared to push her back into her black grief a bit.

Caleb spoke up quickly. "We had some fun in the snow. You and Cindy Sue will too with a pup to play with."

Ma was frowning again. "That dog will track up the house if you keep letting it in."

"I have a mop. Rusty is fine company. Better than some." Jenny Sue gave Caleb a sideways glance. "Kept my feet warm all night."

"You surely didn't let that mongrel sleep on your bed." Ma sounded more than a little shocked. "You should have fixed it a pallet. Out on the porch."

When Jenny Sue merely smiled, Ma shook her head. "Dogs belong outside."

"But poor puppy. He was scared and Caleb said he might run away. I didn't want that to happen before the kids got to see him." Jenny Sue looked straight at Ma with no smile this time. "Besides, it is my bed."

"So it is." Ma pressed her lips together as if to keep from saying whatever she was thinking before she blew out her breath. "Well, you're the one that'll get fleas."

Now as Caleb rode toward Booneville, he had to wonder at the power of one long-eared pup to make such a difference in a house. But the pup wasn't what made the difference. It was Jenny Sue. Somehow Rusty had given her the strength to stand up to his mother and take control of her own house again. While his ma surely hadn't intended to usurp Jenny Sue's place as mother, she had. Sometimes what a person did with loving intent turned out to be all wrong.

He hoped letting Tansy know she had always been the object of his affections didn't turn out to be one of those all-wrong things. He shook away the thought. A man couldn't forever sit back and hope to be noticed. Although he might be wrong, he had the feeling that Tansy wouldn't have minded had he taken advantage of the moment and kissed her on Sunday. If not for noting that flicker of uncertainty, he would have.

But what if that uncertainty was only in his mind with the fear she would push him away? Perhaps he should wait for a sign that she was looking on his attentions with favor. Give her time to get used to the idea that he loved her, had always loved her, not Hilda, before he pushed her for more.

After he left Phoebe at the livery stable, he headed down the street to the school superintendent's office. He'd take care of that first and then go see Tansy's library.

"Barton, wait up!"

Damien Felding. The last man Caleb wanted to see. He'd hoped the man would be long gone by now, but here he was still in Booneville. Nothing for it but to stop and see what he wanted.

"Whew." Felding bent over to catch his breath. "I didn't think I was going to catch you."

"Now you have." Caleb didn't bother smiling. No need pretending something he didn't feel.

"I have." A smile twitched up one side of Felding's mouth as he straightened up. "Aren't you curious as to why?"

"I figure you'll tell me if you want to."

"You mountain men are all so stoic." His smile got bigger as if he knew a joke but didn't plan to share it.

"That's us. Look, I've got things to do. So if you've got something to say, say it."

"Stoic and direct." Felding's smile disappeared as his eyes narrowed at Caleb. "I thought it might be time the two of us had a talk. Without Tansy around."

"I don't have anything to say to you about Tansy." Caleb kept his eyes steady on Felding's face. "Or anything else."

"Stoic, direct, and unfriendly." When Caleb didn't say anything, he went on. "But don't you think you should consider what's best for Tansy? A bright girl like her needs to try her wings away from the mountains. Find out how people live out in the world. That world she's read about in books."

"With you?" Caleb clenched his fists. He couldn't knock him down right in the middle of Main Street, but he wanted to.

"That would be one way." Felding shrugged.

"I guess that's for Tansy to decide."

"It is, but why would she choose a life of hard work and never having enough when she could do so much better?" Felding looked sure of his words. "Let her go, Barton."

"I'm not holding her here."

"But you'd like to."

"You've said your piece. Now I've got things to do."

"So does she. If she gets the chance."

Caleb didn't bother answering as he turned and walked on down the street. He went past the superintendent's office and around the block. He needed time to calm down. He should have knocked Felding flat, but that wouldn't have changed the truth in his words. Tansy might have a better future away from the hills, but not with Felding. He wasn't who Tansy needed, but then it could be Caleb wasn't either.

He pushed that thought out of his mind. First things first. He was here to see the school superintendent.

Mr. Jefferson was more than glad to let Caleb take the test even though he said the state had come up with some new qualifications. "You know how those men in Frankfort are. Making laws that match up to those flatland schools with no sense at all about us up here in the eastern part of the state."

Mr. Jefferson was a tall stick of a man who kept pushing his black-rimmed glasses back up on his nose. He had been a teacher for years before taking the superintendent job when the state required the county to have one. Now he pushed his glasses up yet again and studied the application form Caleb filled out.

"Looks as if you attended primary school at Manders and then went to high school here in town to get your diploma. State says you need a few weeks of training but that can wait till summer. We can consider the rest of the school year something of an emergency situation and overlook a few rules for the good of the children."

Mr. Jefferson laid down the paper and peered over his glasses that had slipped down on his nose again. "Our youngsters lack some of the advantages of children in other parts of the state, but they do want to learn. The way the boys and girls clamor for the books the packhorse librarians carry out to them proves how much. Do you know about the book women?"

"Yes, sir. My nephew and niece love it when the book woman comes."

"You're on Tansy Calhoun's route, I think."

"Yes." Just the mention of her name made his heart beat faster.

"Tansy was a little different pick for the packhorse librarians. The program gives women jobs to keep them off the dole. The others all have children. Tansy is the only one not married, but after her father left, the family needed help." Mr. Jefferson shook his head slowly. "You'd think a man wouldn't take off and leave his family, but it's happening all the time right now."

"I hear he went hunting work."

"With none to find, most probably." Mr. Jefferson sighed. "Seems to me a man should stay with his family, but I suppose some men can't handle seeing that family go hungry." He pushed his glasses up on his nose and picked up the paper again. "After you take the test, you can fill out the rest of the school year at Manders."

Caleb didn't find the test too hard. By the time he left the superintendent's office, he could almost hear schoolchildren calling him Mr. Barton. The thought made him smile and his steps were light as he headed up the street to see Tansy.

Then Felding's words came back to mind. *Don't you think you should consider what's best for Tansy?* Was he thinking what was best for her or only what he wanted? He shook his head. He would never push Tansy to do anything she didn't want to do, but right then, he was more than sure she would want to show him her library. He looked at the sun sinking low in the west and walked a little faster.

He was almost to the building Mr. Jefferson said housed the library when an older-model Chrysler coupe pulled out of a side street, slid sideways, and headed up the street toward him. The car slowed as it came closer. Behind the wheel, Felding had a wide smile as he waved at Caleb. Tansy waved too, but she didn't appear to be smiling. For sure, Caleb wasn't smiling.

Felding pressed on the gas and threw up slush as the car fishtailed and then straightened out as it sped away.

Tansy managed not to gasp when the car started sliding, but only barely. Damien was showing off the way mountain boys did sometimes by jumping fences. But she'd never seen Caleb showing off that way. He wouldn't want to chance hurting his horse. He was too sensible. Boring, Hilda had said.

But Caleb wasn't boring. He simply thought things out and did the right thing. She wasn't sure she'd done the same, getting in the car with Damien.

She glanced back at Caleb, but she couldn't see his face. Then he turned and walked on toward the library. She wished he'd come earlier so she could have been the one to show him around.

At the same time, she'd felt a little breathless when Damien showed up an hour ago to take her for that promised ride. When she told him she had to work until four thirty, he'd paced back and forth by the shelves of books without once pausing to look at the first title. While he hadn't said a word, his pacing was such a distraction that Mrs. Weston suggested Tansy leave early but to be careful.

Tansy hadn't known which Mrs. Weston was cautioning her about—the ride or the man.

Be careful. That appeared to be something Damien didn't think necessary. Tansy gripped the edge of her seat when the car slipped in the snow again.

"Relax, darling." Damien laughed. "Don't you worry your pretty head. I know how to drive this lady in the snow."

Darling? Pretty? Did he mean those words? That was the trouble with Damien. She had no idea whether he meant anything he said.

"I'm not worried. But I'd hate it if you ran over somebody."

"I guess I'd hate that too." He eased off the gas pedal. He

glanced up at his rearview mirror and then over at Tansy. "Your fellow back there didn't look too happy. He must have been going to see you, but lucky me. I grabbed you first."

"He's a friend, not my fellow. I told you that already."

"So you did." He was quiet for a minute. "Don't you want to have a fellow?"

"Sure. Someday." She kept her voice light.

"Aren't you afraid of being an old maid? Seems most of the girls around here marry young."

"Don't young girls get married in the city?"

"Probably, but some of them go to college and wait a while. For the right guy, I guess."

"Girls here in the mountains wait for the right guy too."

"Who's the right guy for you, Tansy Calhoun?" He grinned over at her and went on without giving her time to answer. "Oh, wait. I remember. You're a Jane Austen fan so you must be waiting for a Mr. Darcy. I'm thinking a guy like that might be hard to find up here in the hills. A rich gentleman with an estate. To tell the truth, those guys are pretty hard to find anywhere."

"I'll keep that in mind if I go looking for one," Tansy said. "So how about you? Do you have a girl back in New York?"

"Me? No, I'm free as a bird and poor as a church mouse. But if I were to find a Miss Darcy, I might be ready to give up my freedom."

"A Miss Darcy might be as hard to find as a Mr. Darcy." She peered out the front window. Now that they'd left town, the snow was deeper on the road. Caleb's warning about getting stuck came to mind. Also, if Damien didn't turn around soon, she'd have to ask him to. She needed to start toward home.

"So she would." He grinned over at her again. "But a girl in the mood for adventure might be easier to find."

That wouldn't be a girl anxious to head home. She could wait a few more minutes. Speeding along in a car in the snow was an adventure. Talking about *Pride and Prejudice* made it even better. "But the best thing about Mr. Darcy in that story wasn't his money or his estate."

"Oh? What was it then? His good looks?" Damien asked.

"No, it was how he loved Elizabeth."

Damien laughed. "Ahh, a true romantic, and I took you for the more practical type. But the fact is, the ending of that story wouldn't have been as fine for dear Elizabeth without Mr. Darcy's manor house and fat bank account."

"And it wouldn't have been so fine without the love between them."

"Austen did work some fictional magic in that book. Poor girl. Rich man. Both seeming to dislike the other at first. Complications and misunderstandings. But nothing can stop the course of true love. At least not in a storybook."

"You sound like the practical one, Mr. Felding." Tansy had to laugh. "I would have taken you for the romantic one."

"I can be romantic." He steered the car over to the side of the road and slid to a stop. He put the car out of gear and turned in the seat toward Tansy. "When the occasion calls for it."

Tansy's heart began pounding in her ears. She suddenly felt as out of place as a raccoon in a dog pen. A raccoon would be scurrying out of there and finding a tree to climb, but she didn't have anywhere to go.

She pulled in a breath to calm down. She'd been alone with Damien on Saturday. *On horses*, a little voice whispered. Not shut up in a car. She wasn't shut up. She could open the door and hike back to town if need be. But Damien hadn't done a thing to make her think she needed an escape. He merely said he could be romantic. Not that he wanted to be.

He scooted closer to her and reached to touch her cheek. "Where did your smile disappear to, mountain girl?"

She leaned away from his touch and managed a smile. "I'm smiling."

"You look a little nervous. Surely at your age you've been alone with a guy before." He kept his hand in the air. "Maybe even kissed a few."

"Only those I wanted to kiss." Tansy wasn't about to admit the only boy she'd kissed was Jeremy Simpson, and that was so long ago she'd almost forgot how it felt. They'd both been kids who didn't know much about anything anyway. Damien wasn't a kid, but then, neither was she.

He dropped his hand down on top of hers in her lap. "Do you have a list?"

"A list of what? Books?"

"Forget books, Book Woman." He smiled. "Of those you want to kiss?"

Tansy was wedged against the door. Trapped. Not a feeling she liked. She stared straight into his eyes. "You need to take me back to town."

"Not yet." He traced her knuckles with his finger. "I could teach you some things about kissing. The lessons wouldn't take long, and might be fun."

"Some things don't need lessons." She pulled her hand away from his and pushed him back when he leaned closer.

He laughed then, and she had no doubt he wasn't laughing with her since she wasn't laughing at all. "Are you telling me I'm not on your list?"

"I don't have a list, Mr. Felding, but should I make one and should you be on it, I'll let you know."

"But maybe I won't be in a kissing mood then."

"That's a chance I'll take."

He was still smiling as he scooted back over under the wheel and put the car in gear. Before he turned the car back toward town, he looked over at her. "Say what you want, you do want me on your list. I've kissed enough girls to know when the invitation is in the eyes, if not on the lips."

She didn't know what to say to that. Some part of her did

wonder how it would be to kiss him, but true or not, his saying so rankled. When he laughed, she was mad enough to spit. Not exactly something a lady should do. Spit. Certainly nothing a Jane Austen character would do.

But then she was a mountain girl and not someone in a novel. She might be riding in an automobile with a city man with slicked-back hair and even slicker words, but that didn't change who she really was. She didn't want to change who she was. Not even for a true Mr. Darcy.

In a matter of days she'd had two different men ready to kiss her. Two very different men. She hadn't kissed either of them.

thirty

MORE SNOW FELL AS JANUARY slid into February, with cold winds finding every crack in the cabin. Some mornings, Tansy had to break ice in the water bucket to wash her face before she headed to the barn to break more ice in the watering trough for Shadrach.

February had a way of making her as cranky as Aunt Perdie's cat. Why Prissy had reason to be cranky she couldn't imagine, since the creature spent most of the day in Aunt Perdie's lap or under the blanket on Coralee's bed, all hunkered down. Not that Tansy wanted to do that. She was up every morning before daylight to start out on her book routes. Whenever the weather was particularly unpleasant with bitter winds hitting her face, she would think about the people waiting for a book so they could sit by the fire and read.

Winter, if a person wasn't on a horse riding miles of trails, was a good time for reading by the fire. She made that happen for others, whether or not she had much opportunity to enjoy the same.

Even when she was home, reading time by the fireplace was hard to come by. Since Aunt Perdie and Coralee had come, the house was running over with people. Not only them. Preacher Rowlett started showing up at least three times a week, most always at mealtimes. Ma didn't mind. She said it took her mind off her worries about Tansy's pa. They had food enough, and Coralee was a ready helper in the kitchen. Aunt Perdie even laid aside her sewing to make herself useful peeling potatoes and such. Then she'd go back to piecing a quilt for Coralee's baby. Sometimes she hummed while she stitched the pieces together.

Coralee wasn't far from needing that blanket. She never talked about the baby coming, although sometimes she gently stroked her tummy with a sweet look on her face. She probably talked about the baby in the letters Tansy stuck into whatever book she took her sister. Saralynn always had a letter ready to come back to Coralee. Sometimes the little brothers sent something too, which made Coralee smile and tear up at the same time.

Since Caleb had started up school again at Manders, the boys were there instead of at their house. Saralynn looked a little wistful when she talked about the school.

"I wish I could go with the boys, but somebody needs to tend the fire and such." At fifteen, Saralynn wasn't too old for school, but age wasn't all that mattered sometimes. "I do have these books you bring, and Mr. Barton sends lessons I can do here at home. He's the nicest man. He even brought that dog to school so Sammy could see him. Sammy was practically doing a jig, he was so excited telling me about that."

It sounded funny hearing Caleb called Mr. Barton, but he was the teacher now. Even Arlie called him Mr. Barton.

"Is your pa okay with you doing the lessons?" Mr. Embry hadn't been there when Tansy made a book stop.

"He hasn't said anything against it. Pa's trying to do better. I think how he done Coralee is weighing on his mind and then that with the pup. He told Sammy a boy here in the hills needs to toughen up and not caterwaul over just everything. But I think Pa was feeling some sad in his heart too. Poor little Sammy. He's so little he don't hardly remember Ma. I try to tell him about her, but that ain't as good as having a mama to love on you. Then when Coralee had to leave too . . ." Saralynn wiped a tear away with her apron hem. "Well, things have been hard."

"I'm sorry."

"Oh, I don't mean to sad-mouth you. Pa does care about us. He just got off-kilter when Ma died and then took to drinking some. I reckon he couldn't abide losing her. She had a way of making things good no matter what else was going on. She'd have stood by Coralee, and Pa, I'm thinking he's some ashamed of how he acted instead of doing like Ma would have done." Saralynn fingered the edge of Coralee's letter sticking out of the book Tansy had given her. "Worst thing is that he's the one what brought that jasper around here to sweet-talk Coralee and then light out for who knows where."

"A girl has to be careful about who she gives her heart to." Tansy knew that to be true. But then she wasn't sure she'd ever actually given her heart away. Certainly not to Jeremy Simpson. Could be, she was too careful.

"That's the good Lord's own truth. Sweet words don't mean much unless they lead to standing up in front of a preacher. Someday I hope to find a feller that knows how to do things proper, but I reckon even if that perfect feller came knocking right now, I couldn't answer the door. The boys need me."

"Do you think Coralee will be able to come back home? After the baby comes?" Tansy rushed on. "I mean, we're glad to have her with us. We love her being there." *One a little too much*, she added silently.

"I know. She tells me you're all good to her and how Aunt Perdie treats her like a daughter. The Lord answered our prayers that day in just the right way, but it was too bad about Aunt Perdie's house."

"It's working out."

And it was. Things did have a way of working out. Her own mixed-up thoughts about Caleb would surely work out too. But things hadn't seemed exactly right with him since he'd seen her in Damien's car. Not that Tansy even knew what right might be.

Damien had moved on to another place and she hadn't seen him since the car ride either. She did see Caleb at the school when she dropped off books. Then he came by visiting on the weekends some too to help the boys with getting up wood. He fit right in with the crowd at the house, but Tansy wasn't sure how he was fitting in with her.

At the house, he spent more time helping Arlie with his lessons than talking to her. He hadn't asked her to go riding again. Just the two of them. But the weather had been bad. Not fit for anything but fireside sitting, if a person had a choice. She knew that well enough from being out in it all during the week.

She couldn't exactly put her finger on it, but he seemed different toward her. There had certainly been nothing anywhere near to that moment when she'd held her breath in the middle of the ashes of Aunt Perdie's house and thought he wanted to kiss her. That might have been a one-of-a-kind magical moment, what with finding the Bible and all. But she couldn't keep from thinking Caleb had stepped back because of seeing her with Damien.

The question was, did she want him to step closer instead? She didn't have an answer for that. She was about ready to catch a boat for England and go hunting Mr. Darcy. A manor house might have room for all the folks crowded into their cabin right now and plenty of fireplaces for a person to sit and read a book.

If romance didn't seem to be happening for her, that didn't mean it wasn't happening for others.

One morning toward the end of February when Tansy came down the loft steps, Coralee was up, tending the fire in the cookstove. Sometimes Aunt Perdie would be in her rocker by the fire when Tansy came down, but usually Coralee was curled in bed. Maybe asleep. Maybe not. Sometimes Tansy thought she heard muffled sobs.

Truth was, she was just a girl. Only sixteen and facing going through that hard valley of giving birth. Her mother had died in childbirth. That had to be weighing on Coralee's mind. Tansy had brought her a book about giving birth, but it was more for flatlanders with doctors to help them. That wasn't the way of it up here in the hills. If fetched, Geraldine Abrams would come. She'd helped Tansy's mother when Livvy took time coming.

This morning Coralee had coffee ready and had fried a couple of thick slices of Ma's bread. "I thought you might like a bite of something 'fore you head out. What with it so cold and all." She put the bread on plates and set them on the table.

"That's extra nice of you." Tansy poured some coffee and sat down.

Coralee sat across from Tansy and warmed her fingers on her cup. "I hoped we could talk a minute before the others start stirring. If'n you have time."

The kerosene lamp threw enough light over the table for Tansy to note frown wrinkles between Coralee's eyes.

"I don't have to be in a big rush." Tansy dipped out honey for her bread. "Is something wrong?"

"Not wrong, exactly." She peeked up at Tansy, then stared back at her cup. "But sometimes it's easier to talk without everybody hearing."

"Not much quiet time around here lately."

"I know. It's some like back at my house with the boys full of energy. I'm glad Mr. Barton opened up the school again. Davey said that last feller that quit wasn't much of a teacher anyhow. That some boys gave him trouble. I told Davey I better not hear about trouble from him. But I wasn't really worried. He's a good boy. Tries to take care of little Sammy." She blinked back a few tears. "They miss Ma, and then I up and got in this mess so's I can't be there for them." Coralee sighed.

"You're welcome here for however long you need." Tansy smiled across the table at Coralee.

"That's what your sweet mother says too, but I'm wondering if'n that would hold true were your pa to show up home. Seems womenfolk are more forgiving of a girl who's done wrong."

"No need worrying over that until it happens. We haven't heard anything from Pa. We might never again."

Coralee looked up at Tansy. "I'm not hoping that. I wouldn't ever hope that." She dropped her gaze back to her plate, where her bread was untouched. "Besides, we—me and Aunt Perdie—maybe could find a place with Preacher Rowlett. I think he's sweet on her and I'm nigh on sure she's sweet on him."

"You're kidding." Tansy sat back and stared at Coralee.

"That's how things appear to me. I don't reckon folks ever get too old for some sparking. Aunt Perdie ain't so old that she couldn't do some cooking for Preacher, and it's plain to see he's a man to appreciate a cook." She took a sip of her coffee. When she put the cup down, she glanced toward the mantel clock.

"But that's not what you want to talk about," Tansy said. "Has Josh been worrying you? I'll tell him to stop if you need me to."

Even the dim light didn't hide how her cheeks flushed. "Josh ain't been no bother to me." She pulled in a breath and let it out slowly. "I might as well be out with it. He's done gone and asked me to marry him."

"Oh." Tansy shouldn't have been surprised, but she was. She thought Josh would talk to her again before he did Coralee.

"I know you and your ma might not be happy about that." Coralee looked straight at Tansy. "Him offering to rescue a fallen woman like me."

Tansy took a sip of coffee to have time to think. Josh was so young. But then so was Coralee. The poor girl's fingers were trembling. Tansy reached to touch her hand.

"What Ma or what I think doesn't matter near so much as what you think."

Coralee let out a shaky breath. "I like Josh, but I can't say I'm in love with him. I told him that. I'm not even sure he really loves me the way he thinks. How could he with how I look right now?" Despair colored her words.

"A woman carrying a baby is always beautiful."

"I would think that only true to a man if the woman were carrying his baby. I'm not sure Josh has thought this through. What he might be undertaking by wanting to father another man's child."

"He is young and I don't know how he would provide for you and a baby."

Coralee stared holes in the table. "He says your ma will let us stay. He hasn't asked her, but that she needs him here at least until his pa comes home." She looked up at Tansy. "But you're the one bringing in what the family needs."

"I couldn't do it without Josh." Tansy squeezed Coralee's hand. "I have only one question for you. Do you think you could someday love Josh the way a wife should love her husband?"

"I think I could. Someday." Her voice was barely over a whisper. "But right now, I can't hardly think of anything but this baby I'm carrying. I'd do anything to protect him or her."

"Even marry my brother?"

"I guess that's how it sounds." Coralee looked up at Tansy

and didn't shy away from the question. "And how it might be. Like I said, I might can love Josh right enough someday, but now I can't think about loving anybody in that man-and-wife way. I told Josh as much, but he claims he can wait until I'm ready once the baby comes." She pulled her hand away from Tansy and wiped her eyes. "I thought I was in love before and gave myself to a man who didn't love me back. I'm some scared to give my heart away again."

"This man." Tansy hesitated a second before she went on. "Is he liable to come around to make trouble for you? For Josh? If the two of you marry and Josh takes his child for his own?"

"I don't expect to ever see hide nor hair of him again."

"Did he know about the baby?"

Coralee nodded and then swallowed hard. "He wanted me to figure a way to get rid of it. Called the baby 'it,' as if she or he was something I could just throw away." Her hands slid off the table to cover her baby bulge. "Whatever love I had for him died right then and I knowed he never loved me at all. Just used me for his pleasure."

Tansy went around the table to wrap her arms around Coralee. "Whatever you decide, Coralee, to marry Josh or not, I'll do anything I can to help you. And your baby."

Coralee started sobbing. Tears wet Tansy's cheeks too.

"Shh." Tansy stroked Coralee's head. "Don't cry. It will be okay."

"I'm just so scared," Coralee mumbled. "Of birthing this baby. I saw Ma die trying to have her baby. The baby too."

"That doesn't mean you'll have trouble. We'll pray you have an easy time." Tansy brushed away her tears with her hands before she grabbed a tea towel to give Coralee. "You'll be a great mother."

After Coralee wiped off her face, she looked up at Tansy. "How do you know?"

"I see how loving you are. How good you are to Aunt Perdie and how even her contrary cat likes you."

Right on cue, Prissy padded across the floor to wind between Tansy's and Coralee's legs. The cat tried to jump up into Coralee's lap. Coralee laughed a little as she caught the cat and held her in the little bit of lap she had left. "The baby's done taken up Prissy's spot."

"She'll just have to get used to it." Tansy leaned over and kissed the top of Coralee's head. "Now stop your worrying. Have Josh talk to Ma."

"And I should talk to Aunt Perdie too." Coralee was smiling still. "Maybe we can have a double wedding with her and Preacher Rowlett."

"If you say so." Tansy was still surprised at that thought.

"Or maybe a triple wedding." Coralee peered up at her. "With you and Mr. Barton."

"What in the world would make you say that?" Tansy stepped back from Coralee.

Coralee shrugged. "You aren't promised to nobody and neither is he. Seems to me you'd make a good match. And I'm guessing he'd be willing if'n you were."

"I don't know what makes you think that." Tansy grabbed her cup and gulped down the rest of the coffee.

"A person sits back and watches, they can see all sorts of things. And what I've seen is how he looks at you when he thinks nobody is watching."

"And how's that?" She tried to sound as if she didn't care what Coralee might say, but at the same time her breath was catching in her throat.

"Like you're flowers and stars and a perfect sight for his eyes."

Tansy's face heated up. "Those are pretty words. But you didn't see me looking at him the same way."

"No'm." Coralee shook her head. "The day I first noticed

Mr. Barton's sweet looks at you, you were busy looking at that city feller like as how he was a sparkly rock you'd pulled out the creek and weren't sure whether to put it in your pocket or throw it back."

"That's silly. He was just a guest I was being nice to."

"Mr. Barton was visiting too." Coralee raised her eyebrows at Tansy, then looked down at Prissy. "Don't mind me. I shouldn't ought to have said anything. It's up to you to notice a feller's look. Not me. I sometimes let my lips flap too much. I didn't aim to upset you none."

"I'm not upset. But I'm not sure you were seeing right."

"Maybe not." A floorboard creaked overhead. "Sounds like the boys are stirring. I better poke more wood in the fire to cook something up for them." She dropped Prissy down on the floor and stood up with a quick catch of breath.

"Are you feeling all right?"

"My back's paining me some, but your ma says aches and pains are common when you're carrying a baby." With her hands on the small of her back, she stretched backwards.

"How much longer do you have?"

"Two, maybe three weeks. Soon enough." She smiled a little uncertainly at Tansy. "Thankee for hearing me out this morning. A body needs to speak her worries out loud sometimes."

Coralee might be right about that, but Tansy didn't have a soul to talk out her worries with as she rode Shadrach away from the barn.

"Nobody but you, Shadrach," she said out loud. "But she was right. About me wondering over Damien Felding. About Caleb wondering over me."

Her words got swallowed up in the icy chill of the morning air. A streak of red in the east gave evidence of the sun coming. She pulled in a deep breath and tried to think about the people she'd be taking books to and not Coralee's words.

But she couldn't quite forget that one thing Coralee had said. About being afraid to give her heart away again.

Was that Tansy too? Afraid of being hurt by love? Maybe she was destined to be another Aunt Perdie with nothing but maybes or might-have-beens. Then again, Coralee was matching Aunt Perdie up with Preacher. Who would have thought the girl was such a matchmaker?

thirty-one

PERDITA WAS GLAD when Tansy left. She was tired of shivering by the bedroom door. No heat from the fireplace reached into Eugenia's bedroom. Once out from under the covers, a body needed to waste no time moving toward the fire.

But when Perdita started in that way, she heard Coralee and Tansy talking. Whilst she sometimes had to squint a little to see something in dim light, she could still hear nigh on as good as when she was twenty. Without the slightest compunction, she stepped back, left the door slightly ajar, and put her ear to the crack.

A mother had the right to know what was going on. She smiled at the thought. Coralee might think of her more as a grandmother or kindly aunt, but Perdita felt like a mother. At least how she imagined a mother might feel.

The two girls didn't say anything Perdita didn't expect to hear. At least not right off. Anybody with eyes knew Josh was sweet on Coralee. Perdita wasn't a bit surprised he'd jumped the gun and asked the girl to marry him. He might still be wet behind the ears, but he was old enough for love. Coralee could do worse. Guess she had done worse with the one that helped her make a baby. Good to know there weren't much chance of whoever that was making trouble.

Perdita did admit to being some shocked the girl had noticed her fondness for Hiram. A little smile tickled her lips at the thought that Coralee might be right about him. Maybe he had been showing up here for more than Eugenia's biscuits. Could be Perdita should wade right on out into that pond of possibilities and ask Hiram straight out if'n he had any proper intentions.

A double wedding wouldn't be a bad thing. She weren't as sure about the potential for a triple wedding. That Tansy didn't seem to know how to open her eyes to what was right in front of her. More troubling, the Barton boy had seemed to back off. Afeared of getting waved away like a pesky fly, maybe. But a person couldn't be afeared their whole livelong life. Perdita could tell him a few things about that. Tell Tansy too. A body's years could pass on fast and leave behind a pile of lost chances. It didn't appear Josh aimed for that to happen. Or Coralee either. Both barely out of the tadpole stage.

The two boys bounded down the steps. Arlie grabbed the tin bucket Eugenia had packed up with biscuits and jam the night before and headed for the door like some kind of varmint was after him. Coralee stopped him and made him take a piece of bread to chew on while walking off to school.

Josh poked up the fire and laid more wood on it. Perdita went out of the bedroom to spoil his alone time with Coralee. She hid her smile at the disappointment on the boy's face before he grabbed his coat and headed to the barn. He must have aimed on stealing a kiss if'n Coralee let him.

Yes, indeed. Not much more than tadpoles. She had to think that if they were two tadpoles thinking on marrying, she and Hiram would be two old frogs about ready to croak.

Once she settled in her rocker and got her lap quilt situated, Prissy came running to claim her lap. Then Coralee was there to kiss her cheek.

"You feel about froze, Aunt Perdie. You should have come on out to the fire instead of waiting till Tansy left. We wouldn't have cared if you were out here listening any more than we cared if you were back there listening."

Perdita gave her a look. The girl sometimes tried to almost disappear amongst them, but she did have a way of knowing what was happening. "I reckon I should say I'm sorry for eavesdropping on you."

"No, no. That just saves some time so's I don't have to tell it all again. What do you think about it anyhow?" She stepped away from Perdita to hold her hands out toward the fire. "About Josh, I mean."

"Appears he needs to be looking for a place to build a new cabin. Once he tells Eugenia what he's up to."

"I haven't said yes."

"You ain't said no."

The girl turned back toward Perdita. "I don't want to ruin his life."

"Which way you thinkin' will be the ruining way? The yes or the no?"

Coralee smiled a little at that. "That's the trouble. I ain't exactly sure."

"Then maybe you should quit worrying about ruining his life and think on what your heart is tellin' you."

"I think my heart is froze up right now." Coralee blinked a few times as if fighting back tears. "That ain't much of a way to start out married life."

"You won't get no argument from me on that." Prissy's fur crackled when Perdita stroked the cat's back.

Coralee turned to get her backside warmed by the fire. She pulled her skirt tight against her to make sure no bit of cloth draped into the flames. She looked toward the bedroom door.

"Is Miss Eugenia all right? She seems to be staying abed later each morning."

"She's got you to be up and fixing for the boys." Perdita studied the girl in the firelight. Her belly looked nigh on swelled to popping with baby.

"I don't mind that. I just don't want her to be poorly."

"Eugenia's fine as she ever is, what with the way her hip pains her. She's the kind what can't go to sleep easy at night, then can sleep fine in the morning hours. Me, I'm the other way around. But we manage not to be too worrisome to one another." She rocked back and forth. "Could be I'll have to nap out here in my rocker should Joshua find his way home." She peeked up at Coralee. "Less'n you is right about that Hiram Rowlett. Then I might be resting in a place I never expected to be. In a missus spot in a bed." That made her chuckle.

"I'm thinking it could happen." Coralee started to laugh too but then gasped a little when she moved away from the fire.

"Are you having pains, Coralee?"

Eugenia came out of the bedroom. "Pains? Are you punishin' some?"

Coralee stepped away from the fire and put her hands on the small of her back. "I've had some discomfort, but there's not been any rhythm to it. You said aches and pains are normal, didn't you?"

"So's having the punishin' birthing pains." Eugenia looked worried. "Nothing to do but wait and see. Times are when the pains don't mean nothing." She got a soft look on her face for the girl then. "But the birthing has to happen sometime."

"I know." A tremble sounded in Coralee's voice.

Perdita dumped Prissy out on the floor and got up to wrap her in a hug. "We'll get you through it. Eugenia here knows about babies."

"I do." Eugenia came over to hug Coralee too. "It ain't the

easiest thing in the world, but a sweet baby is worth the suffering. I'll send Josh after Granny Abrams if'n your pains keep on."

Livvy came down the steps in her nightgown. She rubbed her eyes. "I want to hug Coralee too."

That made Coralee laugh as she held out her arms to the little girl. Perdita sat back down in her rocker. There wasn't nothing to do but wait to see what happened next.

"But don't you want to keep it until the next time I come so you can find out how the story ends?" Tansy asked Jenny Sue when she said she hadn't finished the book she was handing back in.

Jenny Sue shook her head. "No need. I know how it'll turn out. The hero will save the day and win the girl and everything will be hunky-dory." She looked like that was the last thing she wanted to read.

Tansy looked down at the Zane Grey book. She hadn't read it, but she was sure Jenny Sue was right. The hero would get the girl and things would end the way a reader would want. "I like a happy ending."

"I'm guessing most folks do, but they don't always happen outside of books." Jenny Sue pointed toward the fireplace. "Come warm by the fire. Cindy Sue will raise a fuss if she misses you."

Tansy looked over at the little girl curled in the middle of the bed. The pup was stretched out by the bed. "She does look to be having a good nap."

"That child can sleep through a thunder boomer, but I'll rouse her if she don't perk up in a minute."

"I'm glad Junie is up to going to school with Caleb." Tansy sat down in one of the chairs. She had some minutes to spare before she moved on to Manders School. "I'll be seeing him there in a bit, but I'll leave him a book here too. Do you know if Ma Vesta wants a book?"

"Who knows what Ma Vesta wants?" Then Jenny Sue shook her head. "Or then again, she's not shy about makin' sure a person knows exactly what she wants. More than wants. Expects."

"I can go ask her." Tansy didn't want to tiptoe into a fuss between Jenny Sue and Ma Vesta.

"Just leave something here for her. She was gonna rest her head a while. Aimed to get away from Rusty. She don't like me letting him be inside here." Jenny Sue smiled. At the sound of his name, the dog flapped his tail before he came to let her fondle his ears. "She don't see Rusty's good points the way I do. But she said if you had that quilting book you'uns put together down at the library, she wouldn't mind taking a look-see at it."

"All right. I do have it with me." Tansy started to get up to fetch it from her saddlebags.

"You can give it to me when you go." Jenny Sue reached to stop her. When Tansy sat back, Jenny Sue stroked the dog's ears again. "It can get powerful lonesome around here, but I'm trying to do better. Be ma to Junie and Cindy Sue again. I gave that over to Ma Vesta a while, but they are mine and not hers."

Tansy didn't know what to say to that, so she simply nodded.

Jenny Sue kept her gaze on the dog. "Rusty here helped me get back on my feet. And me on my feet got Junie on his. A mother has a powerful lot of responsibility to see that her young'uns do right."

"I guess so."

"I know so." Jenny Sue looked over at Tansy. "You have a dog?"

"I pet on Josh's hound some."

"That's how it was with me too. I've been around dogs aplenty, but they were Pa's dogs or my brother's dogs or Reuben's dog. I know Caleb fetched this pup home for the kids, but Rusty here is my dog." When she looked back down at the pup, he licked her hand. "He looks at me like I know the answers. That can be a fine thing. Nobody has ever had much regard for my ability to come up with answers."

When Tansy started to say something, Jenny Sue stopped her. "I ain't saying you. You were just a little kid when I was hanging around with Hilda, but she never had much good to say about me."

"She was jealous."

Jenny Sue smiled. "I knew she was struck on Reuben. But then most all the girls were struck on Reuben. Were you?"

"He was a great guy, but no."

"That's right. You were always off at the edge of the porch or out in the woods climbing trees or whatever with Caleb, best I remember."

"I was young."

"I reckon we were all young." Jenny Sue sighed. "You said you liked happy endings. What is your happy ending, Tansy? Are you thinking you can live a storybook life?"

"I don't know. I guess all lives have ups and downs. Even in storybooks."

"Sometimes you don't get to write your own endings in life. They're writ by somebody else." She sounded sad. "I won't ever get over losing Reuben, but I'm only twenty-four."

"Nobody expects you to stay a widow forever."

"I'll always be Reuben's widow, but that don't mean I won't marry again." Jenny Sue gave Tansy a look. "Ma Vesta has done figured out how to keep me a Barton. Keep it all in the family."

"Oh?" Tansy should have told Jenny Sue she'd read to Cindy Sue next time and left.

"You ain't hearing what I'm saying." Jenny Sue had a tired smile. "I don't know but one other Barton available for marrying."

"Caleb?" Tansy felt like a hand was squeezing her heart. Jenny Sue must be why Caleb had stopped acting interested in her. He had other plans. She had to pull in a breath before she could speak. "So, are congratulations in order?" She was relieved her voice didn't sound as shaky as she felt.

"Not yet. It wouldn't be respectful to Reuben's memory marrying anybody this soon, though I will tell you it feels like it's been two years since Reuben passed instead of six months."

"Caleb will make a wonderful father for your children." The words almost choked Tansy.

"He would, wouldn't he? This ain't none of mine or none of Caleb's thinking. It's all Ma Vesta, but that woman does get what she wants. Especially from her boys. I don't reckon Reuben ever did nothing without asking his ma first. Way more than he ever asked me. Caleb's some the same."

"He left the mountains to find work. I heard Ma Vesta wasn't so happy about that." She didn't know why she was trying to prove Jenny Sue wrong.

"She wasn't all that upset. She still had Reuben to do for her then. If his ma told Reuben to jump, he just asked how high."

"I'm sorry."

Jenny Sue smiled. "I am making it sound bad, ain't I? But it weren't all bad. I was happy so long as Reuben came to my table at night. He loved me right."

"And do you love Caleb at all?"

"Like a brother. A nice brother, but a brother."

"And does he love you?"

"Like a sister. Not sure how nice a sister, but a sister." Her smile got a little broader.

"You don't marry your sister."

"It ain't always a good idea, that's for certain."

"Maybe he won't do what Ma Vesta wants. Maybe you won't."

"Ma Vesta's a hard one to buck."

Tansy didn't have anything to say to that. The woman was known for her strong will. It wasn't Tansy's business anyway. She had no claim on Caleb, but from the sinking feeling inside her, she must be wishing she did. "I have to go. I'll read to Cindy Sue next time."

As if she heard her name, Cindy Sue sat up in the bed. "Book woman!" She scrambled out of the covers to run to Tansy. "I want a Clyde story."

Tansy had to smile despite the way her insides churned over what Jenny Sue had said. Junie and Cindy Sue had liked the first Clyde hound dog story she'd written especially for them so much that she'd made up a new one each time she came to their house. Sometimes she couldn't find pictures in the old magazines to match the story, but the children didn't care. Clyde managed to get in some awful fixes, but whatever happened, he was always ready for the next adventure.

Jenny Sue laughed. "I hope you're sharing those stories with the other kids on your book route. Them and this dog here have brung a little sunshine to this dark cabin. Thankee for that, Tansy."

"I'm sorry, Cindy Sue." Tansy handed her the new story. "I need to be riding on down the way, but I left you and Junie a new Clyde story. Your ma will read it to you."

Cindy Sue's lip pouted out. "I want the book woman to read it."

"Then you shouldn't have slept so long, little missy," Jenny Sue said.

Tansy squatted down in front of the child. "If you're sleeping away the next time I come, I'll roar like some big old bear when I come in and that will wake you up for sure."

Her eyes lit up. "You think Clyde could fight a bear?"

"Who knows about that Clyde? He might be that brave."

"Or that crazy." Jenny Sue walked to the barn with Tansy to get the quilting book for Ma Vesta. "Thank you for talking to me, Tansy. And who knows? Maybe this time the hero will save the day and get the right girl just like in a storybook."

"But who's the right girl?" Tansy tried to keep her voice light. "As far as that goes, who's the hero?"

Jenny Sue smiled. "Tell you what. Maybe you can write us all into a story and make things come out right."

"I think I better stick with getting Clyde in and out of messes. People messes aren't as easy to straighten out."

Tansy's head was in a whirl as she headed away from Jenny Sue's house. If Caleb wanted to marry Jenny Sue to please his mother, then so be it. She had no claim on him. He was a friend. Hadn't she told Damien Felding that about Caleb? But it seemed her heart had a different idea.

thirty-two

CALEB HAD BEEN WATCHING for Tansy all day. Since she kept the same schedule for her routes each week, he had no reason to expect her sooner than one o'clock, but he still kept checking out the window every few minutes. His students probably wondered if he was worried about snow.

Teaching filled a spot inside Caleb he hadn't realized needed filling. He'd grown up helping his father work the land to keep the family in what they needed. After his pa died, he and Reuben kept on tilling and planting. Then working with the Civilian Conservation Corps planting trees and carving out trails through the Smokies had been good, but teaching these children felt like a calling. Something the Lord had led him to.

He had a different empty spot inside him. This one in his heart. The spot Tansy needed to fill. He hadn't asked her to fill it. He'd hinted at it, but he hadn't gone down on a knee to ask her to be his wife. What if she said no? He didn't like to think he was a coward, but maybe when it came to love, he was.

Seeing her in that car with Damien Felding had seemed to shred his hopes of winning her. Caleb loved her, but he couldn't offer her the kind of life Felding could. A life where she wouldn't have to pack in every drop of water she used or cook over a fire. Even better, in a city somewhere, Tansy might find a place in some big library with shelves and shelves of her treasured books.

Here in the hills even finding books for school was hard. That's why the children were almost as eager as he was to see the book woman riding up.

When one of the older boys came in from a necessary trip outside to say she was coming, the youngsters rushed outside to greet her. Few of them even had the presence of mind to grab their coats. They were that eager for a new book.

Tansy laughed as the children began shouting, "We want a book."

With a wool cap pulled down over her hair and a scarf around her neck, about all he could see of her were her eyes. That didn't matter. She was beautiful.

As if his thought reached out to her, she looked over the kids' heads at him, but then darted her eyes away. "It's cold out here, kids. Go back inside and I'll read you a story."

Something about the way she looked away from him so quickly worried him, but he pushed that aside. He had to stop imagining problems at every glance, but his smile faded a little as the kids rushed past him, anxious for Tansy's promised story. They had benches for seats and higher benches for desks. Here and there a real school desk was mixed in. Donated by someone who had known a previous teacher. Caleb wished they'd donated books instead.

The children sat on the floor around Tansy while she read. She had such a way of bringing a story to life that even the wiggliest boy sat still. Junie leaned against Caleb. He tended to hang back from joining in with the other children even when they had recess, but at least he was at school and not hiding under covers back at his house.

Tansy read them a Jack story printed out in one of the librarians' handmade books. Felding had wanted to hear some Jack stories. Maybe that's why Tansy had written this one out—so she could give it to Felding. Caleb pushed that thought away. Felding wasn't here. Tansy was. Caleb was.

After she finished reading, she let the children look at the books available while she talked to Freddy Harrison about his brother. "Is Price coming down for books today?"

"No, miss," Freddy said. "With Ella here at school, Ma needs him to watch Pa. Pa can take a wander now and again when the cabin walls close in on him. I wanted to trade out days with him coming to school, but he says I need the learnin' more'n him."

After the boy took the book Tansy gave him for Price to stash with his lunch pail, Caleb said, "I'll work on finding a way to get Price here next year. This term's about over anyhow."

"So, you plan to teach next year too?" Tansy said.

"If they rehire me. You think they will, Junie?" The boy was still shadowing him.

Junie looked up at Caleb. "I can't come if you're not here."

"Sure you could, but I plan to be here," Caleb said.

Tansy smiled at Junie. "I was by your house a while ago and left a book there for you."

"A Clyde story?" Junie's face lit up.

"It was."

Caleb laughed. "I read that last one you left. You wrote it, didn't you?"

Color rose in Tansy's cheeks, making her even prettier. "It's just something silly I made up for them." She flashed a look at Caleb and then glanced away quickly.

He wanted to put his face directly in front of hers and ask

why she seemed afraid to look at him, but instead he said, "No, it was great. Made us laugh and wonder if you'd been following Rusty around and seeing the trouble he can get into."

"Oh dear," Tansy said. "Has Rusty been getting in trouble, Junie?"

"With Granny," Junie said. "She don't like him much. But Ma just laughs at whatever Rusty does." His face changed a little. "I hope he don't go off and not come back."

"No worries there," Caleb said. "That dog knows when he's got it good." He gave Junie a little push. "Go check out what they've picked for us to keep till next time."

After the boy reluctantly moved away, Caleb said, "He can't stop worrying about people going away and not coming back the way his dad did. Poor kid's afraid of losing more. Like his mother or granny or that dog."

"Or you." Tansy finally met his gaze straight on, but he couldn't quite read her look.

"I'm not going anywhere."

"No, I guess not." She looked away from him. "But I better be going. I'll be back by in a couple of weeks."

When she started to step away, he caught her arm. "Two weeks is too long. I want to see you this weekend." When she hesitated, he went on. "I could use help getting lessons prepared for the older students."

"I don't know." Tansy kept her eyes away from him again. "Maybe you should do whatever Ma Vesta needs you to do."

Something about her words gave him a sinking feeling. Caleb frowned. "Did Ma tell you something she needs?"

"I didn't see her. Jenny Sue said she was resting."

"Then what's wrong?" Caleb asked.

"Nothing's wrong. I'm just busy. With the books and all." She gave him a tight smile before she stepped away.

He watched her pack up the books. He didn't know what, but something was definitely wrong. No way to push her to tell him with the children watching, but he would see her before two weeks went by. He could waylay her on the trail home from Booneville tomorrow and demand an explanation. He hoped that explanation wouldn't be that Felding was back in town.

After Tansy left, the older students took turns reading aloud from *The Adventures of Tom Sawyer.* He tried to listen, but he couldn't stop thinking about how Tansy seemed afraid to look at him. Maybe he would just ride to her house tonight. He couldn't keep standing back fretting over seeing her in Felding's car. If Felding had won her heart, so be it, but he would still tell her in plain words how much he loved her.

At half past two, he let school out. Some of the students had a long walk home, and while the days were getting longer, the edge of night still came early. Junie waited at the door while Caleb raked ashes over the live coals in the stove. Then he picked up a library book left on a bench. He stroked the cover while thinking about the packhorse librarian. Maybe he needed to read one of those romantic stories to figure out how to win Tansy's heart. He shook his head at his foolishness and shelved the book.

Caleb got Phoebe out of the corral he'd built for the mare to wait out the day. Pines at one end sheltered her from the snow. With Junie riding behind him, they were headed up Manders Creek toward home when they met Josh Calhoun on his mule.

Caleb reined in Phoebe. "Josh, what's wrong?" Seemed he kept asking that today.

"I aimed to catch up with Tansy here. It's Coralee. She's fixing to have her baby." Josh looked terrified at the thought.

"Take a breath and calm down. Having babies isn't easy, but most all women get through it."

"But she's hurting bad." He looked near tears.

Junie tightened his arms around Caleb's waist. Josh's panic

must be scaring him. Caleb patted Junie's hands and used his calmest voice. "Your mother will help her through it."

"Ma had me go for Granny Abrams. They say things are going like they should, but I couldn't stand seeing her punishin' like that."

Caleb hid his smile. "So they sent you off to find Tansy."

"Coralee's wishing her sister could know, and Ma thought I might catch up with Tansy here so she could tell her. Now I don't know what to do. I ain't exactly sure how far on it is to their house, and Sully here is about tuckered out." Josh stroked the mule's neck.

The mule was hanging its head. "I'll go tell her sister. After I take Junie home," Caleb said.

"That would be neighborly of you. I'm anxious to get on back home to be there with Coralee. I love her."

"I know."

"I asked her to marry me."

"Did she say yes?"

"She wanted to ponder on it. Said she couldn't think of nothing but the baby right now." Josh looked worried again.

"That's understandable."

"I reckon so, but I aim to take her baby for mine if she'll have me."

"You have to let her be the one to choose."

Those words circled in Caleb's head as he took Junie home and then rode back down to the Embry place. By then daylight was ebbing. The baby might have already made an appearance, but he'd promised to carry the message.

Saralynn met him at the door. The boys peeked out behind her. "Is something wrong, Mr. Barton? Have the boys been acting up at school?"

"No, they're doing fine. I just came to bring you word Coralee is having her baby. She wanted you to know."

After a glance back over her shoulder, Saralynn shooed the

boys inside, stepped out on the porch, and shut the door. "Is she doing all right?"

"Josh said she was having pains when he came with the message for you. That's all I know."

"Josh." The worry lines on her face softened a little. "Coralee wrote me about him. Says he wants to marry her, but she's afraid of ruining his life. You know Josh, don't you, Mr. Barton? Would he be ruined if Coralee married him?"

"From the look on his face a while ago, I'd say he'll be ruined if she doesn't. At least until he got over his broken heart."

The door opened behind Saralynn. Frank Embry glowered at Caleb. "What do you want?"

"It's okay, Pa." Saralynn shrank back a step from Caleb. "Mr. Barton's the schoolteacher."

"I knowed him. He's the one that acted like he bought that good-for-nothing pup." Embry narrowed his eyes at Caleb. "You kilt him dead yet?"

Caleb smiled a little. "Rusty can be a trial at times, but my sister-in-law has taken to him. She'd give more than a quarter for him now."

"Hard to believe, but I ain't got no more dogs for you to buy. So what are you doing standin' on my porch?"

Caleb hesitated to see if Saralynn would make up some answer. When she didn't, he told the truth. "I came to tell Saralynn her sister is having her baby and wanted her to know."

The man's face changed. "Is she faring well?"

"I can't tell you that. I can only hope so, since I haven't seen her."

Saralynn sniffed and brushed away a couple of tears. "It's kind of you to come let us know, Mr. Barton." Her lips trembled a little. "If'n you hear how things go for her, you can send word by the boys from school tomorrow."

Embry shut his eyes and drew in a couple of breaths. When he opened his eyes, he looked straight at Saralynn. "That ain't the way of it, Sarielynn. Your sister is needin' you now." He turned to Caleb. "I ain't one to be askin' nobody for favors, but if'n you could take Sarielynn here to her sister, I'd be appreciative."

"What about the boys and supper?" Saralynn said.

"We can make out. If the man here agrees to it, he can take you there and then I'll come get you on the morrow whilst the boys are at school."

She still looked hesitant. "You wouldn't be making trouble, would you, Pa?"

"No, daughter. I'd be doin' what I can to make amends." He stared down at the porch. "I shouldn't of ever put her out." He looked up at Caleb. "Are you willin' to carry Sarielynn there?"

"I can do that," Caleb said.

"Run get your coat and boots, girl." He waited until she went into the house before he turned back to Caleb. "I hear'd Sarielynn say something about the boy up there at the Calhouns wantin' to marry Coralee. Is he a good boy? Somebody that will treat her better'n her own pa did?"

"He's young, but he's good," Caleb said.

"I knowed his pa. Joshua Calhoun. Never heard nothin' bad about him, but they say he's done gone off and left his family."

"He aimed to find work."

"Did he?"

"No way to know. They haven't heard from him since he left."

Embry shook his head. "Nothin' but trouble down in those flatlands."

"Trouble turns up everywhere."

"If that ain't the truth." Embry breathed out a sigh. "Some of it we bring on ourselves like I done with Coralee. But I've give up the moonshine so's I can be a better man for my boys, but it's a hard go without my woman."

Saralynn was back then. “Thankee, Pa. I’ll tell Coralee you let me come.”

“Tell her she don’t need to be afeared of me. That I reckon there ain’t nobody what’s not done something they shouldn’t have now and then.”

thirty-three

EVERY TIME CORALEE GRIMACED, Perdita's insides clenched and her breath came tight, as if she were the one having the punishing birthing pains. She'd been with her mother when Miriam was born, but she'd been too young to know much about what was happening.

Besides, by then her ma was an old hand at birthing babies. It weren't easy for her. Perdita recalled how her ma had held on to the bedposts and grunted, but she didn't remember it taking so awful long. Eugenia said first babies most always took longer to come. Geraldine Abrams claimed it true too. She should know since she'd had a hand in helping plenty of babies come into the world. Most all those mothers came through the pains without a hitch. But Perdita knew of some that hadn't. Some like Coralee's own mother.

Perdita couldn't stop the worry that she might lose Coralee. The girl was so young and, other than the baby bump, lacked any meat on her bones. Perdita didn't know when she'd been so torn up inside. After her ma died, she mostly kept to her lonesome self in her little cabin and barred the door to her heart. If'n you didn't invite anybody in, losing them couldn't rip you apart. By then, she'd long since lost Hiram to Mary

Ellen and couldn't grab back on to any hope of catching his interest even after she died.

Somehow letting Coralee step into her heart had opened up a pathway back to Hiram, and not just him. Perdita found herself ready to smile at everybody in this crowded house. So when Livvy came crying to the bedroom door, Perdita squeezed Coralee's hand and kissed her brow before she went into the kitchen to fix the child something to eat. Much as she wanted to be with Coralee, Eugenia would be the better help.

As soon as Arlie came in from school, she sent him back out to fill up the woodbox, and after Josh showed up some minutes later and about wore a path in the floor pacing back and forth, she shooed him to the barn. No sooner did she have him out the door than Hiram blew in.

He had been making a regular path to Eugenia's supper table, but Eugenia wasn't the only one what could cook a man supper. Perdita opened up a jar of apples Eugenia must have put up last fall, stirred the shucky beans stewing on the back of the stove, and fried some hoe cakes. It was a fine thing having food in the cupboard.

She was nigh on ready to praise the good Lord for her carelessness that burned down her cabin. Coralee was better off here with them that knew what they were about birthing babies than if she'd been alone with Perdita. Like as not, they'd have starved before now anyhow, less'n she'd sent Coralee on her way back home.

Perdita did hope the sister would show up somehow to be a comfort to Coralee. Josh said he'd turned the problem over to Caleb Barton. If'n anybody could get it done, it would be that boy. The only thing he didn't seem good at doing was getting Tansy to pay enough attention to him to know he was the one she needed to settle on.

That girl let her head be turned by too many stories. Anybody could write down whatever words to make things look good on the other side of the hill. Didn't mean this side of the hill wasn't just fine.

But Perdita was glad for the book Tansy brought in about birthing. Whilst it was mostly for flatlanders, it did talk about how important clean was. She and Eugenia had boiled some towels and sheets a week ago to have them ready for the birthing. Today she'd kept water hot on the stove and told even that Geraldine Abrams she'd have to wash her hands with Eugenia's lye soap. The woman gave her a look to let her know who the granny midwife was, but she'd washed her hands proper anyhow.

Perdita even made Hiram wash his hands before he went in to pray over Coralee. That man could speak a fine prayer. 'Tweren't nothing particularly fancy about his words, but they were true and earnest. Coralee seemed comforted for a spell, especially when he quoted from Psalms about how the Lord covered a baby in his mother's womb and fearfully and wonderfully made each and every person.

Hiram went back to wait by the fire. Said he'd sit and read her Bible and keep on praying. She was glad enough to have his prayers added on top of hers.

Tansy was surprised to see Josh still at the barn when she walked Shadrach in for his night feed. He wasn't doing anything. Just sitting with his head in his hands on the milking stool.

"What's the matter, Josh?"

"It's Coralee. She's workin' on having the baby." His pitiful face when he looked up made her hurt for him. "I'm feared she's gonna die."

Tansy's insides clenched. "Are things going bad?"

"I don't know. But she's been punishin' bad all day since not long after you left. Seems she shoulda had it by now if things were right."

"Maybe. Why are you out here?" Tansy looked at the bucket of milk he hadn't carried to the house.

"Aunt Perdie ran me out. I was being too jumpy." He looked down at his hands again.

That made sense. Aunt Perdie ran out of patience easy. And she was probably as nervous as Josh. The two of them loved Coralee. In different ways but equally as fervently. Tansy took a breath to calm her own nerves. She didn't know things weren't good. She wouldn't know until she went inside, but first she had to tend to Shadrach.

"You take that milk on in and then sit down with Preacher. I see his horse tied up at the porch. You can pray with him."

"I've been praying."

"Then pray some more. Pray till you hear that baby crying. Then you'll know everything's all right."

He stood up. "Coralee might not be all right." His voice was trembling.

"But most likely she will be. Did you fetch Granny Abrams?"

"She's here, but she ain't too happy with Aunt Perdie bossing everybody around and making them do what she read about in that book you give her on baby birthing."

"What's that?"

"Mostly making everybody wash their hands time and again. She made me wash mine before I could even peek in through the door at Coralee." The tragic look came back on his face.

"Some babies come hard, but they come and then the mother is fine. You remember how Ma struggled with Livvy."

"No." He shook his head. "Pa made me stay in the barn and watch after Arlie to keep him from doing something stupid to hurt hisself."

"Well, she struggled, but she was okay. And Livvy was okay." Tansy moved over to give him a quick hug. "Now go."

She rushed through taking care of Shadrach. She was putting him in his stall for the night when another horse trotted past the barn. Now who? Preacher was already there. Granny Abrams was already there. Tansy stepped out of the barn, but it was too dark to see the horse well. When the cabin door opened, a woman and a man went in.

Even in the dim light, she knew Caleb at once. Her heart gave that strange funny jerk she'd felt when she saw him at the schoolhouse. Where Junie leaned against Caleb, needing him.

Why did things have to be so complicated? Why couldn't things be like in a story? She smiled a little at the thought as she walked toward the house. Jane Austen's stories always had plenty of complications. When Tansy read *Pride and Prejudice*, she didn't think Elizabeth and Mr. Darcy would ever figure things out. So maybe those stories were more like life than she'd thought.

But in life nobody was simply writing down the words to make things come out right. In the real world you couldn't always count on a happy ending. People went off and didn't come back. Good men got killed when trees fell on them. Houses burned down. Fathers put their daughters out when they didn't do to please them. Children went hungry, and babies weren't always born easy.

She whispered a prayer for Coralee and her baby. While before she'd thought Josh too young to marry, now she hoped Coralee would say yes, even though Tansy had a hard time wrapping her mind around her little brother being a husband and a father.

As she stepped into the house, a muffled cry of distress came from her mother's bedroom. Somehow being muffled made it that much worse. Preacher had his hand on Josh's shoulder. Both their heads were bent. Praying. Arlie was whittling so desperately that she had to bite her lip to keep from warning him not to slice into his thumb. Caleb sat on the floor, leaning up against the wall with Livvy cuddled in his lap. He looked up at Tansy, and her heart did that funny jerk again.

"What's happening?" Tansy asked.

"A baby, we hope," Caleb said.

"Is she all right?"

Josh raised his head to answer her. "We don't know. They won't let us in. Only womenfolk, but you can go find out and tell us."

Livvy spoke up. "You have to wash your hands first. Aunt Perdie says so."

Tansy smiled down at the little girl. "Well, then I will." She let her gaze go back to Caleb. "What are you doing here?"

"A reasonable question, if somewhat rude." He twisted his mouth to hide a smile. "Right now, I'm holding Livvy, but I came to bring Saralynn here."

"It must be getting crowded in there." She looked toward the bedroom, then back at Caleb. "Did she have to slip away from her house?"

"Believe it or not, her father told her to come. Said he'd come get her tomorrow. He promises no trouble."

Another half groan, half scream sounded from behind the door. Josh went stiff and Preacher started praying out loud. Livvy slapped her hands over her ears, and Caleb tightened his arms around the little girl and began humming a tune.

Arlie threw the piece of wood he'd been whittling into the fireplace and said to nobody in particular, "I'm going to pack in some more wood."

Josh stared over at Tansy. "Go see what's happening, Tansy. Please."

Tansy dropped her saddlebags of books on the floor, then shrugged off her coat. She washed her hands as quickly as she could as more sounds of distress spilled out of the bedroom.

Trembling a little, she pulled in a breath for courage and pushed open the door just as Granny Abrams held up a baby. "You have a girl."

Coralee sobbed as she tried to raise her head to see the

baby, who let out a warbled cry. Saralynn was weeping with her as she grasped Coralee's hands.

On the new mother's other side, Aunt Perdie lifted her hands to heaven. "Praise the Lord!" Then she did a few jig steps.

Coralee laughed and sobbed at the same time when she saw that.

Granny Abrams started to hand the baby to Ma, but Ma shook her head. "You take her, Perdie. Show Coralee her sweet child. Then we'll give the babe a bath." She held a towel toward Aunt Perdie.

"You best do it, Eugenia," Aunt Perdie said. "I don't know nothing about holding babies."

"You'll know about holding this'n." She pushed the towel into Aunt Perdie's hands.

"One of you take this baby." Granny Abrams didn't bother hiding her irritation.

Aunt Perdie practically tiptoed over to take the baby, who was letting the world know she was there. Ma helped Aunt Perdie wrap the towel around her.

Aunt Perdie stared down at her little bundle. "If she ain't the purtiest thing I ever laid my eyes on."

At the sound of Aunt Perdie's voice, the baby hushed for a few seconds. The woman's face lit up as though the Lord had shot a special ray of light down through the roof to shine on her. She turned to Coralee. "Here's your little jewel."

Saralynn wiped Coralee's face with a washrag and pushed a pillow behind her to raise her head up higher. More tears coursed down both their faces as Aunt Perdie laid the baby in Coralee's arms.

"Jewell," Coralee echoed Aunt Perdie's word. "That's her name. Maylin after my ma and Jewell. Maylin Jewell."

Granny Abrams stood up to watch mama and baby. "That's a fine name, and truth be told you'll have more than plenty of times to be holdin' her, but right now we need to finish things up. Then you can show your young man your babe."

Silence fell over the room for a second before Coralee said, "I'm not married."

"Oh, I know that." Granny Abrams smiled over the sheets that covered Coralee's legs. "Yours isn't the first baby I've helped bring into the world without a daddy around to own it. Not by a long shot. But Eugenia's boy that came after me appeared more than a little invested in your outcome here. I figured you had an understanding of sorts."

"He wants to," Ma said. "But it's what Coralee wants that matters."

"Some truth to that." Granny Abrams ducked back under the sheets to take care of the birthing business.

Saralynn brushed back Coralee's hair and put her cheek against her sister's. Aunt Perdie lifted the baby away, and Ma held her in her lap on a towel while Aunt Perdie sponged her off. The baby gave them plenty of complaints.

Ma smiled over at Tansy. "Little Maylin has already loudly announced her presence, Tansy, but you go tell the menfolk that all is well with babe and mother."

"Wait, Tansy." Coralee stopped her before she got to the door. "Tell Josh he can come see the baby as soon as we're presentable."

Again, a few seconds of silence fell over the room. New tears wet Coralee's cheeks as she went on. "If'n he's gonna step in as Maylin Jewell's daddy, he needs to be the first feller to see her."

Ma smiled at Coralee. "Josh is young, but he'll do right by you. I'll see to it."

Josh was standing so close to the door when Tansy opened it that she had to push him back to get into the sitting room. He looked ready to shout.

"So you heard."

"That I'm a daddy. Can you believe that, Tansy? I'm a daddy."

"Not sure I can, Josh," Tansy answered truthfully.

His excited look faded into a worried one. "Do you think that means she's gonna marry me?"

Tansy had to smile at him then. He looked so very young, but he was eighteen. Plenty of boys in the mountains started families that young. "Yes, I think that means she's gonna marry you, but you let her tell you when."

"Can I go in now?"

"You have to wait until somebody comes after you."

"I love her, Tansy. So much that I think my heart is gonna explode if I can't let some of it out."

"Shower some down on that baby. That should make room."

"But then I'll look at Coralee and it will swell up full again."

"Congratulations, Daddy." Tansy tiptoed up to kiss his cheek. "I'm proud of you."

She stepped into the middle of the room. "Coralee has a beautiful, perfect baby girl named Maylin Jewell. Mother and baby are fine. Better than fine."

"About time." Arlie laid the block of wood he'd just carried in down on the hearth and dusted off his sleeves. "I'm going to bed." He looked over at Caleb. "See you at school tomorrow, Mr. Barton."

"Eight o'clock sharp," Caleb said with a smile.

Tansy pointed at Livvy, yawning in Caleb's lap. "You too, young miss. Off to bed with you."

"I want to see the baby," Livvy said.

"Tomorrow. Baby Maylin needs her rest tonight. And so do you."

Caleb lifted her out of his lap and set her on her feet. "Good night, princess."

She stuck out her bottom lip in a pout but headed for the loft stairs.

"I need to go." Caleb pushed up from the floor.

"I'm sorry we didn't give you supper."

"I'm not worried about eating." He looked over toward Preacher Rowlett, who appeared to be nodding off in his chair. "But I want to tell you something before I go."

"It's late." Tansy looked over at the table where dirty dishes were stacked. "I need to get things cleaned up."

"This won't take long." He touched her arm and then put his hand against her cheek to turn her face toward him. "Please look at me."

Tansy could have moved away from him, but instead she let him capture her eyes. Her mouth was suddenly too dry to say anything.

"I don't know what Jenny Sue told you today that has you afraid to look at me."

She moistened her lips and managed to say, "I'm not afraid to look at you."

"I hope not. Ever." His hand gentled on her cheek and he ran his fingers along her jaw to her lips. "I love you, Tansy Faith Calhoun. I've loved you since you were a girl in pigtails."

Tansy's heart was beating so hard she was breathless. "I don't know what to say."

He smiled a little. "I can think of a few words that would be good."

I love you rose up and tickled her tongue, but she didn't let the words out. How could she take him away from those children who had already lost so much? "I can't. Not tonight."

He shut his eyes a second, then opened them to stare straight into hers. He hid nothing from her, and she had the feeling he was reading her every thought. He leaned over and kissed her forehead the way one might a child. "I won't push you, but I do ask you to consider what I said."

He stood still, waiting for her to say something, but she merely looked at him. He dropped his hand away from her face and turned toward the door.

"Wait."

He looked back with hope.

"If you still need help with those lessons Saturday, I can come to the school to help you."

"That sounds good." He looked disappointed she hadn't promised more than help with school lessons.

He was putting on his coat when Ma came out of the bedroom. "Oh good. You're still here, Preacher. Granny Abrams needs somebody to see her home now that it's full night. Her horse is out in the barn lot."

Preacher stood up from the rocking chair. "I could do that, Eugenia, but I'm sure Caleb here won't mind making sure Mrs. Abrams gets safely home since the good woman lives just a little ways down from his house."

"I can do that," Caleb said. "Is she ready to go?"

"Nigh on," Ma said. "Tansy, get the lantern and go saddle her horse."

"Yes'm." Nothing for it but to do as Ma asked, even if she suddenly felt too tired to move.

Ma looked at Preacher. "You plannin' on spendin' the night, Preacher?"

Ma must be as worn out as Tansy to ask that so straight out.

"I wouldn't impose like that, Eugenia. I'll be finding my way on home. I was just hoping for a word or two with Perdita before I left."

thirty-four

PERDITA HAD NO IDEA why Hiram wanted to talk to her. She needed to be in the bedroom helping Eugenia clean up whilst keeping an eye on Coralee cuddling her little Maylin Jewell. Josh kneeling by her bed and looking as awkward as a newborn colt was a sight to warm her heart too.

Soon as things got sorted out, Coralee would move out to her cot so she and the baby would be close to the fire. The bedroom had stayed warm enough while the birthing was happening, with the bunch of them crowding in there around the bed. Then some heat might have come from all that hot water Perdita kept bringing into the room. Geraldine Abrams had done some frowning about that, but then before she left, she'd asked Tansy to bring her that book about birthing. Said whilst books couldn't be trusted for every livelong thing, she might find something in the pages worth knowing.

After Geraldine left, Eugenia told Perdita to go on out and talk to Hiram. "He says he has to talk to you afore he'll leave." She gathered up some of the soiled linens. "And I'm speakin' truth when I say we need to get him on out the door before he expects us to cook him breakfast. He could spend the night, but I ain't knowin' where we're going to find beds enough for us, even without adding Preacher into the mix."

"Don't you worry about that, Eugenia. Saralynn can take her rest in here with you. I've slept many a night by the fire in my rocker when my rheumatism was kicking up. That way I can keep the fire burning through the night, and I'd count it a privilege to be close by Coralee to give her a rest if'n the baby needs soothing."

"I can help too," Saralynn spoke up.

"That would be fine." Perdita gave the girl a smile. She was a pretty thing. Looked like her mother, but barely out of girlhood. "I'll come wake you when I get worn out." Of course, that wasn't going to happen.

She gave Coralee and the baby another smile before she went out into the sitting room, empty now except for Hiram. The Barton boy had left with Geraldine Abrams, and Tansy must have gone off to sleep and left Hiram by his lonesome in front of the fire.

"What are you doin' still hangin' around, Hiram?" Perdita said. "The excitement's over, but I sure do appreciate you praying through it. The good Lord blessed Coralee with a sweet baby, and now it appears that little child will have a papa too."

"Answered prayers." Hiram pointed toward the other rocker in front of the fire. They had been pulled close together for when he was praying with Josh. She'd seen them practically head-to-head on one of her trips to get more hot water. "Come, sit with me, Perdita."

When she started to scoot the chair back, he stopped her. "Don't do that. Let's sit hand to hand."

She wasn't sure what he was up to, but she didn't mind sitting practically in his lap. Not that she had much romance left in her bones, but once upon a time being so near Hiram would have had her heart knocking around in her chest like it wanted to break out of her rib cage. In fact, when he reached the few inches between them to put his hand on hers and look her straight in the eyes, her heart did do some jumping around.

"We've known each other a long time, Perdita."

"Nigh on most our lives." Perdita was glad enough that she didn't sound like a breathless teenager. Like poor Josh in there making promises to Coralee to be the best husband ever. At her age she wasn't up to making no promises to anybody or counting on getting any either.

Hiram pulled his hand away and sat back. He reached to pick up the Bible from beside his chair. Her miracle Bible that made it through the fire. She reckoned when he opened it, she might hear some preaching. That was fine. She was more than ready to praise the Lord after Maylin Jewell had come safe into the world.

But then he read, "'Vanity of vanities, saith the Preacher, vanity of vanities, all is vanity.'"

"Ecclesiastes," she murmured. That verse tried to kick in a memory, but she was too tired to make the connection.

"Indeed." He went on. "'The thing that hath been, it is that which shall be; and that which is done is that which shall be done; and there is no new thing under the sun.'"

He had a fine reading voice. Something like his praying voice. She leaned back in the rocker. The man could talk her to sleep. If not for that tickle of needing to remember something, she might doze off.

Hiram turned the page and dipped into another chapter in Ecclesiastes. The familiar place everybody knew. "'To every thing there is a season, and a time to every purpose under the heaven.'"

She nodded and shut her eyes to concentrate on his words as he read through how it was time to do this or time to do that. She liked the time to be born and shuddered at the time to die. Thank the Lord they had the born time and not the die time today. He was down through the weeping and laughing, the mourning and the dancing. She smiled as she remembered the surprised looks on all their faces when she did that jig step.

He went on through the getting and losing, the keeping and the casting away. His voice got a little stronger then as he read, "'A time to keep silence and a time to speak; a time to love and a time to hate; a time of war and a time of peace.'"

Perdita's eyes flew open as all at once what she needed to remember was right there banging on her noggin. That letter she'd stuck at the beginning of Ecclesiastes right where Hiram Rowlett was reading. She ought to have thrown it in the fire the other day instead of sticking it back in the Bible. *Vanity of vanities.*

"This Bible here is a blessing. Making it through the fire the way it did." He picked up the folded paper scribbled with her words years ago. "Appears this letter did too. A letter addressed to me."

He held it up and smiled. The man always had done more than his share of smiling. She considered leaping to her feet to snatch away that scrap of paper and pitch it in the fire. But it had been a good while since she been able to leap anywhere. Besides, burning it up wouldn't make it not have been. She rocked back and forth, almost bumping his knees.

"That letter is old as the hills, Hiram. It don't mean a thing now."

"I'd hate to think that, Perdita. Seems some of it is as true as it ever was." He looked down at the paper and read, "'I figure it ain't no use both of us sitting by our lonesome in two different houses when we could be proper company for one another.'" He looked up. "The two of us are both still sitting by our lonesome in two different houses."

"I expect we're some used to it by now." No need in denying what she wrote. The words were right there in front of his eyes.

He folded the letter and laid it back on the Bible page. "How come you never gave me this, Perdita? Back when you wrote it."

"I reckon I knowed it wasn't something a proper woman ought to write to a man."

"But you saved it."

"I saved it."

She stared at his hands holding the Bible and wished he'd reach across and put his hand over hers again. She was feeling as unsettled as poor young Josh had felt all day while he worried over Coralee.

But Josh was young. She was old. Way too old for her stomach to be flipping and her heart to be banging around. An old woman like her could just fall right out of her rocker on the floor, kilt dead. Then again, she had always hoped to go happy. She peered up at Hiram. Should she be happy? He looked mighty pleased himself as he waited for her to say more.

"Could be I shouldn't of," she said.

His smile stayed steady. "You know the good Lord can work in mysterious ways his aims to accomplish. Back when you first penned this missive, I wouldn't have been ready to read it. And then the Lord, these years later, kept his Word from the fire and your words too. Maybe for this very moment in time for me to find." He leaned closer toward her. "Your words in this letter are right for now. I am tired of a lonely house and you don't even have one to be lonesome in."

"What are you trying to say, Hiram?"

He looked a tad uneasy as he scooted toward the edge of his chair, then shook his head with a sigh. "Sorry, Perdita. I'm too old to go down on my knees. I might never get off them." He took her hands in his. "But I'd count it an honor if you'd do me the pleasure of agreeing to be my wife."

She must have dozed off. She had to be dreaming, but when she shut her eyes and opened them again, there he was right in front of her. Still with that Hiram smile clear up in his eyes and his hands warm on hers.

"Well, has Prissy got your tongue?" he asked.

As if him saying her name summoned the cat, Prissy was there beside them, waving her tail back and forth the way she did when she was ready to bite or scratch something she shouldn't. "You take me, you have to take my cat."

"I didn't expect any different. I figured if you were to say yes, my old hound and me would have to learn some new ways, but a man shouldn't ever stop learning. You up for an adventure, Miss Perdita Sweet?"

"I might be a mite old for adventure." What was the matter with her that yes wasn't coming out of her mouth?

"We're both still on the right side of the grass. I don't think the good Lord is ready for either of us yet. He let you be the next thing to grandma today."

"And a woman headed for the marriage altar tonight. I will marry you, Hiram Rowlett, but I ain't about to let anybody call me Mrs. Preacher."

He laughed at that. "Never had it in my mind for anybody to call me Preacher, and that's a fact."

She shrugged. "And I ain't none of these people's Aunt Perdie."

"Folks call you what they want to call you. Not much way of changing it. But I know your name, Perdita."

"And I know yours, Hiram."

He leaned toward her, and though she hadn't kissed a man in, well, forever, she knew just what to do. Some things must come natural.

He sat back with that smile twinkling in his eyes. "Now I'd better be on my way before Eugenia chases me out the door with her broom. I'll bring an extra horse and come after you in a few days so we can go down and get a license to marry."

"Won't that set the town on their ear." Perdita chuckled.

"I haven't been to town since I took a load of carvings down to the store back in the summer. Don't know whether any of them sold or not."

"Could be you have a pile of money waiting for you."

"Well, if'n you're thinkin' on marrying me for my money, you'd be better served to think again. I ain't got hardly a thing of worth 'cepting my carving knife and that Bible there."

"A miracle Bible is a fine treasure."

"So it is." Who'd have ever believed it would find a way into Hiram's hands and have him open his heart to let her in? A person could be mightily blessed by the Word of God. "If'n the girl and Josh marry, we can make it a double affair. Private like."

"Or maybe a triple one. I overheard, quite by accident, young Barton declaring his love for Tansy."

"And did she have the sense to vow the same back to him?"

"Sadly enough, no. However, I did break my pretense of sleeping to peek over at the two. Tansy's face was much conflicted. I'm thinking she appeared to want to."

Conflicted. That was an interesting word, but *confounded* better described Perdita as she watched Hiram leave. Prissy jumped up in her lap as if reclaiming her territory now that Hiram was gone. The cat started up a good purr. Inside her heart, Perdita purred right along with her.

Tansy wasn't surprised to see Aunt Perdie rocking Maylin Jewell by the fire when she went down to start her day the next morning. Coralee was asleep in the little bed in the corner of the room. Really asleep and not just pretending. Nobody else was stirring as yet.

The dirty dishes stacked on the table poked Tansy's conscience, but she'd been too tired the night before to tend to them. She'd come in from saddling Granny Abrams's horse, told Preacher good night where he appeared to be reading Aunt Perdie's Bible, and gone straight to bed. She never even heard Josh when he came up the loft stairs to turn in.

Tansy peeked at the baby wrapped in the quilt Aunt Perdie made for her. "Is she doing all right?"

"Better than fine."

Tansy had never seen such a smile on Aunt Perdie's face. "Looks like you aren't any the worse for missing some sleeping time. Have you been out here all night?" She stirred up the fire and fed it some more wood.

"I wanted to be close by in case Coralee needed something."

"Baby Maylin Jewell must have needed something first."

"Coralee nursed her a while ago, but I told her I'd do my best to keep the sweet child content awhile so's she could rest."

"Your best must be good enough. She looks nice and content." Tansy smiled at Aunt Perdie. "And you do too. Maybe better than content."

"I am confoundedly content." She looked up at Tansy.

"Confoundedly?"

"The way the Lord can bring blessings out of hard times can confound a body for certain. Poor Coralee being put out of her house and begging me for shelter. Then that shelter burning down, but you coming along to bring us to this good place. You digging my Bible out of the ashes. A Bible the Lord put a shield around to keep from being ash itself. Me getting the gift of this here baby in my lap." She let her gaze go back to the baby. "Then that Bible being in Hiram Rowlett's hands where he come across a note I writ to him long ago and never sent."

"Oh? Sounds like there's more to this story. Why didn't you send it?"

"I've always had a fondness for Hiram, but whilst I once upon a time considered telling him that, I decided it weren't something a woman should do. So I stuck that letter back in my Bible years ago and left it there. A piece of my heart."

"And he found it last night and read it."

"He did. Claimed it was God's timing. Seeing as how he's some lonesome in his house and I don't have one, he's asked me to up and marry him. And that's what has me confounded."

"So did you say yes?"

"It might seem foolish to consider such a thing at my age. I'm some past sixty, you know. But weren't never no other answer on my tongue but yes." She peered up at Tansy again. "Could be you should consider having some the same answer on your tongue when that Barton boy comes calling next time."

Tansy turned to stare at the flames in the fireplace. "There's complications."

"Ain't there always some sort of complications, for sure."

Tansy pulled in a breath. "His mother wants him to marry Jenny Sue."

"His brother's widow?"

"Yes. So he can be a proper daddy to his brother's children. Reuben Junior needs Caleb."

"Hmm." Aunt Perdie rocked a minute before she went on. "Vesta always has wanted to tell everybody what they oughta be doing. Sometimes she's even right, but in this situation, it don't matter near so much what she wants as what Caleb wants."

"And what I want," Tansy whispered.

"Reckon that's something you need to figure out."

"You're right." Tansy looked over at the woman. "I'm happy for you, Aunt Perdie. But I do wonder how Prissy is going to take to Preacher's hound dog."

"It's his hound that better be worried."

That made Tansy laugh and she leaned down to kiss Aunt Perdie's cheek.

"Well, I go for years with nary a kiss and now everybody's wantin' to give me some lovin'."

Tansy pretended a shocked face. "Are you saying you and Preacher were kissing out here in front of the fire last night?"

"Some things is best left to the imagination. Now grab you a biscuit and get on out of here to deliver them books of yours."

thirty-five

CONFOUNDEDLY CONTENT. Aunt Perdie did have a way with words. Tansy couldn't say she herself was at all content, but confounded was right enough. While she rode toward Booneville, she had plenty of time to think. Too much time when she wasn't sure what to think.

Caleb loved her. He'd told her that straight out. The words to say the same to him had been right on her tongue, but she hadn't said them. Did that mean she still had doubts? If she truly loved him, would she have doubts? And what about Junie and little Cindy Sue? Would she be stealing Caleb from them? Would he be happy with Jenny Sue? Would he be happy with her?

The questions circled around in her thoughts like pesky horseflies in the summertime. She didn't know which of them to swat away first. So she pushed them all out of her mind. She wouldn't see Caleb until Saturday. If then. Who knew what might happen by Saturday? Didn't the Bible say that all a person had was the day at hand? That's what she needed to think about right now. This day when she would be repairing books and helping put together scrapbooks of recipes and quilt patterns.

Maybe instead of letting questions she couldn't answer

chase around in her head, she could think up another story about Clyde the hound dog. One where he met a cat like Prissy. Preacher's hound was sure to be in for a scratched nose. Or she could write out another Jack story.

Thinking about Jack stories brought Damien Felding to mind. Another of those questions without answers. She'd been attracted to him. She couldn't deny that. At least at first when she'd thought him a man with aplomb.

Aplomb. She'd seen that word in a book and from the story had decided on a meaning of someone sure of themselves. But when she searched out the definition in the dictionary Mrs. Weston had at the library, she'd found more to the meaning. Aplomb went beyond simple self-confidence to having confidence and composure even in trying situations.

While Damien radiated self-confidence, at the same time he'd been more than a little shaky on Belle at first and worried about staying the night with Preacher. Not that those were exactly trying situations. On the other hand, Caleb had been a rock when Aunt Perdie's house burned and then a picture of steady calm dealing with Mr. Embry about the pup.

Plus, she was absolutely sure Caleb would never try to force her to kiss him. The thought of that ride with Damien had her cheeks burning even now.

The sun was up as she crossed over the last creek before Booneville. With the feel of spring in the air, the creek was near to overflowing from the melting snow, but Shadrach went through the rushing water without hesitation. She stroked the horse's neck when he stepped back on dry ground. At least she didn't have questions about this good horse Preacher had leased her.

Preacher and Aunt Perdie. Imagine that. Thinking on marrying at their age. Not so long ago, Aunt Perdie didn't have a good word to say about anybody. Now she beamed over a newborn baby like a proud granny while ready to say "I do."

People could surprise you. Caleb teaching school. Josh wanting to be daddy to a baby he hadn't fathered. Her pa going away to who knew where. Ma bearing up stronger than Tansy ever imagined she could.

At the library, they all cheered when Mrs. Weston reported a new shipment of books. "We even have a brand-new Mark Twain book, *A Connecticut Yankee in King Arthur's Court.*" She held up the book. "Lena Nofcier, who works with school libraries, sent out a call for books for the packhorse libraries. Donations have been coming in."

Mrs. Weston ran her hand over the front of the book before she handed it to Tansy, who opened it to sniff the new book scent before passing it on.

Imogene riffled through the pages. "My readers love Mark Twain. His characters could be right at home here in the hills."

"I don't know about that one. A Connecticut Yankee." Rayma laughed.

Imogene grinned. "Guess I was thinking more about Tom Sawyer." She looked over at Tansy. "I read that book you wrote about Clyde the hound at Cox's Creek School last week, Tansy. The kids couldn't stop laughing at the fixes that dog got into. You are going to write some more, aren't you? Please."

"Our Tansy. Writing books for us." Mrs. Weston smiled over at Tansy. "But putting together another one might have to wait since I've got another surprise for you. Damien Felding is back in town."

"Oh?" Tansy wasn't sure if she was glad to hear that or not.

"He was in here yesterday pestering the life out of me to let you take him around today. He claims he simply must write about a one-room schoolhouse for this guide he's doing. I said Manders School might work since it's in your area. Mr. Felding was all for it when he found out Caleb Barton was the teacher. Says he's met Caleb."

"Oh," Tansy said again. She was very sure she wasn't glad

to hear this. Damien and Caleb in the same small building, with Damien judging Caleb and Caleb thinking she wanted to be with Damien.

"I hope he has a slicker." Josephine peered toward the window. She was the oldest of them. "Unless I miss my guess, we're in for bad weather."

"You think we're in for more snow?" Mrs. Weston looked out the window. "Clouds are building up."

"Too warm for snow. Rain, I'm thinking," Josephine said. "I wouldn't doubt it being a gully washer."

"Maybe it will hold off till we're all home," Rayma said. "We can hope anyway. Not that rain or snow, cold or heat can stop the intrepid book women." She smiled. "How do you like that new word I came across this week? Intrepid."

Damien Felding came in the door just in time to hear Rayma. "And here, right on cue, is the intrepid reporter ready to hunt out some new stories in the wilds of Owsley County."

Without a doubt, Damien knew how to make an entrance. Even Rayma laughed. He looked at Tansy. "Is my guide ready to take off for the hills?"

"If you want to see a school, there's one right down the street."

"But that's not what I'm after. I want one of those where all ages are in the same room together. Mrs. Weston says your friend Caleb Barton is teaching at a school that sounds like the very thing." He looked amused, as if he guessed her uneasiness.

"It might rain," Tansy said.

"A little rain won't deter an intrepid book woman and her reporter sidekick." He grinned.

She gave up trying to talk him out of it. "Do you have Belle lined up?"

"Ready and waiting. I even got us some sandwiches at the store to have a picnic on the way."

"A picnic," Imogene said. "I don't know when I last went on a picnic."

"I hope in the summer with the sun shining." Josephine looked from Imogene to Damien. "You'd better have that picnic on high ground. With how full the creeks are already, won't take much to bring a tide."

"A tide?" Damien looked puzzled, but then his smile came back. "Right. You mean a flood. Not much chance of that up high in the hills, is there?"

Josephine gave him another look but didn't say anything as she taped up the spine of a book that had seen better days.

"The rain may hold off awhile, but perhaps you should get on your way," Mrs. Weston said.

"I need to change out my books," Tansy said.

"I'll meet you at the livery stable. With that picnic." He gave Imogene a wink and, with a wave, was out the door.

Rayma grinned at Tansy. "I think the intrepid reporter has fallen for our Tansy."

"The question is," Imogene teased, "has Tansy fallen for the handsome, intrepid reporter?"

"It's what's on the inside that matters most," Josephine said.

"Just because the outside looks good doesn't mean the inside isn't good." Mrs. Weston took up for Damien. He had obviously charmed her.

"That can be." Josephine got up to put the book she'd repaired on the shelf. As she passed the window, she looked out. "But good, bad, or otherwise, I'm nigh on sure Tansy is going to find out how he looks wet. Hope you have your slicker with you, girl."

The clouds were thicker, but no rain was falling when they crossed over the first creek. On up the hill, they came to some boulders Damien decided was a perfect spot to eat their sandwiches. "It could be your granny book woman is right and we're going to have a spot of rain."

"Spot of rain?" Tansy shook her head. "I don't think I've ever heard anybody actually say that."

"But you've read it in some of those stories you love, I'm wagering." He laughed.

She had to laugh too. "Very English of you."

He leaned against one of the boulders and unwrapped a sandwich. "Ham, of course. No cucumber sandwiches for the real men and women in the mountains."

"Cucumber sandwiches? Do people really eat those?"

"They do, and I don't know why you look surprised. You people up here eat pickle sandwiches." He took a bite. "Pickles are cucumbers."

"I guess you're right."

"I usually am."

His smile was the one that made had her feel so uncomfortable when they first met, but now she didn't care if he was laughing at her or not. So that answered one of those pesky questions circling her that morning. The attraction she'd felt for him had faded away.

"Tell me, Tansy Calhoun, have you ever thought about leaving this place? Doing something bold?"

"Bold? What do you mean?"

"I don't know. Go see other places? Big cities? The oceans? Real mountains out west?"

"My feet aren't stuck to the ground," she said. "I can travel to those places if I want."

"How about now?" He stepped away from the boulder, closer to her. "I'm leaving Owsley County today. Won't have any reason to come back."

"So you've finished gathering your stories?" She took a bite of her sandwich.

"Finished here. I'm heading to Harlan County next." He

settled his gaze on her. "How about you come with me? Help me find stories there. Then we can head to some of those other places out there in the big wide world."

She swallowed. "What are you asking me?"

"You're a pretty girl. We could have some fun tooling around these hills in my car. Just stay where there are roads. I can find plenty to write about without ever getting on another horse. No more saddle sores for us."

Tansy wasn't sure she was hearing right. "You want to get married?"

He backed up a step. "Well sure, I intend to get married someday, but first let's have some fun."

"You're joshing me."

He moved closer again. "I'm a modern guy. You're a modern girl. We can try things out and then see about marrying."

Tansy stared at him. "I don't think I'm modern enough for you."

"It's your chance to escape these hills."

"Escape?" She had the sudden urge to laugh out loud.

"To a better life." He reached to touch her cheek, but she jerked away from him. His eyes narrowed a little. "I thought you had an adventurous spirit, but maybe I was wrong."

"I get plenty of adventure taking these books around each week. And looks like we're in store for a new adventure." She stuck the remains of her sandwich in her pocket and held out her hand to catch the first raindrops. "I think Josephine was right. We're in for a downpour." The dark clouds coming over the hill were ominous.

"Should we go back?" Damien looked worried.

"Maybe you should, but no need me going that way."

"Can't we take shelter somewhere?" He glanced around.

The rain dashed down harder. They might find a barn nearby, but she didn't particularly want to be trapped somewhere with him during a rainstorm.

"We're not too far from the school." That was where Tansy wanted to go. She was ready to find out the rest of her answers with Caleb. "But town isn't too far either."

The rain dashed down in buckets. Tansy grabbed Shadrach's bridle before he could take off for the trees. She was already wet, so she pulled her slicker out to drape over her saddlebags.

Tansy yelled over at Damien. "Best mount up and hold on. Belle might head for higher ground. Town's that way." She pointed back the way they'd come.

"I can barely see," he yelled back. "I'd lose my way for sure. I better stick with you."

"Suit yourself." He was probably right. Even if she did wish him gone, she didn't want him to be washed away in a tide because he didn't know the creeks. She could assure Caleb that Damien had no place in her heart.

thirty-six

CALEB DIDN'T LIKE THE FEEL of the morning when he went out to feed the animals before he headed for school. When he came in for breakfast, Ma had the same thoughts about bad weather on the way.

"Might be one of those hard rains that come on with spring just around the corner." She looked out the window after dishing up his eggs. "You pay mind to the sky and don't keep those children there if the sky goes to looking mean."

"I'll watch out."

She kept fretting. "You're right on the side of that creek. Don't know what folks were thinking building the school there. Should have put it up on the hill a ways."

"No changing that now." Caleb finished eating and picked up his bag. "I'd better go get Junie."

His mother's frown got darker. "He was sniffling some last night. I'm thinking he shouldn't be out there getting soaked."

"I'll take care of him."

"Will you?" Ma gave him a narrow-eyed look. "You seem some resistant to that idea."

"I'll always be ready to do what I can for Junie, but I'm not marrying Jenny Sue." He looked straight at her.

She didn't back down. "Jenny Sue says you're struck on

that Tansy Calhoun. Couldn't have her sister so you're after second choice."

"Tansy has always been my first choice." No need to mince words with his mother. "I'm going to marry her."

"Jenny Sue didn't seem to think Tansy had agreed to that when she talked to her yesterday when she brung by books." Ma didn't back off. "I've heard rumors she's struck on some city feller."

So it was something Jenny Sue had told Tansy that had her looking everywhere but at Caleb yesterday at the schoolhouse. He pulled in a deep breath and let it out slowly. "Rumors are just rumors, Ma. This is between Tansy and me. You stay out of it."

"I'm your mother."

He cut her off before she could say more. "And I love you, but for the last time, I'm not going to marry my brother's widow."

"Reuben wouldn't treat me like this." Ma looked halfway between tears and anger. "He listened to me."

"He listened and then went off and did whatever he wanted. As well as I remember, you had some other girl picked out for him to marry, but he brought home Jenny Sue."

"He coulda done better," Ma muttered. Then she looked up at him with real tears in her eyes. "I've always just wanted the best for my boys and my grandbabies."

"That's why I know you'll be glad when Tansy marries me." He turned to the door.

"If she does."

His mother's words followed him out, but he pushed away any doubts. Tansy might not have said the words out loud, but she had love in her eyes last night. He would find a way to get past whatever reservations she had, whether because of something Jenny Sue said or something else. He'd given up too soon when he thought he'd lost her to Jeremy Simpson, and he'd been wrong to back off after seeing her with Felding. He was ready to fight for her now.

Ma was right about Junie. When Jenny Sue came to the door, she said, "The onliest one not sneezing and coughing around here is that pup you brought us. Junie better stay home. He can practice his reading with that book Tansy left off here yesterday."

He almost asked what she'd said to Tansy, but he didn't have time for a fuss. Bad enough he had his mother so upset, she was ready to shed tears.

By the time he had the fire built in the potbelly stove at school, his misgivings about the clouds hanging low and heavy grew. Even though it was early in the year for thunderstorms, a flash of lightning streaked through the gray. A few students straggled in, but most parents must have had the good sense to keep their children home.

The Embry boys showed up, but not an hour later, Mr. Embry came after them. "I know weather, Barton, and we're in for a gully washer. I wouldn't trust none of these creeks around about here. Better send the young'uns on home whilst they can still get there."

"What about Saralynn?"

"She'll make out with those Calhouns until I can fetch her." He hesitated before he asked, "Do you know about my other girl?"

"Coralee has a little girl. Both baby and mother appeared to be doing fine when I left last night."

He rubbed his hands across his face, but that didn't hide his look of relief. "I reckon that's good news."

When the Embrys rode off, Caleb looked up at the sky. The first raindrops hit his face as another flash of lightning lit up the clouds. In her pen, Phoebe whinnied. She didn't like storms.

A pony one of the older boys rode to school circled the pen

uneasily. Embry was right. If they did have a gully washer, the children might not be able to safely cross some of the streams to get home.

When he went back inside, he told the children to head home. Without dawdling, the boys and girls grabbed their coats and hurried out the door. They knew about storms. Caleb followed the last one outside where he yelled at the boy with the pony to wait for him to help with the corral gate, but too late. The boy was already pulling it open. That was all Phoebe needed in her storm-panicked state to make an escape.

Caleb called her, but with another crack of lightning, it was useless. The mare disappeared across the hill, headed for her barn. He hoped Ma wouldn't worry over the mare showing up without Caleb the way Reuben's mule had without him last fall. But she knew Phoebe was storm skittish.

Nothing for it except to shank's mare it to the house. By the time he straightened the room, tended the fire, and packed up his teacher bag, rain was pounding down on the roof and hitting the windows in sheets. He sat down at his desk to wait out the worst of the storm.

The minutes ticked by with no letup in the downpour. The creeks would be tiding and he prayed the children all made it home safely. He jumped up when the door crashed open with a rush of wind and water and Tansy stumbled inside. Pushing in behind her was Damien Felding.

"What are you doing here?" he asked.

Tansy wanted to fall into his arms, but no welcome showed on his face. She tried to catch her breath to explain, but Damien spoke first.

"We need shelter. What do you think?"

"I don't know what to think." Caleb peered past them out the door. "Where are your horses?"

Damien gave Tansy a hard look. "She let the horses get away."

He'd been unseated when a limb fell and spooked Belle. The mare surprised them by taking off for high ground. When Tansy dismounted to help Damien, the wind snatched Shadrach's reins away and he raced after Belle. She hoped he didn't lose the saddlebags.

They'd pushed on through the rain to the schoolhouse. She hadn't expected Caleb to still be there when she shoved open the door. By then she could hear the floodwaters roaring past them.

"That doesn't explain why you're here." Caleb frowned at her. "With him."

"No time for that now," she gasped. "The creek's tiding."

No sooner were the words out of her mouth than water rushed in through the door and swirled around their feet. In another instant it was to their knees.

"We need to get out of here," Damien shouted.

"That would have been good a half hour ago." Caleb sounded calm even if he was still frowning. "Now the water's rising too fast."

"I can swim." Damien moved through the water toward the door.

"You'd get washed away." Caleb grabbed the axe by the stove. "Help me stack up some of these benches."

Tansy pushed a bench over toward him. The library book she'd left the day before floated past. She started to catch it, but Caleb stopped her. "We can't save the books." He looked over at Damien still by the door. "Help us, Felding. If you want to live to write another story."

Damien carried a bench over to add to the pile.

"Now steady them." Caleb climbed up bench by bench.

Tansy leaned all her weight against the benches to keep them from toppling over.

"What are you doing?" Damien demanded. "There's nowhere to go up there."

"Just do what he says," Tansy ordered him.

"Yes, ma'am." Damien moved over to brace the other side.

Caleb balanced on the top bench and chopped a hole in the roof. He jumped down into the water. "You first, Tansy. I'll climb up after you to help you out on the roof."

"What about me?" Damien asked.

"First Tansy." Caleb held the benches steady while Tansy climbed up. Then he followed to boost her out on the roof. "Hang on and help Felding out."

"What about you?" Tansy's heart was in her throat.

His eyes gentled on her. "I'll come last. You just hang on."

"I love you, Caleb," she said, but he had already jumped back down.

Damien scrambled through the hole and then crawled up to cling to the roof's ridge. Tansy stayed by the hole to help Caleb. He was almost to the top of the pile of benches when the water swept them out from under his feet. He grabbed the edge of the roof. Tansy caught his wrist.

"Let go," Caleb yelled.

"I can help you up." Tears mixed with rain on her cheeks.

"No." He gripped the edges around the hole and stared up at her. "I love you." Then he pulled away from her and fell back in the water.

"Caleb!" she screamed, but he was gone.

"Grab my hand." Damien scooted down to reach for her.

She let him help her up to the roof ridge. Caleb floated out of the schoolhouse door, clinging to one of the benches.

"Don't look so tragic. He'll make it," Damien said. "He's strong."

She stared at the roiling creek water. Caleb was no longer in sight.

"Do you think so?" She didn't know why she asked Damien,

who wouldn't know anything about tides, but she needed to hear him say yes.

"I know so." Damien leaned close so she could hear him better. "After this is over, the two of you will get married and live happily ever after with shelves and shelves of books and a dozen children."

"Thank you, Damien." She did so want to believe him. "I'm sorry about the horses getting away."

"We can ride this roof like a horse." He smiled.

"Watch out!" Tansy yelled as an uprooted tree slammed into the schoolhouse.

The roof groaned as the building broke free from its foundation and began bouncing along in the water.

"If I live to tell it, this is going to be some story." Damien sounded excited.

The schoolhouse banged into a ledge rock. Tansy lost her grip and slid down the roof. Damien grabbed her coat, but Tansy slipped out of it into the water. She went under but then was tossed back to the surface right under a tree. She grasped a thick overhanging branch and pulled up out of the raging creek.

She inched along the branch until she was next to the trunk. An American chestnut. Although the tree was probably dead, the branch was strong and solid under her.

"Thank you, Lord," she whispered. Then she prayed fervently that Damien's prediction of "happily ever after" years with Caleb would have a chance to come true.

The rain eased to a gentle shower as the storm clouds moved away and the sun peeked through. Somewhere there would be a rainbow.

Caleb fought the current until his arms felt like lead. Still, he couldn't give up. Not with the way he kept seeing Tansy's eyes full of fear for him as he turned loose of the roof. That had to mean she loved him. He had to turn loose. She would have fallen back in the water trying to help him. He hoped Felding was man enough to keep Tansy safe on the roof.

Somebody yelled at him and then a rope hit the water beside him. He caught it and a few minutes later he was on solid ground, coughing up water. The tide had stripped off his boots and shredded his shirt and pants.

"I was feared the creek might wash some folks away," the man said.

Caleb shook his head to clear his eyes. "Herman Clark. Thank you. I'm not sure how much longer I could have lasted."

"Didn't expect to see you floating in the tide, Caleb," the man said. "You know of any others we should keep an eye out for?"

Caleb looked at the creek. "I did leave two hanging onto the schoolhouse roof when I got swept away. That reporter fellow and Tansy Calhoun."

"The book woman? Why didn't she ride her horse to high ground?"

"Her horse must have spooked in the storm and then she had that city greenhorn to deal with."

"Flatlander." Herman spat on the ground. "But they should be able to ride out the storm. Long as the building don't give way."

Caleb's heart squeezed tight at the thought.

Herman must have seen his worry. "I got my boy looking upcreek to see if'n we need to pull anybody else out of the water." Sunshine suddenly split the clouds and sparkled through the last remnants of the rain. Herman pointed. "Looky there. The good Lord's put his rainbow in the sky over yonder."

Dear Lord. Caleb stared at the rainbow. *Please.*

The Clark boy rode a mule out of the trees. "There's a feller stuck on a roof lodged in a logjam up the way."

It might not be the schoolhouse roof, but what if it was? Ignoring his bare feet, he ran up the creek, where he stared out at the battered remains of the schoolhouse. His heart sank at the sight of Felding alone on the roof.

"Where's Tansy?" he yelled.

Felding yelled back, but Caleb couldn't hear him over the sound of the creek.

"We best get him out of the water first off." Herman threw out the rope. When Felding caught it, Herman motioned for him to tie it around his chest. They pulled him toward the bank seconds before the tree jam shifted and the roof broke apart.

"That was close," Felding gasped when he got out of the water.

"Where's Tansy?" Caleb demanded.

"We banged into a rock and she lost her hold. I grabbed her coat, but she slipped out of it."

Caleb balled up his fist and knocked the man flat.

"I tried, man." Felding scrambled to his feet.

Caleb would have knocked him down again, but Herman grabbed his arms and held him back. "Easy, Barton. Ain't helping nothing to pound on this here man."

Felding rubbed his jaw as he edged away from Caleb.

Caleb shook off Herman's hands and breathed in and out as he stared at the water rushing by. Tansy could be in that water. Under that water. He shut his eyes.

"She loves you, Barton," Felding said.

"She tell you that?" Caleb could hardly push out any words. He just wanted to hit something again. Anything.

"Not in words, but no mistaking the look in her eyes when you washed down the creek."

He swallowed hard but had no words.

"Don't give up on that girl," Felding said. "When she was afraid for you, I told her you were strong. That you'd make it, and you did. Tansy will too."

Felding was right. Tansy could have found a way to save herself. He stared at the man and uncurled his fists. The flood wasn't Felding's fault. "Look, I'm sorry I hit you."

"Don't let it worry you." The man actually smiled. "A writer is always ready for a new experience, and getting socked by a mountain man felt pretty new." He worked his jaw up and down.

That sounded nutty to Caleb, but he couldn't waste time trying to figure out Felding. He turned to Herman. "I need to borrow your mule."

Caleb didn't wait for an answer. He yanked the reins from the boy and jumped up on the mule's back.

"We'll keep an eye out down here," Herman called after him.

The water receded as Caleb rode along the creek. The worst was over. Gully washers came and went fast. A calf floated by. Bits of the schoolhouse swirled past. Tansy's library book was caught in some brambles.

Now the sun was shining full. The storm was over, but would it ever be over in his heart?

"Tansy!" he called over and over. Each time silence answered him, his heart sank lower.

"Hello!"

The word was so faint he thought he might be imagining the whisper on the wind, but then after he called her name again, the next shout was loud and clear.

"I'm down here."

He kicked the mule to move faster through the debris to where Tansy huddled in the crook of an American chestnut nobody had harvested.

"Tansy." He blinked back tears as he looked up at her. "If you aren't evermore one sight for sore eyes."

"Is that really you, Caleb?" She sounded as though she couldn't believe he was actually there.

"It really is."

"I'm stuck up here like a treed raccoon." She laughed. "Couldn't figure how to get down without breaking my neck."

He felt like laughing too. Not because she was stuck, but because they were both alive. "How did you get up so high?"

"Wasn't near this high before the water went down." She looked out at the creek. "Sorry about the schoolhouse. Did Damien make it?"

"Do you care?"

She frowned down at him. "Of course I care. You should too."

She was right. Jealousy was messing with his thinking. "We pulled him out of the water, Herman Clark and me. Then I punched him."

"Why would you do that?"

"Because you weren't on that roof with him."

"Poor Damien. Did he punch back?"

"He was smarter than that. He got up off the ground and told me you loved me." Caleb stared up at her. "Do you?"

"That's not a question you should ask while I'm stuck up here in this tree, shivering so much I might fall out any minute."

Even from the ground, Caleb could see her eyes flashing sparks at him.

"Right. Let me get this mule in position and then you can swing down where I can catch you."

"Will you catch me?"

"I'll catch you, Tansy Faith Calhoun, and never let you go."

He was glad the Clarks' mule was cooperative. He stood steady while Caleb reached for Tansy. She grasped the branch and swung down into his arms. He cradled her against his chest, his cheek resting on the top of her head.

After a minute, she pushed back from him and stared straight into his eyes. "Now you can ask me anything you want."

He smiled a little. "Then I'll get right to the question I should have asked you years ago. Will you marry me?"

A worried look clouded her eyes. "I don't want to steal you from Junie and Cindy Sue."

"If you don't say yes, you'll break my heart and I won't be good for anybody."

One corner of her mouth twitched up in a smile. "Then yes, I'll marry you, Caleb Barton."

"Tomorrow?"

"Of course not. I have books to deliver to people on my book route tomorrow. If Shadrach hasn't lost them all by now." She smiled, but he knew she was serious. "But come spring, we can find an American chestnut still pushing out leaves and get a preacher willing to go into the woods to marry a couple of tree lovers."

"Lovers. I like the sound of that." He feasted his eyes on her face. "May I kiss the future Mrs. Barton?"

"Oh, I do wish you would."

When Caleb touched his lips to hers, she slipped her arms around his neck and let him know without words that yes indeed, she did love him.

thirty-seven

PERDITA ROCKED BACK AND FORTH in the May sunshine as she whittled on a piece of chestnut Hiram had brought in from the woods. The wood hadn't spoken to her yet, but it surely would eventually. Something happy, for it was a purely pleasurable thing to sit on the front porch, her in one rocking chair and her husband in another.

They weren't talking. Often as not, they didn't have a great need to fill the air with words, although sometimes Hiram did take a talking spell. That was fine with Perdita. She didn't mind listening. But much of the time, Hiram would be reading. Sometimes his Bible. Sometimes one of those books Tansy brung him.

Perdita was still some disappointed they hadn't had that triple wedding. Folks in the hills would have talked about that till the cows came home. But Coralee and Josh hadn't wanted to do any waiting to get married. Leastways, Josh hadn't. Coralee was still some tender about the heart, but she was warming to Josh. And what a fine little mother she was. Couldn't nobody nowhere love her baby more'n Coralee. Well, unless it was Perdita herself.

She did think Maylin Jewell hung the moon and polished some stars whilst she was up there. What a difference a season could make. When winter was coming on back in December, Perdita the same as thought she wouldn't see the whole of another year come around. She didn't have much to live for anyhow and no way to make that living possible with no food in the cupboard excepting what she got by the charity of her neighbors. Poor Hanley. She still grieved over that good man dying before his time.

But Perdita had said that prayer for family, and the Lord had thrown open heaven's doors and showered blessings down on her. Coralee was teaching her sweet baby to call Perdita Granny.

Of course, the baby wouldn't do any talking for some time, but she grinned like it was Christmas morning every last time she saw Perdita. Eugenia said that was just the reflection of Perdita's own joy. And maybe it was. Eugenia was happy enough to be a granny too, and even Coralee's ornery pa had come around to being mostly decent.

Hiram said it was time Perdita forgave and forgot what the man had done to his own daughter, kicking her out in a snowstorm to maybe freeze, but sometimes forgiving took a while and forgetting even longer.

Joshua hadn't come home to Eugenia, but he had sent an envelope with some dollar bills wrapped in a letter some other man had writ for him. Could be someday he'd make it home. Poor man would be some surprised at the changes that had come about whilst he was gone. He didn't even know he could claim being a grandpa. Through love. In time Coralee would carry more babies fathered by young Josh, but Perdita knew in her heart that none would ever be as dear to her as Maylin Jewell.

"Have you come up with what you're carving over there, wife?" Hiram looked up from his book.

"It ain't come to mind yet. I've been doing too much woolgathering." She let her hands rest. Prissy was stretched out on the porch railing and Hiram's hound skulked behind the chair. He aimed not to get too close to the cat. At times, Perdita was right sorry for the critter, but then he'd go to howling at the moon in the middle of the night and she'd be ready to sic Prissy on him.

"Thinking about that Maylin Jewell, I'm guessing." Hiram smiled over at her. He gave her some teasing over how she loved the babe, but he was taken with her himself.

"Some. But more still thinking how we missed our chance to be a mountain legend by having that triple wedding."

He reached his hand over toward her, and she laid down her whittling to put her hand in his. Another fine thing. Having a man want to hold her hand now and again. Being a married woman had made her feel ten years younger. She had half a notion to stand up and do a jig, but Hiram wasn't much on jigging. All she'd do was disturb the animals.

"Some of us were in too much of a hurry to wait for Tansy to tie the knot with Caleb. Our ceremony down at the Reverend Reid's house in Booneville felt proper enough to me." He squeezed her hand a little.

"We done fine, Hiram. And I'm glad you agreed to say the vows so Josh and Coralee could tie the knot. I didn't know you hadn't never helped nobody get married afore that."

"I guess it's time to let the title 'preacher' be more than name only. Somehow reading that Bible of yours that came through the fire settled a calling in my heart that I hadn't ever felt before. I knew the words but didn't always have the feeling."

"And now you do?"

"Now I do."

"I do." She echoed his words. "Funny how two little words can change so much. And whether all three of us couples stood up together at the same time or not, we all got mightily changed by those two little words. It was good you tying the knot for Tansy and that Barton boy too."

"Even if I keep marrying folks, I don't know that I'll ever do another wedding like that one." Hiram chuckled as he shook his head and pulled his hand back to smooth down the Bible page. "That was something today, the way the two of them stood under that tree to promise to love and honor each other all the days of their lives."

"Seemed the height of foolishness to me." Perdita picked up her whittling. "Having all of us troop out there with them to stand around in the woods with nary a place to sit. And then that dog Jenny Sue brung with her running crazy around everybody. I was plumb worried about him knocking over Coralee whilst she was holding Maylin Jewell."

"But you saw how Coralee spoke him down. That girl has a way with animals."

"And with old ladies like me too, I reckon," Perdita said. "Vesta didn't look overly happy about the whole proceeding. I hear she had things planned out different for Caleb."

"Maybe so, but there wasn't ever any other choice for Caleb. He's been in love with Tansy before he hardly knew what it meant to be in love."

"Sort of like young Josh," Perdita said.

"Some, but Caleb thought he'd lost Tansy back when. Josh and Coralee are starting at their young ages."

"Tansy ain't so old. Merely twenty. Not nearly so old as this bride here." This time Perdita reached over toward Hiram and he took her hand.

"Sometimes love knows no age." Hiram looked out toward where the sun was going down. "But didn't it do your heart good to see those two standing under that chestnut tree somehow putting out leaves in spite of the killing air that's took down most of the mighty ones? If love can keep that tree alive, then it will stay tall and strong forever."

"I do like a story that has a fine ending," Perdita said. "Why don't you tell me one from the Bible that does?" She guessed he was in a talking mood and she was ready to listen. It was funny how in just a few weeks together she could already read his aims and wants.

"There's the story about Ruth. You ought to like it, with Ruth claiming Naomi for her mother and finding a good husband in Boaz."

She rocked and listened as he made the old story she'd read who knew how many times come to life again. The wood in her hand began to speak to her. Without a doubt it was going to be a twist of entwined love.

She smiled as she imagined that Tansy girl and the Barton boy loving one another as they entwined their lives. She did owe a debt to the two of them with how they pulled her Bible out of the ashes. They'd pulled her life out of the ashes right along with it.

While Hiram kept talking, she sent up a silent prayer of thanks for such a season of weddings. Then she prayed blessings down on Mr. and Mrs. Caleb Barton wherever they were. Knowing them, she wouldn't wonder at all that they were spending their first night together as man and wife right under that chestnut tree where they'd promised their love forever.

"Have you ever thought about how many things a person might wonder about, Caleb?" Tansy stared up at the tree branches over them.

"You mean the wonder of how I got such a beautiful woman to marry me?" He raised up on one elbow to stare down at her in the moonlight. "Or is the wonder that I let you talk me into spending our honeymoon night under this tree?"

"Seems perfect to me. No chance of anyone finding us to do a shivaree." She pulled him down to kiss her and then cuddled against him for warmth. The May night was pleasant enough, but she liked feeling his strength wrapped around her.

"We can praise the saints for that, but don't many people do that these days."

"They might for the schoolteacher."

"And the book woman."

"But I was talking about natural wonders. The stars. The soft feel of the night air. This tree that's still putting out leaves instead of dying. Do you think our children will be able to climb around in its branches?"

"I don't know, but I like the sound of 'our children.'" He pulled her closer. "I love you, Mrs. Barton."

"I know."

She could hear his heart and felt her own matching his beat for beat. Maybe he was right. Maybe the best wonder of all was how an intrepid book woman had found a man with aplomb to marry. No writer, not even Jane Austen, could come up with a better ending. Or not an ending at all. A beginning. Only a beginning.

LOVED THIS STORY?

Turn the page for a preview of another historical tale from Ann H. Gabhart!

Available Now

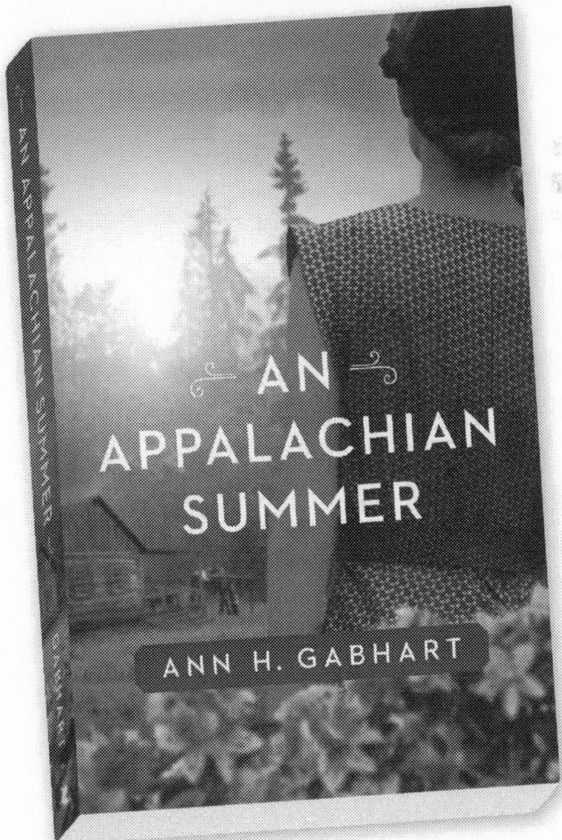

Discover what happens when one intrepid young woman steps away from the past into a beautiful, wide-open future . . .

one

May 20, 1933

Piper Danson's cheeks hurt from smiling for what seemed like hours with no relief in sight. More people waited in line to take their turn in front of her and pretend happiness over her debut into society. Then again, their smiles might be sincere. Piper was the one feigning excitement as she repeated socially appropriate words of welcome.

Her ridiculous gauzy white dress looked made for a sixteen-year-old instead of a woman with two years of advanced studies at Brawner Women's College. If only she could fiddle with the neckline where it chafed her skin, but a debutante didn't adjust her clothing in public. She pretended everything was wonderful and that she loved all the flowers presented to her in celebration of her coming-out party. But the cloying odor of so many flower arrangements made her feel as if she were at a wake. Perhaps she was. The funeral of her freedom. Time to pick a man and marry.

That wasn't exactly right. More like time to accept the man her parents had chosen for her and settle down into a proper life, the way her sister Leona had done after her debutante season four years prior.

Where was Jamie Russell when she needed him? She quickly scanned the room before the next person stepped in front of her. Jamie was nowhere to be seen. His absence was disappointing, but hardly surprising. Not now. Not after his family had lost everything in the stock market crash. While debutante balls had surely waned in importance for him in the face of such misfortune, she still expected him to come to hers. If only someone would open the ballroom's balcony doors to let in some air. She considered fainting simply for the novelty of it, but her mother would never forgive her. Besides, fainting was for fragile girls. Piper was anything but fragile. Tall, willowy for sure, but strong enough to rein in the most fractious of horses.

"How beautiful you look." One of her mother's friends took Piper's hand.

Piper held on to her smile and tried to remember the woman's name.

"You are so lucky to have this lovely ball with so many families struggling right now."

Piper didn't know whether to keep smiling or look sad. Perhaps this woman's family had fallen on hard times. Her dress did look like last year's fashion. Piper glanced over at her mother for a clue. Not only had her mother's smile not wavered, it looked genuine, as if produced specifically for this very woman.

Piper murmured something polite and continued to smile too, although she thought having such an elaborate event at a time when men stood in soup lines out on the streets of Louisville was reprehensible. Were it not for her mother, Piper would have flouted Emily Post's guide to proper etiquette for a debutante and escaped to somewhere. Anywhere away from this receiving line.

But she couldn't disappoint her mother, who had worked tirelessly to organize this ball as though Piper's future depended entirely upon a successful debut. Piper had managed to put her off for two years, until now at twenty she was a bit old for a first-time debutante. She told her mother that, but she would have none of it.

"We may have been wise to wait a few years in between Leona's and your debuts. Especially with the situation as it is," her mother had said.

That was the closest her mother ever came to speaking about the depressed economic state of the country. She chose to sail above it, as though money were the least of her worries. She had been a debutante in better financial times and married the man her parents thought she should.

When Piper had asked if she loved her father when they married, her mother avoided a straight answer. "My parents had my best interests in mind. Love grows with time."

Whether love had grown or not, her parents were comfortable with their union. Her mother maintained appearances and ensured their two daughters and son had the advantages of an upper-society life. Her father supplied the necessary funds to make that possible through his partnership in a prestigious law firm, although the family had made some adjustments due to a reduction in clients able to pay the firm's fees.

The guest list was shorter than when Leona had her debutante ball, the hotel ballroom smaller. Piper didn't care. She had tried to convince her mother to simply have a tea and forget the ball. Her mother was aghast.

"What would people think?" She had actually turned pale at the thought. "Appearances are important. Vitally important for your father's firm. If clients thought we were affected by the situation, then they might fear bringing their concerns to Danson and Harbridge."

"But a tea would be so much more sensible." Piper paused and then added, "Considering the situation." When her mother's eyes narrowed on Piper, she knew she had made a mistake using her mother's word for the depression.

While her mother, who had the look of a hothouse lily, might be several inches shorter than Piper, she could be hard as nails when crossed. "That's enough, Piper. You will have a ball. Leona had a ball. A very successful one where she captured a perfect husband in Thomas Harper. Now it is your turn."

When Piper opened her mouth to continue her protest, her mother held her palm out toward her. "Not another word."

Now Piper looked out to where Leona sat with her perfect husband. She looked absolutely miserable, but that could be because she was well along with her first baby. Leona was petite like their mother, and no matter how flowing her dress, her condition couldn't be hidden. Some of the ladies were no doubt whispering behind their hands that Leona should have stayed home, started her confinement. A proper lady didn't parade her expectant body around for the whole world to see.

From the look on Leona's perfect husband's face, he wished they were both home. Or at least, Leona at home and he in his accounting office making sums add up. The man was ten years older than Leona and continually looked as though his cheeks might crack if he smiled.

Piper again resisted the urge to massage her own cheeks before they did crack. After tonight, she was not going to smile for a week. Maybe two. What was there to smile about anyway, with Jamie not showing up at this mockery of a party? He had to know about it, even if he had moved to Danville with his mother after his father's fatal heart attack. Brought on by the collapse of the Russell fortunes. Or so people said when they weren't gossiping that perhaps he hadn't had a heart attack at all but had taken an overdose of some sort.

People did like to gossip in the social arena. As if they had little else to do but find fault with one another. Piper held in a sigh. She was definitely fodder for the gossips with her late debut. A girl of twenty should already be married or at least promised to someone. Piper could almost hear the whispers. *If that girl doesn't watch out, she'll end up the same as Truda Danson.* Alone. With no prospects.

As if she'd beckoned her with the thought, her aunt Truda stepped in front of her.

"You look like you just swallowed a raw fish, my dear." Truda took both Piper's hands in hers and gave them a shake. "A very slimy one at that."

Fortunately, Piper's mother had turned to signal the musicians since all the guests had been greeted. She either didn't hear Truda or chose to pretend she hadn't. Piper's mother often turned a deaf ear to her sister-in-law. That made for a more peaceful family life.

On the other hand, Piper's father failed to follow her mother's example. He and Truda often had very animated discussions about matters of politics or money. Truda, who held a position in the bank their father had founded, was the main reason Piper's family hadn't lost everything in the slump. She feared a crash was coming and talked her brother into selling the family stocks that were next to worthless a mere month later.

Millions of dollars of investments disappeared into thin air. But Truda wisely socked away the Danson money in a fail-proof account. Piper's mother wasn't at all sure she hadn't stuffed her mattress with it, but wherever it was, they had avoided ruin.

"I'm smiling." For the first time that evening, a real smile sneaked out on her face.

"That's better." Truda gave Piper's hands another shake before she turned them loose. She lowered her voice. "I know you would rather be jumping your horse recklessly across fences or curled in a corner with a book, as would I. The book for me, not the horses. But instead, here we are, making your mother happy. A daughter has to do that at times."

"Did you?" When Truda gave her a puzzled look, Piper went on. "Make your mother happy."

"Oh heavens, no. Poor dear had to give up on me." Truda laughed. "Although I did indeed wear the white dress that looked as atrocious on me as this one does on you. Green is your color. No pale sickly green either. Vibrant green to make your eyes shine like the emeralds they are." She winked. "That would have set those Emily Post readers on their ears. Who decided Post was the expert on everything anyway? Why not Truda Danson's Rules of Etiquette?"

Piper's mother gave Truda a strained smile. "Really, Truda, you promised not to upset Piper's evening."

"Not to worry, Wanda Mae." Truda's face went solemn, but her eyes continued to sparkle with amusement. "I will refrain from speaking any more truth the entire evening and speak only words that will tickle my listeners' ears."

"Piper and I will appreciate your restraint." Her mother motioned Piper toward the dance floor. "Now, go. Braxton is waiting to usher you out for the first dance. I hear he is an excellent dancer. Do try not to step on his toes."

"But he has such big feet, Mother dear." Truda whispered the words near Piper's ear as they moved away from her mother.

Piper stifled a laugh.

"Does the young man indeed have big feet?" Truda peered out at the guests as though checking shoe sizes.

"Braxton's feet are fine. My feet are the clumsy ones." Piper sighed, her giggle gone. She'd taken dancing lessons. Her mother insisted on it, but though she learned the steps, the smoothness of their movement escaped her.

"Then you should do something else. Something better." Truda turned her gaze back to Piper.

"But a debutante must dance to the tune played for her."

"Perhaps for this evening, but come tomorrow you can pick your own tune. It is 1933, dear girl. We are no longer in the dark ages where a woman has no say in the choices she makes."

Truda gave Piper's arm a squeeze. "Marry if you must, but only do so for love."

"Mother says love will grow."

"So it can. Properly nourished." Truda raised her eyebrows. "But a good seed well planted in the rich loamy soil of romance puts down the strongest roots and grows best."

"Did you ever plant such seeds?" Piper had never heard of Truda having a suitor.

Truda shook her head without losing her smile. "I was born before my time. Independence in a woman was not admired twenty years ago. Nor was I beautiful enough to encourage young men to court me in spite of that stubborn lack of coyness. Or perhaps I never met the right man to tempt me to court him."

"They say I look more like your daughter than my mother's." Piper smiled. "So perhaps I will be in the same situation."

"Come now, child. You are much lovelier than I ever hoped to be. Didn't your mother just say this Braxton was waiting to sweep you off your feet? One of the Crandalls, isn't he?"

"Yes."

"Your excitement at the prospect sounds a bit lacking. I can't remember which one he is. Point him out." Truda looked out at the guests again.

"He's beside Thomas." Piper didn't look his way. She could feel him waiting for her. A nice man. Already established in his family's business. Something to do with railroads. Her father claimed him a good match. Love would grow.

"Hmm. A pleasant-looking fellow. Tall enough so you won't have to worry about towering over him if you wear a shoe with a heel. That's good. Men don't like to feel short. That's why they sometimes prefer those petite girls, but I say be glad you're tall enough to reach the high shelves in a cabinet. A useful ability." Truda let her gaze wander around the rest of the room. "But where is that curly headed boy with the burnished brown eyes who was always trailing you around before you went off to college? Jamie Russell, wasn't it?"

"He must be otherwise occupied this evening." Piper pretended she didn't care.

"Or uninvited. We do close ranks against the less fortunate, don't we? Such a shame about his father. I hear his brother is trying to revive their business. Manufacturing washing machines, I think. Or was it stoves? Either way, no one can afford new things right now."

"Yes." Piper looked around at the ornate room, the flowers, the plates of food. "Unless one is a debutante."

"Try not to sound so thrilled." Truda laughed softly. "Or so much like me." She gave Piper a little shove. "Go. Dance with Braxton of the Crandall railroad fortune. Tomorrow you can take a vow of poverty and walk a different path. But for tonight, be your mother's daughter. A blushing debutante."

A flush did climb up into Piper's cheeks as she turned toward Braxton but stayed where she was. Surely a blushing debutante should wait for the man to approach her.

She scarcely knew him. Since he was five years older than her, he'd been away at Harvard while she and her friends first tasted the freedom of stepping out. Then she'd been away at school except for holidays or summers when she spent every moment possible with Jamie.

She wanted to glance around again to see if perhaps, invitation or no invitation, Jamie had come. But instead, she kept her gaze on Braxton Crandall. One might consider him handsome. A strong chin line, a nose not too big, neatly coiffed brown hair parted on the side. He excused himself from the group around him and came toward her. She had to wait until he stepped nearer to see that his eyes were a grayish blue. He was clean-shaven. That was a plus. Piper had never cared for mustaches.

She almost laughed aloud as she imagined Truda's voice in

her head. *"Well, I should say not. A mustache never looks good on a lady. That's why some wise person invented tweezers."*

The nonsensical thought did help. Her smile was genuine and whether it was meant for Braxton Crandall or not little mattered. His own smile got wider.

"Miss Danson." He reached for her hand. "I do think you, as the lovely lady of the hour, are expected to lead off the dancing. Would you grant me the pleasure?"

Piper inclined her head and let him take her hand. As they walked toward the dance area, she hoped for a slow waltz where she could count her steps, even as she remembered the last time she had danced with Jamie. A fast Charleston that had them laughing and leaning on one another in exhaustion when the music stopped.

With Jamie, she never had to count steps.

two

A BALCONY DOOR OPENED and music floated out to where Jamie Russell leaned against the brick wall around the Grand Hotel's prized rose garden. The hotel's brochures spoke glowingly of the beauty and peace it afforded all their guests.

Jamie felt none of that peace. He shouldn't be here. He had told himself not to come. Better to stay in Danville where his mother had found refuge on her brother's estate.

Uncle Wyatt was a physician. While well respected in his town, he was not rich. He claimed any doctor worth his salt could never get rich. Too many needed his services without the coin to pay. Especially now. But he was thrifty and had preserved his inheritance from his much more ambitious father. Part of that inheritance was the family house and acreage in Danville. Jamie's mother had inherited a like amount of money, along with a second house in Louisville.

All was lost when Jamie's father's loans were called in after the crash. He had so wanted to be rich. None of them knew how deeply he went into debt to buy stocks. It seemed a failsafe prospect, with how the market kept booming. For a while it had worked. Profits mushroomed. His father bragged about doubling his money. He repaid the loans but turned around and borrowed more. The gains were there to be grabbed by those brave enough to play the market or foolhardy enough to think stocks would continue rising instead of the bottom dropping out. The crash took it all.

Not only from his father. Others ended up in the same sorrowful position after the ticking of the stock market tape on Black Tuesday.

Jamie had never cared much for numbers. He liked words. Hated the hours he spent in the family business, figuring supply and demand. Supply had overwhelmed demand and now nothing was worth anything. Certainly not Jamie himself, if money were the measure of worth.

Money did seem to be the measure at events like the one playing out in the ballroom above him. He could go in. He was appropriately dressed. The creditors hadn't taken their clothes. Only their self-respect. And his father.

Financial ruin had been more than his father's heart could stand. A stronger man might have fought through. Come back from nothing. Jamie's older brother was that kind of man. Simon was working to revive the family fortune by finding investors to finance a new manufacturing venture. He claimed the economy had to improve and people would again want to spend money.

Perhaps they would, but now all commerce moved at a snail's pace. Still, a new president seemed ready to bring the country out of the depression. President Roosevelt's fireside chat had come through the radio to bolster the courage of men like Simon. So much so that Simon was thinking of changing from manufacturing washing machines to making radios. Even during this downturn in fortunes, people still wanted their radios. That was the future. Simon was every bit as ambitious as their father had been but with a more conservative bent. No loans to gamble on the market. Only on his business future.

Jamie, at twenty-two, was five years younger than Simon and

five years older than their baby sister, Marianne, who would never have an elaborate debut party like the one going on in the hotel. That worried their mother, who feared their loss of fortune would keep Marianne and Jamie from finding a good match. Simon was already married with two children. Fortunately, he had made a good match, a lovely lady. An inheritance from her grandmother kept them from losing their house.

Simon and Estelle could have been on the guest list for Piper's party. If so, Piper's parents probably hoped Jamie wouldn't ride Simon's coattails through the door and mess up their plans to match Piper with a more likely husband candidate.

Not that Jamie and Piper had ever mentioned marriage back when they were forever together. Before the crash changed everything. Jamie had been able to continue his education. Uncle Wyatt made sure of that. Jamie had just graduated from Centre College in Danville. A fine college that had tried to prepare him for the future, if he only knew what that future was.

Simon said he could work for him as soon as he got the new business up and going, but Jamie hated the thought of being stuck behind a desk, adding up figures. Uncle Wyatt said he could consider medicine, but the sight of blood made Jamie queasy. Teaching was a possibility, although the idea didn't excite him. Nor would it excite a debutante's parents.

He looked toward the balcony and wished Piper would step outside. He hadn't seen her for months, but at one time they could almost converse without words. Guessing each other's thoughts had been a game they played. She was better at it than him, always knowing when he was thinking blue instead of red or yellow. At church, sitting on opposite sides of the aisle, if he looked toward her, she was always turning to look at him at the same time.

He wondered now why he had never told her he loved her. Why he hadn't made her promise to marry him when they came of age. She would have kept her promise whether her parents thought she should or not.

Perhaps he should climb up the trellis to the balcony. Be a Romeo to his Juliet. But then that story hadn't ended so well for Romeo or Juliet.

That didn't mean he couldn't still ask. Step up to her and ask for a dance. Any dance she wanted to do. A waltz. A Charleston. A dance for life.

Piper, I am here. In the rose garden. He pushed the thought toward her and stepped out of the shadows. He felt foolish, but he couldn't tamp down the hope, making his heart beat faster. If she came outside, that would mean their special connection hadn't been broken by his change in fortune.

The music stopped. The balcony doors opened, and Piper stepped out. Jamie's smile faded when a man followed her. They were obviously together. He recognized the man. Braxton Crandall. The son of the man Simon hoped would invest in his radio factory. The Crandalls' railroad money hadn't disappeared in the crash.

Jamie moved back into the shadows. Money did matter. In so many ways. Perhaps not for love but for all those practical things a person needed. Love wasn't practical.

What if the two had slipped out on the balcony for a kiss? Jamie could not bear watching that. Better to leave without anyone knowing he was there. He pulled in a quick breath when he brushed against a bush. If they heard the rattle of leaves, he hoped they would think it was the wind.

He resisted the urge to look back toward the balcony as he went out the gate. Nor did he think *goodbye.* Instead he thought the words he should have said when he was sixteen or nineteen or twenty. *I love you, Piper Danson.*

Acknowledgments

I LOVE WRITING STORIES and I'm happy each time I hear from a reader who has enjoyed one of my books. If you are one of those readers—and since you're reading this, I assume you are—I thank you. I love the way you welcome my characters into your hearts and continually ask me to write more stories about what might happen next.

I'm forever grateful to the many people on my team at Revell Books and Baker Publishing Group who work to make my stories the best they can. Thank you, Rachel McRae, for your careful editing and support of my story vision in this book. Thanks also to the art department that wraps my stories in such beautiful covers. I've never been to a book event where someone didn't comment on them. I'm blessed to have Michele Misiak and Karen Steele on my team and always ready to help with any questions I have. Even better, they find great ways to get my books in front of readers. Barb Barnes has a way of making me a better writer as she helps with those final edits to be sure I've told the story in the best way possible. I also owe gratitude to many other people behind the scenes at

Baker and Revell who help my imagined stories become the best they can be. Thank you all.

Many thanks also to my wonderful agent, Wendy Lawton, who is ever encouraging, even when I'm wondering if I'll ever find my way from chapter 1 to "the end." She always assures me I can, and somehow the story comes to life and gets written.

I'm blessed to have a family who loves stories and even reads mine. Through our many years together, my husband has been unfailingly supportive of my writing dream. Last but certainly not least, I am continually grateful and blessed the Lord granted me the desire of my heart to write stories to share with you, my readers.

Ann H. Gabhart is the bestselling author of several Shaker novels—*The Refuge, The Outsider, The Believer, The Seeker, The Blessed*, and *The Gifted*—as well as other historical novels, including *Angel Sister, These Healing Hills, River to Redemption*, and *An Appalachian Summer*. She and her husband live on a farm a mile from where she was born in rural Kentucky. Ann enjoys discovering the everyday wonders of nature while hiking in her farm's fields and woods with her grandchildren and her dogs, Frankie and Marley. Learn more at www.annhgabhart.com.

"What a jewel of a story. Reminded me of *To Kill a Mockingbird.*"

—Lauraine Snelling,
award-winning and bestselling author

Ann H. Gabhart tells a story of hope and love in the face of desperate circumstances during the Great Depression.

Revell
a division of Baker Publishing Group
www.RevellBooks.com

Available wherever books and ebooks are sold.

Meet

Ann H. Gabhart

Find out more about Ann's newest releases, read blog posts, and follow her on social media at

AnnHGabhart.com

Printed in the United States
by Baker & Taylor Publisher Services